Carbonate Depositional Systems: Assessing Dimensions and Controlling Parameters

Hildegard Westphal · Bernhard Riegl
Gregor P. Eberli

Editors

Carbonate Depositional Systems: Assessing Dimensions and Controlling Parameters

The Bahamas, Belize
and the Persian/Arabian Gulf

Springer

Editors
Hildegard Westphal
University of Bremen
Germany
hildegard.westphal@uni-bremen.de

Bernhard Riegl
Nova Southeastern University
Dania, FL
USA
rieglb@nova.edu

Gregor P. Eberli
University of Miami
Miami, FL
geberli@rsmas.miami.edu
USA

ISBN 978-90-481-9363-9   e-ISBN 978-90-481-9364-6
DOI 10.1007/978-90-481-9364-6
Springer Dordrecht Heidelberg London New York

Library of Congress Control Number: 2010932327

Printed on acid-free paper

Springer is part of Springer Science+Business Media (www.springer.com)

Preface

This book was initiated by a study conducted at Rosenstiel School for Marine and Atmospheric Sciences for Shell Research. For the present book this study has been considerably enlarged and modified.

The book has benefited from discussions with numerous colleagues, and numerous colleagues have provided us with unpublished data and information. Without this support, this book would not have been possible.

Numerous colleagues have reviewed parts of the manuscript. Gerald Friedman has reviewed the concept of the book. The Bahamas chapter has been reviewed by Paul Enos, Pascal Kindler, André Strasser, and Peter Swart. The Belize chapter has been reviewed by Ian MacIntyre, Ed Purdy, and Eberhard Gischler. The Persian Gulf chapter has been reviewed by Tony Lomando and Christopher Kendall. Further reviews of various parts of the book were done by Paul Wright, Fred Read, Mitch Harris, and Bob Ginsburg. To all these colleagues we would like to extend our gratefulness. Sonja Felder and Anastasios Stathakopoulos are acknowledged for proof-reading and formatting the manuscript.

We would also like to gratefully acknowledge the support by the late Wolfgang Engel of Springer Verlag who accompanied the first preparation of this book and who did not live to see it finalized. Suzanne Mekking and Martine van Bezooijen then accompanied the project from the manuscript to the book you hold in your hands now.

January 2010

Hildegard Westphal
Bernhard Riegl
Gregor P. Eberli

Contents

1 **Parameters Controlling Modern Carbonate Depositional Environments: Approach** 1
Hildegard Westphal, Gregor P. Eberli, and Bernhard Riegl

2 **Controlling Parameters on Facies Geometries of the Bahamas, an Isolated Carbonate Platform Environment** 5
Kelly L. Bergman, Hildegard Westphal, Xavier Janson,
Anthony Poiriez, and Gregor P. Eberli

3 **Belize: A Modern Example of a Mixed Carbonate-Siliciclastic Shelf** 81
Donald F. McNeill, Xavier Janson, Kelly L. Bergman,
and Gregor P. Eberli

4 **The Gulf: Facies Belts, Physical, Chemical, and Biological Parameters of Sedimentation on a Carbonate Ramp** 145
Bernhard Riegl, Anthony Poiriez, Xavier Janson,
and Kelly L. Bergman

5 **Summary: The Depositional Systems of the Bahamas, Belize Lagoon and The Gulf Compared** 215
Gregor P. Eberli and Hildegard Westphal

Index 231

Chapter 1
Parameters Controlling Modern Carbonate Depositional Environments: Approach

Hildegard Westphal, Gregor P. Eberli, and Bernhard Riegl

First research on carbonate depositional environments dates back to the middle of the nineteenth century, when Nelson (1853) described the general morphology of the Bahamas and realized the origin of calcareous eolianites. However, systematic studies on carbonate sediments and particularly their modern analogues remained scarce until the 1950th and 1960th. Then, pioneer work on the modern (sub-) tropical carbonate depositional environment, that was triggered by research groups of several large petroleum companies, ignited a boom in carbonate research (among others: Ginsburg 1956, 1957; Ginsburg and Lloyd 1956; Lowenstam and Epstein 1957; Newell and Rigby 1957; Wells 1957; Purdy 1961, 1963; Imbrie and Purdy 1962).

Despite a great increase in knowledge on carbonate depositional environments, understanding ancient carbonate rocks still remains challenging for the simple reason that the carbonate depositional environment is a dynamic system that responds to a variety of parameters such as climate (humidity, temperature), nutrient availability, productivity, sea level changes, tectonic movements, changes in water and wind energy, and biological determinants. Within this dynamic system, the multitude of parameters is interdependent and superimposed upon each other, and individual processes cannot be easily separated in the rock record. As a result, although extensively studied, the relative importance of the various parameters on the stratigraphic architecture of carbonates is still poorly understood.

The goal of this book is to add to the understanding of the carbonate depositional environment and to close some of the still existing gaps in our understanding of the

H. Westphal (✉)
MARUM and Department of Geosciences, Universität Bremen, Germany
e-mail: hildegard.westphal@uni-bremen.de

G.P. Eberli
Rosenstiel School for Marine and Atmospheric Sciences, University of Miami, Miami, Florida, USA
e-mail: geberli@rsmas.miami.edu

B. Riegl
National Coral Reef Institute, Nova Southeastern University, Dania Beach, Florida, USA
e-mail: rieglb@nova.edu

H. Westphal et al. (eds.), *Carbonate Depositional Systems: Assessing Dimensions and Controlling Parameters*, DOI 10.1007/978-90-481-9364-6_1,
© Springer Science+Business Media B.V. 2010

1

influence and interplay of individual parameters. It is written by geologists for geologists in order to provide an easily accessible overview over the large amount of relevant information provided by the neighboring sciences. Therefore, our view is strongly biased towards what we considered helpful for the interpretation of the sedimentary record. We do not claim the book to be a complete oceanographic, biological and geological review. For our task we concentrate on modern carbonate deposition, because separating the individual parameters is easiest in the modern environment where physical and biological parameters can be measured and directly compared with the actual properties of a depositional body. The approach of this book is to construct an image of modern depositional environments of three classical areas of carbonate deposition, in order to assess both, the range of physical, biological and chemical parameters, and their sedimentary response. This book presents a comprehensive compilation based on data from published work and from unpublished theses, and the integration of these data in order to extract previously undiscovered relationships between the discussed parameters and carbonate deposition.

This book concentrates on classical (sub-)tropical carbonate and mixed carbonate-siliciclastic depositional environments. Today there is growing awareness of the presence of other types of carbonate producing environments such as the cool- to cold-water realm (e.g., Lees and Buller 1972; Nelson et al. 1988; James and Clarke 1997; Freiwald 1998; Hageman et al. 2000; James et al. 2005) and the deep sea (e.g., Lazier et al. 1999; Freiwald and Roberts 2005), but also of the influence of nutrient levels on the development of different types of carbonate platforms (e.g., Pomar 2001). The (sub-)tropical environments presented here are understood as end-members of the wide range of carbonate platform types of the present-day world and the rock record. To expand this approach to the other carbonate depositional environments remains for the future.

The three study areas described here were chosen to represent different environmental settings of (sub-)tropical carbonate deposition:

1. The **Bahamian Archipelago** consists of several isolated carbonate platforms on a passive continental margin in a humid climate.
2. The **Belize Lagoon** is a rimmed carbonate shelf with various amounts of siliciclastics admixed. It is located in a strike-slip tectonic regime in a humid climate.
3. The **Gulf** is a flooded foreland basin, but is often considered a classical carbonate ramp. It is situated in an arid climate, contains evaporites and is influenced by siliciclastics.

For each study area the following parameters are investigated:

- **Tectonic setting**: structural constraints, terrestrial influence
- **Physical environmental parameters**: wave energy and direction, tidal range, currents, wind, water temperature, water clarity, depth of photic zone
- **Chemical parameters**: nutrient supply, ocean chemistry, carbonate saturation state, salinity

- **Biological parameters**: richness of calcifying fauna, ecological reaction to physical and chemical parameters
- **Geometries** of various facies belts: width, length, height, grain size, distribution, dominant organisms, slope gradients.

In addition to assessing the influence of these parameters within each of the three depositional environments, a comparison of the three study areas provides new insights in their relevance for carbonates in general. This comparison reveals interesting similarities and differences between the three sites.

References

Freiwald A (1998) Modern nearshore cold-temperate calcareous sediments in the Troms district, northern Norway. J Sed Res 68:763–776

Freiwald A, Roberts JM (eds) (2005) Cold-water corals and ecosystems. Springer Verlag, Berlin, Heidelberg

Ginsburg RN (1956) Environmental relationships of grain size and constituent particles in some South Florida carbonate sediments. AAPG Bull 40:2384–2427

Ginsburg RN (1957) Early diagenesis and lithification of shallow water carbonate sediments in South Florida: In: Le Blanc RJ, Breeding JG (eds) Regional aspects of carbonate sedimentation. SEPM Spec Publ 5:80–100

Ginsburg RN, Lloyd RM (1956) A manual piston coring device for use in shallow water. J Sed Petrol 26:64–66

Hageman SJ, James NP, Bone Y (2000) Cool-water carbonate production from epizoic bryozoans on ephemeral substrates. Palaios 15:33–48

Imbrie J, Purdy EG (1962) Classification of modern Bahamian sediments. In: Ham WE (ed) Classification of carbonate rocks. AAPG Mem 1:253–272

James NP, Clarke J (eds) (1997) Cool-water carbonates. SEPM Spec Publ 56:440

James NP, Bone Y, Kyser TK (2005) Where has all the aragonite gone? Mineralogy of Holocene neritic cool-water carbonates, Southern Australia. J Sed Res 75:454–463

Lazier AV, Smith JE, Risk MJ, Schwarcz HP (1999) The skeletal structure of *Desmophyllum cristagalli*: The use of deep-water corals in sclerochronology. Lethaia 32:119–130

Lees A, Buller AT (1972) Modern temperate water and warm water shelf carbonate sediments contrasted. Mar Geol 13:M67–M73

Lowenstam HA, Epstein S (1957) On the origin of sedimentary aragonite needles of the Great Bahama Bank. J Geol 65:364–375

Nelson RJ (1853) On the geology of the Bahamas and on coral formation generally. Geol Soc London Quart J 9:200–215

Nelson CS, Keane SL, Head PS (1988) Non-tropical carbonate deposits on the modern New Zealand shelf. Sediment Geol 60:71–94

Newell ND, Rigby JK (1957) Geological studies in the Great Bahama Bank. In: Le Blanc RJ, Breeding JG (eds) Regional aspects of carbonate sedimentation. SEPM Spec Publ 5:15–79

Pomar L (2001) Types of carbonate platforms: a genetic approach. Basin Res 13:313–334

Purdy EG (1961) Bahamian oolite shoals. In: Peterson JA, Osmond JC (eds) Geochemistry of sandstone bodies. AAPG Spec Vol: 53–63

Purdy EG (1963) Recent calcium carbonate facies of the Great Bahama-Bank, I and II. J Geol 71:334–355

Wells JW (1957) Coral reefs. GSA Mem 67:609–631

Chapter 2
Controlling Parameters on Facies Geometries of the Bahamas, an Isolated Carbonate Platform Environment

Kelly L. Bergman, Hildegard Westphal, Xavier Janson, Anthony Poiriez, and Gregor P. Eberli

2.1 Introduction and Research History

The Bahamas are among the most extensively studied carbonate regions in the world, and a number of phenomena typical of calcareous environments have been first observed in the Bahamas. Early geological research in the Bahamas was undertaken by Nelson (1853) who surveyed their geography and topography. He noticed the "remarkable lowness of profile" and the dynamics of construction and destruction of the islands, outlined the biota and lithologies, described the formation of the carbonate rocks, and noticed the eolian origin of many Bahamian islands. Forty years later, the examination of modern carbonate environments rapidly progressed with the expedition of L. and A. Agassiz in 1893 (Agassiz 1894). Their explorations focused mainly on the fringing reefs of GE Great Bahama Bank. Research on abiotic carbonate components followed, by Vaughan (1914) who emphasized that

K.L. Bergman (✉)
ETC Chevron Corporation, San Ramon, California, USA
and
Rosenstiel School for Marine and Atmospheric Sciences, University of Miami, Miami, Florida, USA
e-mail: KBergman@chevron.com

A. Poiriez and G.P. Eberli
Rosenstiel School for Marine and Atmospheric Sciences, University of Miami, Miami, Florida, USA
e-mail: geberli@rsmas.miami.edu

H. Westphal
Department of Geosciences, Universität Bremen, Germany
e-mail: hildegard.westphal@uni-bremen.de

X. Janson
Bureau of Economic Geology, Austin, Texas, USA
and
Rosenstiel School for Marine and Atmospheric Sciences, University of Miami, Miami, Florida, USA
e-mail: Xavier.Janson@beg.utexas.edu

H. Westphal et al. (eds.), *Carbonate Depositional Systems: Assessing Dimensions and Controlling Parameters*, DOI 10.1007/978-90-481-9364-6_2,
© Springer Science+Business Media B.V. 2010

carbonate constituents can originate from both skeletal secretion and chemical precipitation, and introduced the terms "organic" and "inorganic" limestones. Black (1933) first characterized the sedimentary facies on Great Bahama Bank and noted the significance of the widespread aragonitic mud. The sand-sized calcareous components of the Bahamas and their origin, including ooid sands, were described in detail in the classic papers by Illing (1954) and Newell et al. (1960).

In the 1960s, the Shell research group led by Robert Ginsburg significantly advanced our understanding of carbonate systems. Important work coming out of this research group included publications by Dunham (1962) who established the first classification of carbonate rocks, and Ball (1967) who classified sand bodies according to their geometry and setting. At the same time, Purdy (1963a, b) made a thorough petrological study of the carbonate facies on GBB addressing the distribution, origin and composition of carbonate sediments. The paleoclimatic significance of humid versus arid tidal flats was recognized by Shinn et al. (1969) and Ginsburg (1976). Enos (1974) compiled available data to produce a facies classification and map of surface sediments of the banks (see Section 2.5).

Work in the 1970s and 1980s incorporated the slope and basins surrounding the shallow banks and the effect of sea-level on platform growth. Hine and Neumann (1977) characterized the energy balance associated with platform margins and used subsurface data to reveal the internal structure and growth of the banks. Mullins and Neumann (1979) classified bank margins and characteristics of leeward and windward margins. Schlager and Chermak (1979) described the platform-basin transition while Schlager and Ginsburg (1981) elaborated on slope and deep-water trough evolution surrounding the Bahamas. Mullins et al. (1984) used geophysical and coring tools to capture comprehensively the anatomy of the open-ocean platform slope north of Little Bahama Bank.

Many questions surrounding the long-term evolution of Great Bahama Bank were revealed on the first deep-penetrating seismic profiles. Eberli and Ginsburg (1987, 1989) recognized the internal structure of the Bahamas and the role of lateral progradation in their growth. Subsequent drilling of the margins documented the influence of relative sea-level on the prograding pulses and the diagenesis in the shallow subsurface (Schlager et al. 1994; Eberli et al. 1997a; Ginsburg 2001).

The well-studied Bahamian platforms have long been considered the type setting of platform carbonates and are the basis of now classical depositional models for carbonate deposition in general. However, there still is surprisingly little process-oriented understanding about the effect and interaction of different parameters influencing carbonate production and deposition. Here a comprehensive compilation of data is attempted in order to unravel some of the potential interrelationships between the different parameters and to provide a useful database for further research.

2.1.1 Morphology

The remarkable flat and shallow morphology of the Bahamas was noticed by Spanish explorers. Herrera (1601; in the English translation by Stevens 1726) wrote about the

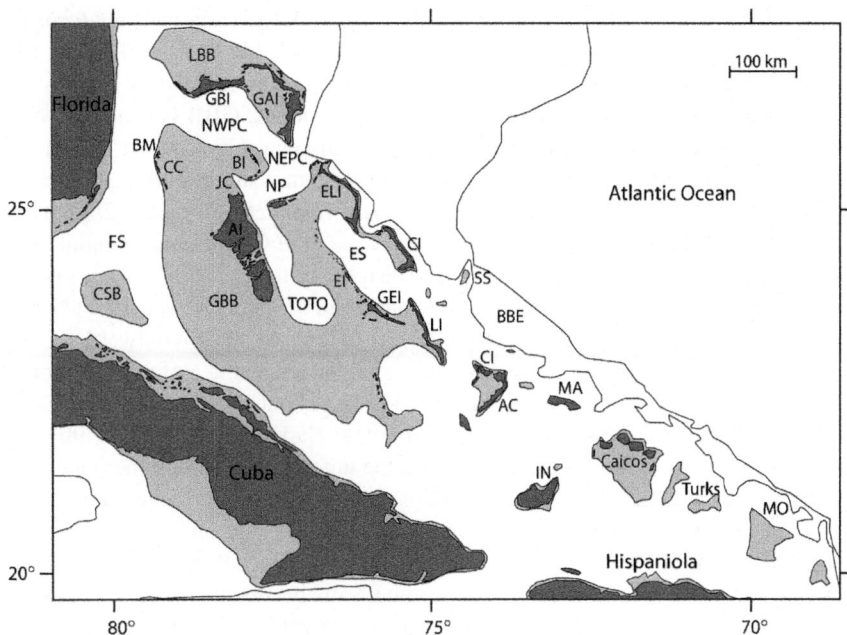

Fig. 2.1 Setting of the modern Bahamas at the western margin of the tropical North Atlantic, to the north of the Caribbean. LBB = Little Bahama Bank and GBB = Great Bahama Bank are the major platforms of the Bahamian system that also includes the smaller Turks and Caicos platforms. Other islands and platforms: AC = Acklins; AI = Andros Island; BI = Berry Islands; BM = Bimini; CC = Cat Cay; CI = Cat Island; CR = Crooked Island; CSB = Cay Sal Bank; ELI = Eleuthera Island; EI = Exuma Islands; GAI = Great Abaco Island; GBI = Great Bahama Island; IN = Great and Little Inagua; JC = Joulters Cays; LI = Long Island; MA = Mayaguana; MO = Mouchoir; NP = New Providence; SS = San Salvador. Oceanographic features: BBE = Blake-Bahama Escarpment; ES = Exuma Sound; FS = Florida Straits; NEPC = Northeast Providence Channel; NWPC = Northwest Providence Channel; OBC = Old Bahama Channel; SC = Santaren Channel; TOTO = Tongue of the Ocean; WP = Windward Passage

reconnaissance of Ponce de Leon in 1513: "… they went out from the islets (…), navigating among some islands he took to be overflowed and found it to be Bahama". Craton (1986) remarks that "it is an interesting fact that bajamar means shallow (strictly "low tide") in Spanish. Herrera was probably quoting the name given to the islands by the Spanish between 1513 and 1601". The name of the Bahamian archipelago thus refers to a morphological description.

The Bahamian archipelago extends from the Straits of Florida to the Puerto Rico Trench (Fig. 2.1). It is located between 20° and 28° N at the southeastern continental margin of the North American plate, which formed during the Jurassic when Laurasia broke up and the North Atlantic started to form. The Bahamas presently receive virtually no siliciclastic input, except in the form of windblown dust, due to the deep channels surrounding the platforms, resulting in a pure carbonate system that is considered the classical example for an isolated carbonate environment.

The Bahamian archipelago consists of several isolated carbonate platforms of which the largest are Great Bahama Bank (GBB) and Little Bahama Bank (LBB) (Fig. 2.1). The modern topography of the Bahamas is characterized by two distinct realms; the shallow-water banks and the deep-water areas. GBB is separated from LBB by the Providence Channel and is further dissected by Exuma Sound and Tongue of the Ocean. The shallow-water areas (<200 m water depth) are composed of several flat-topped, steep-sided carbonate platforms that sum up to 125,000 km^2 (Meyerhoff and Hatten 1974). The shallow-water realm of GBB extends continuously over more than 400 km from north to south. The relief of the Bahama banks is low and most submerged areas are covered with less than 10 m of water (Newell 1955; Newell and Imbrie 1955). Small islands cap the banks mainly on their windward, eastern margins. They are mainly composed of lithified Pleistocene and Holocene carbonate sand, largely eolian, and some reefal material and cover about 11,400 km^2 in area (Doran 1955; Milliman 1967; Meyerhoff and Hatten 1974; Harris, 1979, 1983). With 60 m above sea-level, Mt. Alvernia on Cat Island is the highest elevation on the Bahamas.

Channels and re-entrants, together with the periplatform deep ocean, make up the deep-water realm. In Exuma Sound and Tongue of the Ocean (TOTO), the two embayments within GBB (Fig. 2.1), water depths exceed 2,000 m. These deep-water areas are separated from the shallow-water platforms by steeply dipping slopes that generally show higher angles on the eastern (windward) side, and lower angles on the western (leeward) side of the platforms. Below the platform edge at 25–60 m below sea-level, the slopes dip almost vertically down to depths of around 135–145 m below sea-level (Ginsburg et al. 1991; Grammer and Ginsburg 1992; Grammer et al. 1993), where the slope angle decreases again. Where facing the open Atlantic Ocean (Bahama Escarpment), the slopes reach depths greater than 4,000 m below sea-level. With angles exceeding 40°, these slopes belong to the steepest sustained modern continental slopes world-wide (Emiliani 1965; Freeman-Lynde et al. 1981).

2.1.2   Origin and Tectonic Setting

Until recently the tectonic setting of the Bahamas was controversial (Meyerhoff and Hatten 1974; Mullins 1975; Mullins and Lynts 1977). Early workers assumed a continental basement. Nelson (1853) thought the Bahamas to be based on a huge delta formed by The Gulf Stream. Hilgard (1871, 1881), Gabb (1873), and Suess (1885–1909) proposed the Bahamas to be isolated fragments of The Gulf of Mexico-Florida crust. The hypothesis of a continental basement of the Bahamas was questioned by a plate-tectonic reconstruction of the continents prior to rifting (Bullard et al. 1965), wherein the area of the Bahamas entirely overlaps the African continent. Based on this observation the hypothesis formed that the Bahamas were underlain by oceanic crust (Dietz et al. 1970; Le Pichon and Fox 1971; Glockhoff 1973; Sheridan 1974).

Meyerhoff and Hatten (1974) provide strong evidence that the underlying crust was continental based on seismic, gravimetric, and magnetic data and the geological similarity to the adjacent areas of Yucatán and Florida. Their concept was corroborated by Mullins and Lynts (1977) who resolved the overlap problem by a pre-rift recon-struction involving rotation of the region, thus obtaining a perfect fit.

Connected to the question of the nature of the underlying crust, the striking present-day morphology of the Bahamas was subject to a second long-lasting controversy. Three general concepts were proposed in the literature: (1) the concept of an inherited tectonic structure, (2) the concept of an inherited erosional morphol-ogy, and (3) the concept of a constructional Cretaceous carbonate system:

1. It was long believed that the recent morphology reflects an underlying, buried relief inherited from basement structures like folds or faults (e.g. Talwani et al. 1960; Ball et al. 1969; Lynts 1970; Sheridan 1971, 1974, 1976; Uchupi et al. 1971; Glockhoff 1973; Mullins and Lynts 1977). These fault structures were thought to be related to the Jurassic rifting (Mullins and Lynts 1977) or a volcanic precursor topography (Schuchert 1935). Rifting would have produced a horst and graben topography whereby the horsts formed the foundation for the platforms.
2. Hess (1933, 1960) assumed a subaerial erosional relief like a drainage pattern as precursor morphology. Ericsson et al. (1952), Gibson and Schlee (1967), and Andrews et al. (1970) suggested the deep Bahama channels to be essentially val-leys eroded in the submarine environment by slumping and turbidity currents.
3. The concept of a constructional origin of the morphology was based on a com-parison of the steep marginal profiles to the vertical growth of Pacific atolls (Newell 1955). Dietz et al. (1970), Paulus (1972), and Dietz and Holden (1973) thought the present-day morphology is inherited from reef development in the early Late Cretaceous. The so-called "megabank" concept states that the modern patterns result from disintegration of a much larger carbonate platform in the Cretaceous (Paulus 1972; Schlager and Ginsburg 1981; Sheridan et al. 1981). The megabank is thought to have drowned during the Mid-Cretaceous. According to this concept, the modern platform configuration are isolated remanents on the drowned platform (Schlager and Ginsburg 1981).

In contrast to the three aforementioned concepts that assume an upward growth after the establishment of the platform without considerable lateral migration of the margins, Ball (1967, 1972) suggested that the basement structure probably did not contribute much to vertical relief. He stated that any precursor structure most likely has been significantly modified. The existence of a Cretaceous megabank underlying LBB was corroborated by ODP Leg 101 but also showed that the Providence Channel was a long-lived deep-water re-entrant (Leg 101 Scientific Party 1988). Additionally, the results of ODP Leg 101 and a re-evaluation of core Great Isaac-1 revealed that in contrast to the more static concepts, some margins of the Northwest Bahamas have migrated laterally over 10 km since the Miocene (Schlager et al. 1985; Leg 101 Scientific Party 1988; Schlager et al. 1988).

In the mid-eighties, a high-quality seismic reflection profile across GBB, showed that the modern topography is the result of much more dynamic processes even than inferred from ODP Leg 101, including significant bank migration (Fig. 2.2; Eberli and Ginsburg (1987, 1989). The modern topography thus reflects neither inherited rift-graben structures nor the morphology of a Cretaceous megabank.

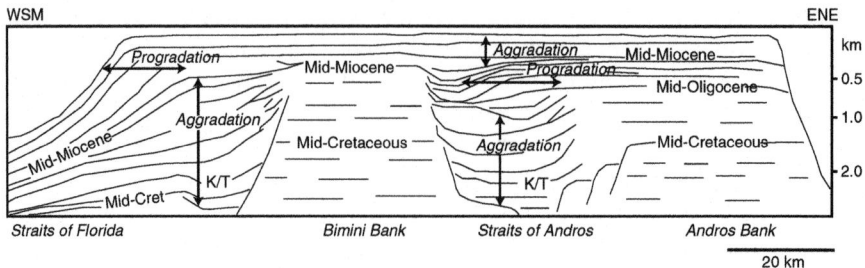

Fig. 2.2 Interpretation of seismic section through Great Bahama Bank displaying the complicated internal architecture of the bank. Two nuclear banks, Andros and Bimini Banks, coalesced by the infilling of an intraplatform seaway, the Straits of Andros. Progradation of the western margin of the platform during the Neogene expanded the bank more than 25 km into the Straits of Florida (Modified after Eberli et al. 1994)

The modern northwest GBB evolved from a process of repeated tectonic segmentation, related to rifting and reactivation of tectonic movements associated with the collision of the North American plate with Cuba, and subsequent coalescence by progradation (Eberli and Ginsburg 1987; Masaferro and Eberli 1999; Masaferro et al. 1999). For example, a former deep channel termed the Straits of Andros by Eberli and Ginsburg (1987) had similar dimensions as the Tongue of the Ocean and was completely filled and incorporated into the GBB. Similar infilled seaways exist in the southern GBB where a left-lateral strike-slip fault system had created several depressions in the Late Cretaceous and Early Tertiary (Masaferro and Eberli 1999). In addition, lateral growth on the leeward side of GBB led to progradation of more than 25 km westward into the Straits of Florida since the Miocene. Since the Cretaceous, the repeated tectonic segmentation and subsequent coalescence, induced by the high productivity of the shallow-water carbonate factory, led to the progressive modification of the GBB. Thus, coalescence and progradation are the most striking features governing the development of this carbonate platform.

Offbank sedimentation, which is a basic process of progradational deposition, is observed on the present-day Bahamas. Even though the environmental conditions on the Bahamas have not been the same for the entire history of the archipelago, an understanding of the present-day system is crucial for an understanding of the past. In the following, parameters controlling carbonate production and deposition in the present-day environment are described and discussed.

2.2 Physical Parameters

Physical parameters that directly or indirectly affect the carbonate depositional system include wave energy, direction of wave propagation, tidal range, currents, wind, and temperature. Generally, in the Bahamas, the physical parameters that influence water energy are most influential at the bank margins (Illing 1954). There, the highest degree of kinetic energy interacts with existing topography and biological barriers (e.g. sand shoals, reefs). Variability in exposure to water energy along the bank margins is responsible for the differences in margin architecture and the distribution of facies belts, especially carbonate sand bodies (Ball 1967; Hine and Neumann 1977; Hine et al. 1981b).

2.2.1   Climate

The Bahamas have a subtropical, humid, marine climate. The annual average air temperature is 25°C (Table 2.1). The annual temperature range is larger in the northern Bahamas, with a deviation of almost 8°C, than in the South with only 4°C. The north-south gradient disappears during midsummer, when the 28°C isotherm extends from within The Gulf of Mexico and envelops the region entirely (Isemer and Hasse 1985).

The average yearly rainfall is 85.7 cm (Table 2.2; Bosart and Schwartz 1979). Regional variations in average yearly rainfall range from 10 to 150 cm per year (Newell et al. 1959). Average seasonal rainfall increases from south to the north (Table 2.2). Seasons are characterized by warm, wet summers and cooler, dry winters. Most precipitation occurs between May and October (Table 2.3 and Fig. 2.3; Gebelein 1974).

2.2.2   Prevailing Currents

The current system around the Bahamas is dominated by the North Atlantic gyre to the east of the Bahamas, and the Florida Current to the west. The Florida Current is composed of waters from the Caribbean Sea and Gulf of Mexico that flow into

Table 2.1 Monthly temperature recorded at the Nassau International Meteorological station, at 25°03N and 77°28W from 1973 to 1993 (Data from www.hpc.ncep.noaaa.gov)

Temperature (°C)	Maximum	Minimum	Average
January	25	18	22
February	25	18	22
March	26	19	23
April	27	21	24
May	29	22	26
June	31	24	28
July	32	25	28
August	32	25	28
September	31	24	28
October	29	23	27
November	28	22	25
December	26	19	23
Yearly average	28	22	25

Table 2.2 Seasonal precipitation (cm) from various locations in the Bahamas recorded from 1935–1967 and 1968–1975 (From Bosart and Schwartz 1979)

Station	Winter	Spring	Summer	Fall	Yearly total
Mangrove Cay (Abacos, E LBB)	3.2	11.2	10.3	14.8	129.7
Nassau (N GBB)	4.3	8.2	15.0	18.6	138.2
Dunmoretown Harbour (Eleuthera, NE GBB)	4.2	7.5	10.7	13.0	106.1
The Bight (Cat Island, E GBB)	4.1	7.1	8.3	12.2	95.4
Georgetown (Exumas, SE GBB)	2.4	7.0	9.8	11.3	91.4
Duncan Town (Ragged Island)	2.3	5.0	5.0	11.7	71.7
Riding Rock (San Salvador)	2.6	4.6	4.4	7.6	57.6
Albert Town (Acklins)	3.3	4.9	6.3	9.9	73.2
Matthew Town (Great Inagua)	1.4	2.6	2.5	4.5	33.1
Grand Turk (Turks Islands)	5.6	4.3	5.0	10.4	76.4
All stations (average)	3.1	6.6	7.6	11.3	85.7

Table 2.3 Total precipitation recorded at the Nassau International Meteorological station, at 25°03N and 77°28W from 1855 to 1989 (Data from www.hpc.ncep.noaaa.gov)

	Total Precipitation (cm)		
	Maximum	Minimum	Average
January	29.7	0	4.8
February	18.3	0	1.3
March	19.3	0	4.1
April	42.9	0	6.6
May	45.5	0	13.2
June	50.8	0	17.8
July	37.1	0.8	15.2
August	27.7	0.3	17
September	55.9	0.8	18
October	59.4	0	17
November	26.4	0	7.1
December	27.2	0	4.3
Yearly total	232.4	52.1	128.5

the Straits of Florida and, north of LBB, become a major component of The Gulf Stream, the surface-flowing limb of the North Atlantic gyre (Fig. 2.4). The mean total transport through the Florida Straits is around 32 Sverdrups (SV; 1 SV = 10^6 m^3/s) and total transport in the upper 200 m is between 15.5 and 16.5 SV (Table 2.4; Johns et al. 1999). With the decreasing cross-sectional area of the Florida Straits in the northward direction, current velocities increase (Fig. 2.5). Thus, the surface velocities of the Florida Current accelerate from 180 cm/s in the Southern Florida Straits to up to 200 cm/s in the Northern Florida Straits (Wang and Mooers 1997). As the Florida Current wraps around Florida, the highest flow velocities of about 200 cm/s are found along the western side of the straits (Richardson et al. 1969).

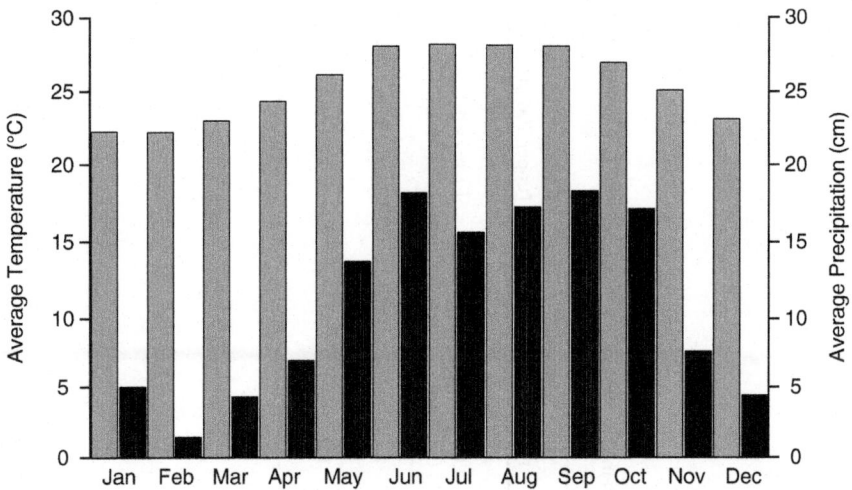

Fig. 2.3 Average temperature (*gray columns*) and precipitation data (*black columns*) from Nassau, New Providence, GBB (Tables 2.1 and 2.3) indicates pronounced seasonality of precipitation (Data from www.hpc.ncep.noaa.gov)

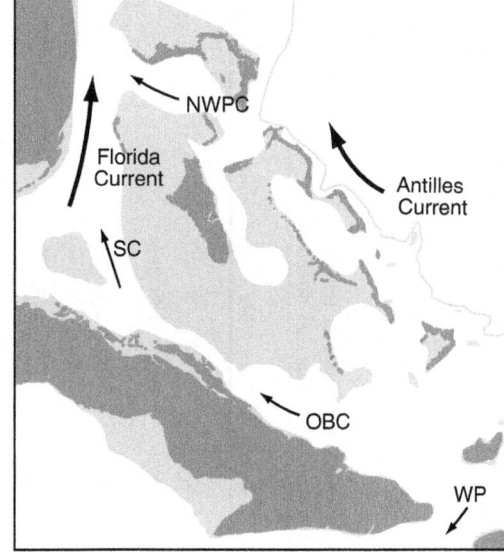

Fig. 2.4 The ocean currents around the Bahama Islands flow into the Florida Straits where they join the Florida Current. North of the Bahamas, the Florida Current joins the Antilles Current to become The Gulf Stream. NWPC = Northwest Providence Channel; SC = Santaren Channel; OBC = Old Bahama Channel; GIP = Great Inagua Passage; WP = Windward Passage

Surface flow velocities along the eastern side of the channel are considerably lower at 20–60 cm/s (Richardson et al 1969; Wang and Mooers 1997).

Overall flow from the Santaren Channel and Northwest Providence Channel is directed toward the Florida Current. Flow in the upper 200 m through the Northwest Providence Channel is primarily westward with a maximum flow of 30 cm/s and measuring 0.4 SV in volume (Johns et al. 1999). The Santaren Channel transports

Table 2.4 Flow direction, velocity and volume of Bahamian currents (From Atkinson et al. 1995; Leaman et al. 1995; Johns et al. 1999). Negative flow volume and velocity indicates reversal of mean flow direction

Current or passageway	Mean flow direction	Mean flow velocity	Maximum surface flow velocity	Mean flow volume (upper 200 m) (SV)	Flow volume range (upper 200 m) (SV)
Florida Current at 26°N	NNW	–	>170 cm/s	15.5 ±1.4	13.7–17.6
Florida Current at 27°N	N	–	>160 cm/s	16.5 ± 2.4	13.1–19.3
NW Providence Channel	W	>30 cm/s	–	0.4 ± 0.8	1.7–(–0.5)
Great Inagua Passage	SW	–	>20 cm/s	2.2 ± 1.5	0.5–5.2
Santaren Channel	N	–10 to 20 cm/s	20 cm/s (300–400 m depth)	1.8	–
Old Bahamas Channel	NW	~26 cm/s	193 cm/s (250 m depth)	1.9	6.6–(–2.4)

SV = Sverdrup

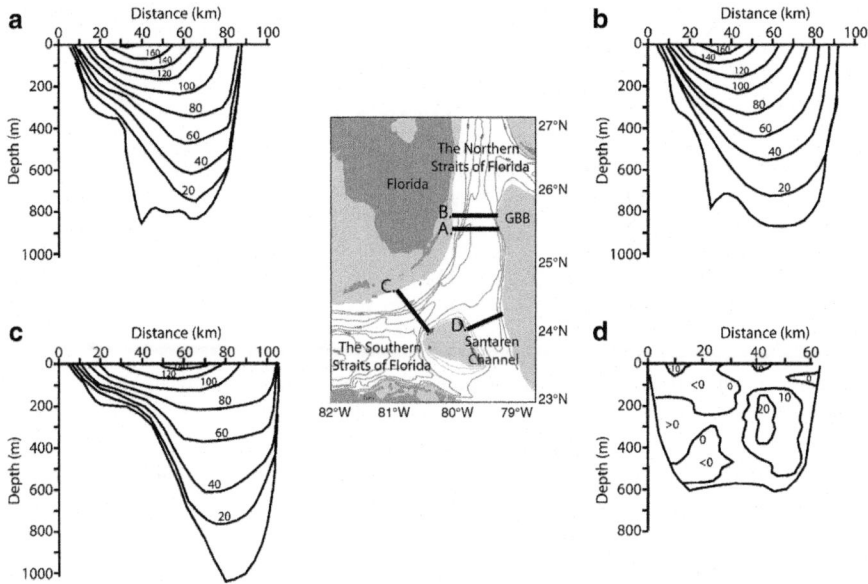

Fig. 2.5 Cross-sections and downstream current velocity contours in cm/s of the Florida Straits ((**a–c**); Richardson et al. 1969) and Santaren Channel ((**d**); Leaman et al. 1995). For locations see map

about 1.8 Sv of water northward at a maximum velocity of 20 cm/s (Leaman et al. 1995). Flow velocities in the Old Bahama Channel are on average 26 cm/s with transport of 1.9 SV northwestward (Atkinson et al. 1995). Flow through the Great Inagua Passage is to the southwest totaling 2.2 SV in the uppermost 200 m and has a maximum velocity of 30–40 cm/s (Johns et al. 1999). A summary of current flow direction, velocity, and water volume around the Bahamas is listed in Table 2.4.

2.2.3   Wind Energy

The northeasterly tradewinds of the region are modified by the North American low-pressure system which results in a wind regime dominated by southeasterly winds (Fig. 2.6). The easterly wind regime of the Bahamas crosses the region at an average speed of 6–7 m/s (Table 2.5). Easterly winds (from NE, E, and SE) account for 77% of the wind frequency in August, and for 46% in February, averaging 63% over the whole year (Fig. 2.6; Table 2.6; Sealey 1994). Additionally, in winter (February) cold northwesterly winds account for 30% of the wind, whereas warm southerly winds occur at 9% of the time (Fig. 2.6; Sealey 1994). This general wind pattern is fairly consistent throughout the Bahamas, with the exception that the northwest wind accompanying cold fronts affects the northern Bahamas more than the southern Bahamas (Sealey 1994).

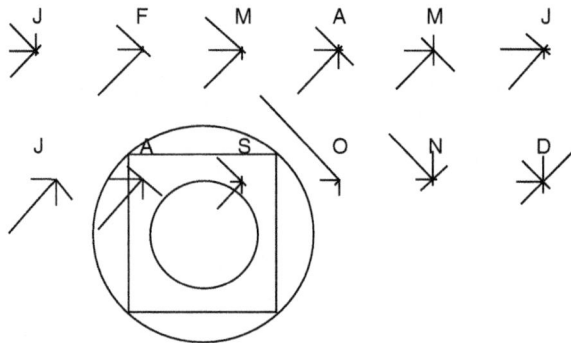

Fig. 2.6 Wind direction in days per month recorded in 1939 from the Nassau Meteorological Station records (see Table 2.6 for quantitative data from Smith 1940)

Table 2.5 Mean monthly wind speed in m/s (±1.25) (From Young and Holland 1996)

Month	Wind speed		Exceedence (%)	Wind speed
January	9.0		10	11.7
February	8.5		50	7.0
March	8.0		90	2.5
April	6.0			
May	6.8			
June	4.0			
July	3.8			
August	4.0			
September	8.5			
October	7.0			
November	9.0			
December	9.5			

Table 2.6 Wind direction in days per month recorded in 1939 from the Nassau Meteorological Station records (From Smith 1940)

Month	Wind direction							
	N	NE	E	SE	S	SW	W	NW
January	3	6	5	7	1	1	0	2
February	1	7	4	12	1	2	0	0
March	1	9	7	10	2	1	0	0
April	1	5	5	11	3	3	1	1
May	3	3	6	10	3	5	0	0
June	0	4	8	10	2	1	1	1
July	0	0	5	14	5	5	0	0
August	0	3	7	12	3	4	0	0
September	1	7	2	7	2	1	0	0
October	0	21	4	1	2	0	0	0
November	6	12	3	3	1	0	1	4
December	5	5	5	5	2	2	0	7

These persistent wind patterns influence sediment production and accumulation in the long term while tropical storms and hurricanes may also influence sediment redistribution in the short term. Depending on intensity, duration, and circulation pattern, storms can erode or deposit sediment (Perkins and Enos 1968). The greatest influence of storms is on non-vegetated, unstabilized environments such as sand bodies. The Tongue of the Ocean is subject to an average of two tropical storms per year, whereas approximately eight tropical storms pass over central LBB during 10 years (Cry 1965). Of all tropical disturbances affecting LBB, 67% pass to the north and east, traveling in a northerly, northwesterly, or westerly direction. They generate dominant winds from the northwest, north, and northeast. The result is a net bankward energy and sediment flux along the northeast bank margin, and a net offbank transport along the leeward margin on LBB (See Section 2.5.1.2; Crutcher and Quayle 1974; Hine and Neumann 1977). Some sand belts might only be active during hurricanes (e.g., tidal-bar belts in the Lily Bank area, Hine 1977; Hine et al. 1981b).

Yet storm influence on facies distribution are generally limited to the superficial and the short-term as ambient energy conditions reestablish pre-storm facies configurations (Perkins and Enos 1968; Boss and Neumann 1993; Major et al. 1996; Rankey et al. 2004). A re-examination of deposits on northern GBB after the passage of Hurricane Andrew in 1992 showed minor effects on sand bodies, hard bottom communities, and low-lying islands (Boss and Neumann 1993). In addition, bioturbation by organisms in vulnerable facies such as lagoonal sediments may destroy any record of winnowed deposits left behind by storms (Perkins and Enos 1968; Boss and Neumann 1993). Shinn et al. (1993) observed that muddy storm layers were preserved in high-energy channel sediments, because migrating ooids prevented obliteration by bioturbation.

Geologists disagree on the degree of impact storms have on the off-bank transport of sediment, because background (not storm-related) currents rarely exceed sediment transport threshold velocity (Cry 1965; Smith and Hopkins 1972). Hine et al. (1981a) suggest that storms are the primary transport mechanism of shallow-water sands found in 200–400 m water depth off the leeward margin of GBB and LBB and cite evidence of offbank oriented sand waves along the bank margins that are only active during storms. They concluded that normal tidal and current flow could not be responsible for carrying the sand to the deep (Hine et al. 1981a). Yet, observations after the passage of Hurricane Andrew over GBB in 1992 and Frances and Jeanne over LBB in 2004 showed that surprisingly little sediment transport was induced by the storm (Major et al. 1996; Reeder and Rankey, 2008).

2.2.4 Wave Energy

Wave energy in the Bahamas is highest along the windward (eastern) margins of the platforms, e.g. at eastern Abaco on LBB, and at Long Island on GBB (Table 2.7). Wave and swell energy arrive along these margins unimpeded from the open Atlantic Ocean. The rest of the Bahamas are exposed to lower wave energy due to either

Table 2.7 Percent frequency of wave heights at east Abaco (EA), east Long Island (LI), and southwest Great Bahama Bank (GB) (From National Buoy Data Center 1973)

		January	February	March	April	March	June	July	August	September	October	November	December
(%) Frequency >1.5 m	EA	43	43	41	35	22	17	17	24	34	40	43	42
	LI	41	37	35	41	34	26	33	28	30	33	42	47
	GB	34	32	28	28	23	21	29	19	20	21	32	39
(%) Frequency >2.4 m	EA	7	11	10	7	3	2	2	3	10	12	11	7
	LI	8	7	6	9	4	2	3	3	4	7	13	15
	GB	3	3	3	4	2	2	2	1	2	2	7	5
(%) Frequency >4 m	EA	2	2	3	2	1	1	0	1	4	4	3	1
	LI	1	2	1	1	1	0	0	0	1	2	3	4
	GB	0	0	1	0	0	1	0	0	0	0	0	0

limited fetch, infrequency of onshore wind (leeward margin, e.g. Bimini) or because they are shielded by energy-absorbing environments, such as the platform interior of southwest GBB. Wave energy also shows a seasonal trend with the highest waves occurring during winter months (Table 2.7). The easterly tradewinds result in a generally westward movement of water in summer (Smith 1940). A decrease in tradewind influence in the winter results in a southward movement of water (Smith 1940).

Wave-energy dissipates rapidly on the platform edges. A study of wave height changes in a shallow reef-crest environment indicates that energy loss from fore-reef to back-reef averaged through a tidal cycle is approximately 67%. Correspondingly, coral species associations vary from fore-reef to back-reef according to their tolerance to wave-energy exposure (Roberts 1979; Lugo-Fernandez 1989).

Wave energy greatly influences sediment composition and accumulation. Wave-dominated shorelines are characterized by grainy facies, whereas the most protected areas are sites of mud accumulation. Where shorelines are exposed to even a limited amount of wave or storm influence and under conditions of low sediment supply, the coastline will undergo erosion; this is the case along the coast of northwest Andros Island. In contrast, the protected southwest coast of the island is prograding (Shinn et al. 1969). The direction of wave propagation is an important factor in sediment distribution because transport direction is dependent upon direction of wave incidence, which is essentially a response to wind direction. Seasonal shifts in wind patterns from the east-northeast in the winter to east-southeast in the summer cause variations in the direction of swell-propagation (see Table 2.6) (Sealey 1994). Under extreme conditions such as a hurricane, waves can arrive from any direction and may reach heights of 15 m, even on the western banks (Abel et al. 1989).

2.2.5   Tides and Tidal Currents

Tides in the Bahamas are semidiurnal and microtidal with a mean range of about 0.7 m (Table 2.8). The tides are amplified where resonance takes place, for example in the deep-water embayments of the Tongue of the Ocean and Exuma Sound (Ball 1967), resulting in tidal currents reaching up to 200 cm/s (Halley et al 1983). On GBB, tidal currents generally are radially oriented and sweep bankward (0.32 m/s) during flood tide and offbank during ebb tide (Sealey 1994). On LBB, radial near-bottom currents rarely exceed 0.2 m/s. Here, the strongest currents are oriented alongbank, as opposed to the on-bank – off-bank direction on GBB (Hine et al. 1981b).

2.2.6   Energy, Water Depth, and Distance from Platform Edge

In the shallow-water environment, wave energy, tidal force, and currents dissipate from the platform margin to the platform interior with little to no significant change in water depth. Consequently, circulation is poor in the platform interior as indicated by the increase in salinity from 36–37 to 46.5 ppt from the western margin of GBB

Table 2.8 Predicted mean and spring tidal range and mean tide level for 2004 (From http://co-ops.nos.noaa.gov/tides04/tab2ec4.html)

Location	Position		Mean tide range (m)	Spring tide range (m)	Mean tide level (m)
	Latitude	Longitude			
Guinchos Cay	22°45'	78°07'	0.64	0.79	0.37
Elbow Cay, Cay Sal Bank	23°57'	80°28'	0.64	0.79	0.37
Fresh Creek, Andros Island	24°44'	77°48'	0.73	0.88	0.40
North Cat Cay	25°33'	79°17'	0.70	0.85	0.40
North Bimini	25°44'	79°18'	0.73	0.88	0.40
Memory Rock	26°57'	79°07'	0.70	0.82	0.40
Settlement Point, Grand Bahamas Island	26°42'	78°59'	0.82	0.94	0.43
Pelican Harbor	26°23'	76°58'	0.79	0.94	0.43
Nassau, New Providence Island	25°05'	77°21'	0.79	0.94	0.58
Eleuthera Island, West coast	25°15'	76°19'	0.73	0.88	0.40
Eleuthera Island, East coast	24°56'	76°09'	0.67	0.79	0.37
The Bight, Cat Island	24°19'	75°26'	0.79	0.94	0.43
San Salvador	24°03'	74°33'	0.70	0.85	0.40
Clarence Harbor, Long Island	23°06'	74°59'	0.79	0.94	0.43
Nurse Channel	22°31'	75°51'	0.64	0.79	0.34
Datum Bay, Acklin Island	22°10'	74°18'	0.61	0.79	0.34
Mathew Town, Great Inagua Island	20°57'	73°41'	0.64	0.79	0.37
Abraham Bay, Mayaguana Island	22°22'	73°00'	0.61	0.76	0.34
Hawks Nest Anchorage, Turks Island	21°26'	71°07'	0.64	0.79	0.34

to the interior near Andros Island (Cloud 1962). This has consequences for the distribution and prediction of sediment composition and grain size. In contrast to carbonate depositional environments elsewhere, in the Bahamian shallow water environment, facies is not related to depth but is controlled to a large extent by energy level. For example, ooid sediments are deposited in the high-energy setting at Joulters Cays, whereas carbonate mud is deposited in the same water depth on the tidal flats of Andros Island.

Based on the assumption that the distribution of facies is energy-controlled and energy decreases towards the platform interior, Purdy (1963b) and Wilson (1974) proposed a generalized facies zonation from the margin to the platform interior. These assumptions are generally valid for cases where no restrictions and no antecedent topography exist at the platform margin. Islands at the platform margin and reefs significantly influence the energy distribution and facies belts can widen along strike. Consequently, distance from the platform margin is not always a good indicator of either energy or facies. In the Joulters Cays area, for example, ooids are found at the same distance from the margin as some kilometers further south mud accumulates near Andros Island (see Section 2.5.1.1).

2.2.7   Sea-Surface Temperatures

Generally, the sea-surface temperatures in the Bahamas are warm and stable, in part due to the prevailing surface current regime. The Antilles Current and the Florida Current dominate the regional flow pattern and serve as a source of seawater exchange between the shallow banks and surrounding seaways (Fig. 2.4; Mullins et al. 1980).

Sea-surface temperatures west of Andros Island average 28.5°C in late spring (maximum 29.5°C; minimum 27°C) (Cloud 1962). The seasonal temperature maximum and minimum in Middle Bight, Andros Island is 29.7°C and 21.6°C, respectively (Smith 1940). Temperatures as high as 36°C have been recorded on the bank near Bimini (Table 2.9; Shore and Beach 1972). During the passage of cold fronts, platform interiors occasionally exhibit large fluctuations in temperature (Wilson and Roberts 1992).

Fluctuations of sea-surface temperatures affect the carbonate system in two ways. Most tropical organisms found in the Bahamas (especially zooxanthellate corals) are sensitive to temperature changes (see Section 2.4.1). Corals will die in just a few days if exposed to water colder than 14°C (Roberts et al. 1982). Corals are also

Table 2.9 Sea-surface temperatures of four Bahama localities (From Shore and Beach 1972)

Location	Maximum (°C)	Minimum (°C)	Mean (°C)
North Bimini	28–36	18–23	26.8–27.3
Gold Rock Creek, G.B.I.	32	18	–
North Riding Point, G.B.I.	32–33	16–17	–
San Salvador	29–30	22–24	–

G.B.I. = Grand Bahama Island

adversely affected by water warmer than 31°C, however, with significant local variability (Reaka-Kulda et al. 1994). The other major effect of fluctuating sea-surface temperatures is the production of hyperpycnal density flows. Both evaporation in summer (see Section 2.3.1) and rapid cooling of platform top water during the passage of winter storms result in denser bank-top water relative to adjacent surface water. The denser water flows off bank and contains entrained sediment (Wilson and Roberts 1995). These density-driven flows are potentially an important mechanism for offbank transport of fine-grained sediment and slope erosion. They can also affect platform-edge biota, where lethally heated or cooled bankwater can cause bleaching and death of reef corals at variable depths (see Section 2.5.1.2; Lang et al. 1988; Wilson and Roberts 1992; Riegl and Piller 2003).

2.2.8   *Light and Radiation*

Due to its low latitude position, the Bahamas experience little variation in daylight during the year (11–13 h/day; Hidore and Oliver 1993). The 1% light penetration level is at around 90 m in winter and 70 m in summer (Liddell et al. 1997). Certain photosynthetic organisms such as *Halimeda* and even hermatypic corals occur even below this depth (Liddell et al. 1997).

2.3   Chemical Parameters

The large-scale and shallow-water nature of the Bahama banks has a significant impact on water chemistry. The size of the banks causes circulation to be sluggish, particularly in the calmer summer months. Waters experience a high residence time that, combined with the high evaporation rate in the subtropical climate, controls salinity, alkalinity, and carbonate chemistry.

2.3.1   *Water Chemistry*

Bahama waters originate in the deep channels surrounding the platforms. Waters flow from the Florida Straits, TOTO and Exuma Sound onto the shallow banks (Fig. 2.1). Salinities on GBB typically range from 36–46.5 ppt (Cloud 1962; Traverse and Ginsburg 1966; Fig. 2.7). Salinities exceeding 80 ppt have been reported from restricted regions close to the shore of Andros (Fig. 2.7; Queen 1978; Bourrouilh-le-Jan 1980). Highest salinities occur during the summer months when evaporation is at a maximum. Salinity increases with distance from the bank edge, and salinity on GBB exhibits a linear relationship with residence time (Fig. 2.8). This relationship was observed early on by Maurice Black during his 1930 expedition (as cited by Jeans and Rawson 1980). The longest residence time on the bank

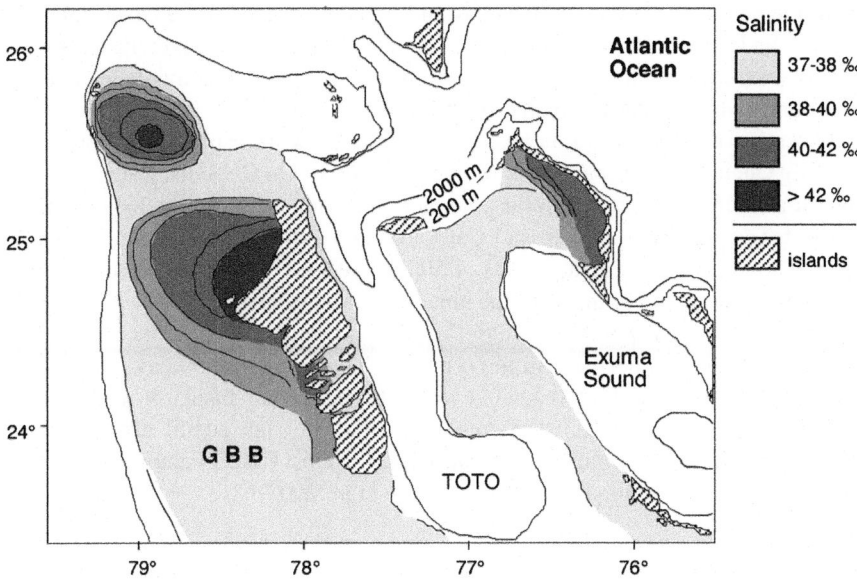

Fig. 2.7 Salinities on Great Bahama Bank, summer 1955 and 1956. Highest salinity waters are found in the lee of Andros Island where the highest residence time of bank waters occurs (Redrawn from Traverse and Ginsburg (1966))

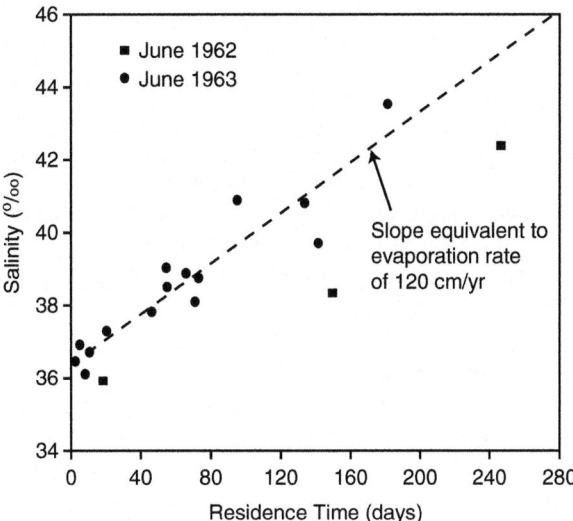

Fig. 2.8 Relation between salinity and residence time (From Broecker and Takahashi 1966)

has been estimated at 240 days in waters with the highest salinities (Broecker and Takahashi 1966; Morse et al. 1984). However, salinities do not reach a concentration that would result in the precipitation of evaporites except in the drier tidal flat environments of the southern Bahamas and in saline ponds of the island interiors (such as on Lee Stocking Island, central Exumas; Dix et al. 1999). In the Turks and Caicos, evaporites such as dolomite, gypsum and halite may precipitate in enclosed tidal ponds where water is isolated (Perkins et al. 1994).

Waters on the Bahamas Banks are supersaturated with respect to aragonite and high-Mg calcite (Morse et al. 1984; Morse et al. 2003). On GBB, the waters with the highest residence time and salinities, i.e. near the bank center in the lee of Andros Island, have had a higher net loss of $CaCO_3$ (Cloud 1962; Broecker and Takahashi, 1966). Cloud (1962) estimated a loss of 50 mg $CaCO_3$/L from the bank edge to the inner bank. As a result of this loss, the inner bank waters have lower levels of $CaCO_3$ supersaturation and lower rates of $CaCO_3$ removal (Smith 1940; Morse et al. 1984). $CaCO_3$ removal rates decrease from 60 mg $CaCO_3$/cm^2/year in 50 day old water to 15 mg $CaCO_3$/cm^2/year in 200 day old water (Broecker and Takahashi 1966). Inner bank waters also show a decline in total inorganic CO_2 and specific alkalinity with increasing salinity as the oldest waters have had the most $CaCO_3$ removed (Broecker and Takahashi 1966; Morse et al. 1984; Morse et al. 2003).

2.3.2   *Inorganic Precipitation and Whitings*

Mud sediment composed of aragonite needles is abundant on the Bahama banks, particularly on GBB. The debate over the inorganic versus organic origin of this mud is not resolved. Neumann and Land (1975) estimate that calcareous algae could produce 1.5–3 times as much mud as has accumulated on the bank top. However, two lines of evidence suggest the mud does not have only an algal source. First, the strontium content of the mud is different from that derived from codiacean algae and is similar to other inorganically precipitated sediment (Milliman et al. 1993). Second, algally-produced aragonite needles are morphologically different compared to GBB mud aragonite (Macintyre and Reid 1992). Aragonite precipitation on resuspended sediment may be the primary source of aragonite mud (Morse et al. 2003). Milliman et al. (1993) estimated using Sr concentrations of sediment that chemically precipitated aragonite contributes 55–78% of the clay-sized aragonite.

A phenomenon relevant to the origin of aragonite mud is whitings, also known as "fish muds" by local fisherman. Whitings are white clouds of suspended fine-grained carbonate mud that drift across shallow platforms. Bahamian whitings are generally elongate and can exist for days to weeks (Shinn et al. 1989). They tend to be most abundant northwest and west of Andros Island concentrating around 25°N and 78°50'W, occupying around 1.24% of the area on GBB and occurring over the mud and pellet-mud facies on GBB (Table 2.10; Shinn et al. 1989; Robbins et al. 1997; Morse et al. 2003). Typical whiting material consists of 86% aragonite, 12% high-Mg calcite, and 2% low-Mg calcite (Shinn et al. 1989). Space shuttle and satellite image analysis by Robbins et al. (1997) found the highest frequency of whitings in the months of April and October with a frequency index of 72 to 75 km^2/day and the lowest in summer and winter months ranging from 15 to 30 km^2/day.

Table 2.10 Monthly variation of whiting frequency calculated and normalized from photographs taken during the NASA manned spacecraft program from 1965 to 1993 (From Robbins et al. 1997)

Whitings on Great Bahama Bank	
Month	Frequency index (km^2/day)
January	29
February	n/a
March	28
April	75
May	28
June	32
July	33
August	31
September	16
October	72
November	50
December	52

The origin of whitings is controversial and three main hypotheses have been proposed: (1) sediment stir-up by biologic activity, (2) direct chemical precipitation on resuspended sediment, and (3) biologically-induced precipitation. Some geochemical measurements have suggested material in whitings may be stirred up bottom sediment. To determine whether whiting sediment was in equilibrium with the surrounding water Shinn et al. (1989) estimated from stable isotopes that 90% of whiting $CaCO_3$ could be resuspended sediment. Using the partial pressure of CO_2 as a measure of potential precipitation and C^{14} age estimates, Broecker and Takahashi (1966) estimated that only 25% of the whiting material could be derived from precipitation within the whiting itself and the average age of whiting material was ~200 year. Similarly, Broecker et al. (2000) demonstrated that whiting sediment is not in equilibrium with the ^{14}C content of the water, but is similar to that of older bottom sediment indicating the whiting material did not precipitate *in situ*. In addition, there is no apparent mechanism for resuspending sediment, as no fish have been observed associated with whitings (Shinn et al. 1989).

The above hypothesis leaves room for a second possibility in that direct chemical precipitation may occur on resuspended pre-existing carbonate nuclei which is supported by geochemical evidence of ongoing precipitation of $CaCO_3$ in bank waters (Shinn et al. 1989; Morse and Mackenzie 1990). Morse et al. (1984) observed that whitings form outside high-salinity waters where precipitation of calcium carbonate is 1.5 times higher than within the high-salinity waters. The whiting area is also the site of higher rates of specific alkalinity removal in comparison to the highest salinity waters (Fig. 2.9; Morse et al. 1984). In addition, due to high precipitation rates all over the bank, bank waters are less supersaturated with respect to calcium carbonate than adjacent Gulf Stream and bank-edge waters (Morse and He 1993). Yet, no significant differences in pH, total alkalinity and total CO_2 between water inside and adjacent to whitings has been observed (Broecker and Takahashi 1966; Morse et al. 1984).

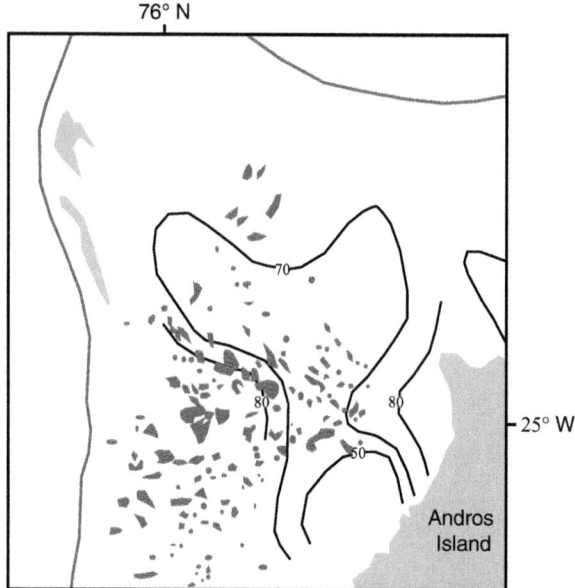

Fig. 2.9 Precipitation rate of calcium carbonate in µmols of $CaCO_3$ per kg of seawater and whiting distribution on northern Great Bahama Bank based on 1975 cumulative satellite photographs compiled by Harris (unpublished) (From Morse et al. 1984)

Morse et al. (2003) attempted to resolve this paradox using calculated reaction kinetics between calcite, bank waters and resuspended sediment. They concluded whitings represent precipitation on resuspended sediments, a major portion of which, however, can also take place outside the visually dramatic whitings. They estimate that single aragonite needles may be resuspended in a whiting about 80 times over a 30-year period during which they experience repeated overgrowth.

A third hypothesis suggested by Robbins and Blackwelder (1992) and Thompson (2000) is that whitings are biologically induced precipitation by algal, picoplankton or bacterial blooms. These authors found larger amounts of organic components within whitings than outside whitings. Yet, bank waters are nutrient poor and no detectable difference in chlorophyll levels inside and outside of whitings, indicating algal productivity, have been observed (Morse et al. 2003).

2.4 Fauna and Flora

This section focuses on basic parameters controlling eight major groups of organisms that contribute directly or indirectly to carbonate sediment production. These groups include: (1) corals, (2) red algae, (3) green algae, (4) foraminifers, (5) sponges, (6) molluscs (including bivalves and gastropods), (7) echinoderms, and (8) microbial mats. For each of these groups an attempt is made to characterize their growth rate, their contribution to calcareous sediment production and destruction, and the factors controlling those parameters. Temperature, salinity, light (depth/turbidity), nutrient level (nitrate and phosphate), and water motion all play a significant role in the distribution of the various organisms. An additional important source of fine-grained sediment is bioerosion that destroys framework-building skeletons (compare Hassan 1998).

2.4.1 Hermatypic Corals

Compared to the size of Great Bahama Bank, and the amount of apparently suitable habitat, hermatypic corals and reefs are not developed to their full, theoretical potential. Even though volumetrically not among the most important groups, zoox-anthellate hermatypic corals are the best-studied carbonate-producing organisms in the Bahamas. Typical Caribbean reef types that occur in the Bahamas are: (1) shelf-edge reefs, framework buildups situated close to the shelf edge, typically between 15 and 30 m (2) mid-shelf reefs, more or less well-developed framework buildups in around 10 m depth (3) fringing reefs, i.e. frameworks close to shore, usually *Acropora palmata* dominated, between the water level and about 5 m depth (4) barrier reefs, i.e. similar framework type as fringing reefs, but forming a lagoon, usually >1 km to coastline (5) patch reefs, i.e. small, circular or elongate frameworks on the bank, and (6) coral covered hardgrounds, i.e. more or less dense coral growth without framework development (Riegl and Piller 2003). Chiappone et al. (1997a) cite four morphological categories: (1) patch reefs, (2) channel reefs, (3) windward hard-bottom reefs, and (4) fringing reefs (Table 2.11). Relatively few fringing reefs and barrier reefs occur along edges of platforms, predominantly along windward margins and in front of islands where they are protected from bank waters (Ginsburg and Shinn 1964, 1994) e.g., on the eastern side of GBB, the northeastern side of LBB, and the western margin of TOTO. Well-developed fringing reefs are mainly found in the southern Bahamas (Mayaguana, Inagua) and Turks and Caicos; Riegl et al. 2003). Very few patch reefs in relation to the amount of available habitat are scattered within the platform interiors and in the lee of fringing reefs. Hardbottom reefs, i.e., hard-bottom areas that are scattered with coral cover, are found in windward shelf areas (Chiappone et al. 1997a). Linear channel reefs occur in tidal channels primarily in the Exuma Cays.

The major reef building corals are dominated by 18 species. Most abundant are the shallow water species *Acropora palmata* (with a maximum at about 3 m water depth) that are now increasingly rare at several sites due to significant, disease-related mortality (Kramer 2003) and the deeper water species *Montastrea annularis* (with a maximum at about 10 m water depth) (Linton et al. 2002). *A. palmata* has since been significantly reduced in abundance and is missing in much of its former range (Kramer et al. 2003). Other abundant species include *Acropora cervicornis*, *Diploria sp.*, *Porites porites, Porites astreoides*, *Agaricia agaricites*, *Montastrea*

Table 2.11 Results of a large-scale survey of reef corals and reef morphology of the Exuma Cays (Summarized from Chiappone et al. (1997a, b)

	Patch reef	Channel reef	Windward hard-bottom reef	Fringing reef
Morphology	Circular	Ridges oriented along strike of channel; spur and groove	Low-relief, hard-bottom areas with occasional patches of sand	1. Fringing platform, along platform edge with low relief 2. Reef ridge, parallel to shore, 3–5 m relief 3. Deep spur and groove, elongate spurs perpendicular to shore, 6–7 m relief
Dimensions	1. Leeward, nearshore: 20–30 m 2. Windward bank: <200–1,000 m²	<10,000–30,000 m²	–	1. Reef ridge, 40 × 100's m 2. Deep spur and groove, spurs 100 m in length
Water depth	1–6 m	2–11 m	3–13 m	1. Fringing platform, 1–7 m 2. Reef ridge, 1–11 m 3. Deep spur and groove, 10–20 m
Coral species total	27	38	33	38
Common coral species	M. annularis, P. astreoides, P. porites, Dichocoenia stokesi, Favia fragum	P. porites, Siderastrea siderea, M. annularis, A. agaricites, Millepora alcicornis, M. complanata, P. astreoides	A. agaricites, M. annularis, P. astreoides, S. sidereal, Diploria clivosa, P. porites	M. annularis, P. astreoides, P. porites, S. siderea
Coral cover	9–50%	2–44%	1–21%	8–37%
Occurrence	Leeward and lagoonal environments	Exuma Island tidal channels	Windward margins	Along edges of platforms, nearshore, along forereef escarpment

cavernosa, and *Siderastrea siderea* (Smith 1948; Tucker and Wright 1990; Chiappone et al. 1996, 1997a). On the fringing reef along the windward side of Andros Island, the so-called Andros Barrier Reef, corals exhibit a clear zonation according to energy. The reef crest is dominated by *A. palmata*, seaward to 3 m depth *A. palmata*, *A. cervicornis* and *Diploria* dominate, and below 3 m *Montastrea, A. palmata* and *P. porites* dominate (Bathurst 1975; Kramer et al. 2003).

The latitudinal distribution of coral species in the Bahamas is homogenous and does not express a latitudinal zonation (Chiappone et al. 1997b). Live coral cover on coralgal reefs is highly variable ranging from 4% near San Salvador to over 30% on the Turks and Caicos Banks (Linton et al. 2002).

Corals are warm-water oligotrophic organisms adapted to normal marine salinities. The mean maximum summer temperature of the Bahamas, 27°C, corresponds to the optimum temperature for coral growth (Table 2.12). Due to the stable surface-current regime, water temperatures in the Bahamas rarely exceed the upper temperature limit of coral growth (between 30°C and 34°C). Corals show optimum growth rates in normal open-ocean salinities of 35–36 ppt, but can tolerate a range of 25–40 ppt (Table 2.12; Wood 1999). Most areas on the Bahamas platform do not exceed this upper salinity limit (Fig. 2.7). Reasons why the platform areas of the Bahamas are relatively poorly settled by corals, are recurrent cold and warm anomalies. In winter, due to their position close to the North American continent, the Bahamas are under the influence of northerly cold-fronts. These cause rapid cooling of the shallow bank waters as well as increase in salinity, due to evaporative moisture loss to the dry, cold air (Roberts et al. 1982). In summer, during doldrums conditions (which can be mediated by ENSO) bank waters heat up and also become hyperpycnal due to moisture loss, this time to the hot atmosphere. Both in summer and winter, the cooled or heated, but in both cases hyperpycnal waters, form density-cascades flowing over the bank margin (Wilson and Roberts 1992, 1995). Thus, not only reef growth on the banks, but also on their edges is negatively influenced (Riegl and Piller 2003), which is the reason why the banks are relatively poor in coral buildups and well-developed fringing-, barrier-, and shelf-edge reefs only occur where the margins are protected from bank waters by islands (Ginsburg and Shinn 1994).

Coral reefs are among the most active producers of calcium carbonate in the present-day world. Vertical accretion rates of Holocene reefs are on the order of 10 mm/year in the Caribbean (Macintyre et al. 1977). Calcification rates vary for different coral species. In the Caribbean the growth rate of the massive *Montastrea annularis* is around 0.6 mm/year, whereas *Acropora cervicornis* exceeds 100 mm/year (Enos 1991; for growth rates of other species see Table 2.13). Coral-reef growth is preferentially initiated on topographic highs or open settings where wave energy and circulation are high. High water energy equates better oxygen availabiliy to the corals and removes sediment and waste products that can damage the corals and decrease light penetration. A steady sedimentation rate exceeding 0.2 g/cm^2/day is lethal for certain coral species (Stafford-Smith 1992). The minimum water velocity that prevents sediment deposition onto corals is about 0.6 m/s. The upper limit of water motion for sustained coral growth is 5–7 m/s, beyond which algal pavements or

Table 2.12 Relationships between growth rates of corals, red algae, and green algae and chemical and physical parameters

	Corals	Red algae	Green algae
Growth rate	0.85–150 mm/year (Enos 1991)	0.3–22 mm/year vertical growth, 10 to 40 mm/ky rhodolith growth (Littler et al. 1991; Payri 1997)	0.41–0.46 m/year (Freile et al. 1995)
Contribution in $CaCO_3$	–	2–3,600 g/m^2/year (Payri 1997)	0.06–2.4 kg/m^2/year (*Halimeda*) (Freile et al. 1995)
Light	Threshold depth = 1% of light irradiance (Chalker et al. 1988)		Threshold depth = 0.05% of light irradiance (Hillis 1997)
Depth	0–75 m (Liddell et al. 1997)	>290 m (Littler et al. 1991)	At least 0–80 m, possibly 150 m (Freile et al. 1995)
Salinity	25–50 ppt (Wilkinson and Buddemeier 1994)		
Temperature	20 to 32–34°C (Jokiel and Coles 1977)		

Table 2.13 Coral growth rates for the Caribbean, Florida and the Bahamas (From Enos 1991)

Individual coral growth rates (mm/y)	
Massive corals	4.0
Montastrea annularis (Caribbean)	3.0–12.0
Acropora palmata (Caribbean)	100–120
Porites porites (Caribbean)	21–36
Porites astreoides (Florida, Bahamas)	5.7–13

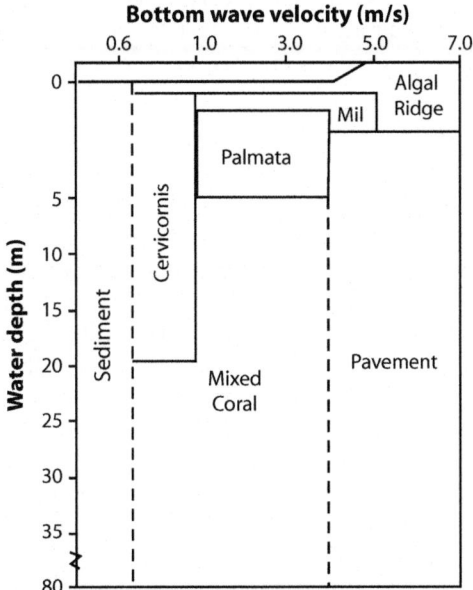

Fig. 2.10 Relationship between water depth and bottom water velocity and their influence on reef composition. Mil = *Millepora* (From Graus et al. 1984)

ridges begin to dominate (Fig. 2.10; Graus et al. 1984). Although the calcification rate decreases with depth, rates of injury to the colonies and bioerosion are much higher in shallow water.

2.4.2   Red Algae

Red algae (Rhodophyta) are important primary sediment producers, sediment binders, and reef builders in carbonate systems in general (Steneck and Testa 1997) and in the Bahamas in particular (Woelkerling 1976). The growth rate of red algae ranges from 0.3 to 22 mm/year, and $CaCO_3$ production reaches up to 3,600 g/m²/year (Table 2.12; Payri 1997). Because red algae are phototrophic, their growth rates depend on light. Red algae cover only a few percent of seafloor in the Bahamas to a depth of 50 m (Littler et al. 1991). On GBB, red algae contribute up to 2% to the sediment in skeletal grainstone facies (Purdy 1963b). Below 50 m red algae become locally dominant and are found to 290 m water depth reflecting their oligophotic nature (Littler et al. 1991). In the tropical Atlantic, most red algae belong to the genera *Neogoniolithon*, *Porolithon* and *Lithophyllum* (Adey and Macintyre 1973). Red algae favor high water energy and nutrient-rich waters (Table 2.12; Littler and Littler 1984). Under laboratory conditions, growth of red algae has been promoted by either, high temperatures and normal UV conditions, or low temperatures and high UV conditions. For the same experimental conditions, corals experience strong bleaching (expulsion of their algal symbionts) (Reaka-Kulda et al. 1994).

2.4.3   Green Algae

Green algae (Chlorophyta) are among the most important groups of primary calcium carbonate producers in modern coral reefs, lagoons, fore reefs and on slopes (e.g., Neumann and Land 1975). Udoteaceae, including the genera *Halimeda*, *Penicillus*, and *Udotea*, are the most common green algae in the Bahamas. On GBB, *Halimeda* remains compose up to 10% of the sediment in the skeletal grainstone facies and around 1–3% elsewhere on the bank (Purdy 1963b). The light limit for *Halimeda* is 0.05% of surface irradiance (Hillis 1997; Freile et al. 1995). On the leeward side of GBB, *Halimeda* occurs to 150 m water depth but shows highest abundance between 20 and 50 m where it represents 30–65% of the sediment cover (Freile et al. 1995). Similarly, on the San Salvador platform, *Halimeda* reaches 150 m depth (Blair and Norris 1988). *Halimeda* contributes 2.4 $g/m^2/year$ total to Bahamian sediments (Table 2.12; Freile et al. 1995).

   Whereas *Halimeda* produces predominantly sand fraction sediment (e.g., Hoskin et al. 1986), *Penicillus* is a major producer of lime mud. In the inner reef tract of Florida, *Penicillus* reaches a production rate of 25 $g/m^2/year$, whereas in Florida Bay values of only 3 $g/m^2/year$ are common (Stockman et al. 1967). In Bahamian lagoons, the algal mud production from the green algae *Penicillus*, *Halimeda*, *Ripocephalus*, and *Udotea* amounts to a total of 90 $g/m^2/year$ which exceeds the observed sediment accumulation rates in these lagoons (Neumann and Land 1975). The difference is balanced by off-bank transport (see Section 2.5.2.1). Large amounts of algal debris are exported off-bank and become an important component of mud- and sand-size sediment on the slope and in the deep sea (Neumann and Land 1975; Freile et al. 1995).

2.4.4   Foraminifers

Foraminifers are a ubiquitous group in the carbonate system of the Bahamas. About 50 species make up the bulk of the benthic foraminifers present in the Bahamas (Rose and Lidz 1977). Foraminifers belonging to the families Soritidae and

Miliolidae compose 60–80% of the total foraminifer fauna (Rose and Lidz 1977). Other important foraminifer families include Peneroplidae and Homotrematidae, in particular the genus *Homotrema*. Although the ecological distribution of Homotrematidae is well known for the modern environment (Rose and Lidz 1977), few studies have looked at their contribution as sediment producers. On GBB, they make up 1–7% of the sediment, whereas peneroplids amount to 0.5–3.5% (Purdy 1963b). At southern GBB near Ragged Island, 2–5% of the sediment consists of benthic foraminifers with a higher percentage in vegetated and reefal areas (Illing 1952). Whereas no estimates for the Bahamas are available, benthic foraminifer production rates of southern Florida are estimated at 60 g of $CaCO_3/m^2/year$ corresponding to 4.9% of outer margin sands and 6.6% of inner margin sand (Enos and Perkins 1977; Hallock et al. 1986). Planktonic foraminifers amount to 20% of the sediment at periplatform settings of Northwest Providence Channel in the Bahamas (Pilskaln et al. 1989).

2.4.5 Sponges

Sponges are present in low numbers to depths exceeding 400 m (Maldonado and Young 1996). While zooxanthellate corals decrease in abundance and diversity below 50 m water depth, at the same depth sponge abundance and diversity increases (Liddell et al. 1997). Sponges are conspicuous as primary colonizers of the deeper slope to more than 500 m water depth (Maldonado and Young 1996). In the Caribbean, sponges of the class Sclerospongiae are the most active calcium carbonate framework constructors below depths of 70 m (e.g., Lang 1974).

In the shallow-water environment, some endolithic sponge species, especially of the family Clionidae (class Demospongiae) are important bioeroders especially in reef systems. The product of sponge erosion ("*Cliona* chips") can represent up to 30% of the total sediment in lagoonal carbonate environments in general (Fütterer 1974). In a subtidal setting on Grand Cayman Island, the species *Cliona caribbaea* makes up 5% of the total bottom areal cover between 1 and 10 m water depth and is responsible for destruction of the coral carbonate framework, producing 8 $kg/m^2/$ year of sediment (Acker and Risk 1985).

2.4.6 Molluscs and Echinoderms

Molluscs and echinoderms are moderate contributors to calcium carbonate sediment production but are important bioeroders (see Section 2.4.8; Moore 1972). Abundance in the sediment varies strongly. In nearshore areas of Smuggler's Cove and Rod Bay, St. Croix, they make up around 1.2% of the sediment fraction; similarly, 1.2% of the >2 mm fraction near Andros Island are composed of molluscs and echinoderms (Greenstein and Meyer 1990; Greenstein 1993). In the platform interior near Ragged

Table 2.14 Most common occurrence of sea urchin and sand dollar species and their density in a high energy lagoon, Fernandez Bay, San Salvador Island (From Greenstein 1993) and the intertidal zone of Black Rock, Little Bahama Bank (Hoskin and Reed 1985)

	Species	Shore zone	Distance from shore (m)	Density (individuals/m²)
Fernandez Bay, San Salvador Island	Sea urchins			
	All sea urchins	Reef/near-reef	–	60
	Echinometra lucunter	Nearshore	0–18	35.6
	Tripneustes ventricosus	Nearshore	0–18	5.4
	Diadema antillarum	Offshore	28–138	0.3
	Eucidaris tribuloides	Offshore	28–138	1.2
	Sand dollars			
	Mellita quinquesperforata	Farshore	46–156	–
	Leodia sexiesperforata	Farshore	46–156	–
Black Rock, LBB	Sea urchins			
	Echinometra lucunter	Intertidal to 7 m water depth		37

Island and southwest of New Providence Island, they make up 6% of the sediment fraction and 18% on the bank edge near Ragged Island (Illing 1954). A study of their sensitivity to light and temperature shows that both groups are tolerant to a temperature range of 27–31°C associated with a 20% increase in incoming ambient UV light (Reaka-Kulda et al. 1994). Echinoid distribution is partly dependent on substrate. For example, in the high-energy lagoon of Fernandez Bay, San Salvador Island, regular echinoid species (sea urchins) prefer hard rubbly substrates, whereas irregular species (sand dollars) prefer soft sand in which they burrow (Greenstein 1993). The density of living sea urchins is strongly dependent on the distance from shore and the environment (Table 2.14).

2.4.7 Microbial Mats

Microbial mats containing cyanobacteria are found on sediment surfaces throughout the Bahamas. Microbial activity is thought to be responsible for stabilizing, altering and even creating sediment grains (i.e. ooids, grapestones; Fabricius 1977). Two types of microbial mats are distinguished: (1) the lithified, mostly subtidal stromatolites made up of sandy sediments (that are focused on below); (2) the *Scytonema-Schizothrix* mats of supratidal to subtidal mud (see, e.g. Monty 1976; Hardie 1977; Shinn et al. 1969; Neumann et al. 1970) that are effective sediment binders, but are not lithified.

Microbial deposits can take on unlayered thrombolitic and massive forms or can be layered to form stromatolites (Monty 1976). Bahamian microbial mats contain a bacterial bioassemblage dominated by the cyanobacteria *Schizothrix* and *Scytonema* (Reid et al. 2000). The bioassemblage produces biofilms, which trap

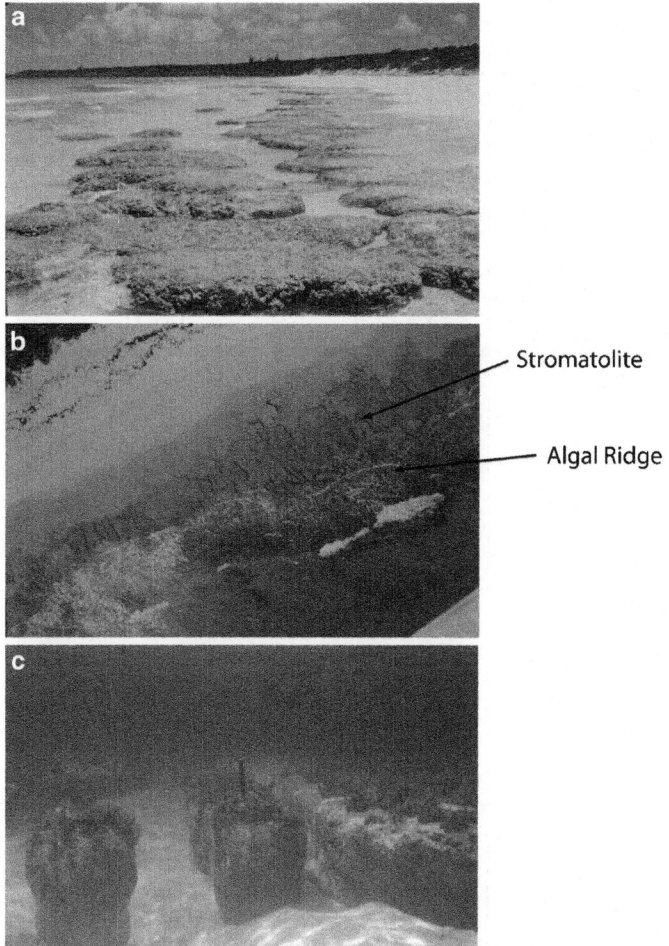

Fig. 2.11 (a) Intertidal thrombolites; Highborne Cay, Exuma Cays; (b) Aerial view of intertidal fringing reef consisting of a red algal ridge and backreef intertidal stromatolites, Stocking Island, Exuma Cays; (c) Subtidal columnar stromatolites around 5 m in height. Little Darby Island, Exuma Cays (Photographs courtesy of R.P. Reid, RSMAS, University of Miami)

and bind sediment and are the site of carbonate precipitation leading to stromatolite lithification (Macintyre et al. 2000; Reid et al. 2000). Microbial mats can adopt various non-planar morphologies to form attached buildups, such as stromatolites, and unattached nodules, such as oncolites (Fig. 2.11; Table 2.15; Monty 1976).

Before the discovery of modern stromatolites in the Bahamas, stromatolites were thought to represent anactualistic, mostly Precambrian buildups that only survived in extreme environments such as hypersaline lagoons; e.g. Shark Bay, Australia (Logan 1961). Modern stromatolites living in normal-saline marine waters were first described in the Bahamas by Dravis (1982) (Schooner Cays in Eleuthera Bight).

Table 2.15 The three types of environments where stromatolites have been described in the Bahamas (From Reid et al. 1995)

Stromatolite environment	Energy	Depth	Height	Morphology	Location
Subtidal in confined tidal pass	High energy, confined currents, up to 1.5 m/s	<10 m	0.5–2.5 m	Asymmetric columns, form walls perpendicular to tidal flow; linear ridges form perpendicular and parallel to tidal flow	Channels in the vicinity of Lee Stocking Island, Bell Island, and Wardrick Wells
Subtidal in sandy embayment	Unknown	0.5–2 m	0.5 m	Pillars; vase-shaped columns for ridges perpendicular to beach	Little Darby Island
Subtidal along open sandy beach	High wave energy, up to 1.5 m/s	< 1.5 m	Several cm to 1 m	Columnar; tabular coalesce to form columns	Stocking Island; Highborne Cay

In the Bahamas, stromatolites occur where high stress from high sediment flux in high-energy environments inhibits the colonization of competing eukaryotic algae (Reid et al. 1995). Even though stromatolites are geologically important sediment binders and reef builders in the modern environment of the Bahamas, they are an insignificant part of the total carbonate system and their contribution to the sediment budget is minor.

Modern stromatolites occur in depths ranging from 0 to 10 m in the intertidal and subtidal zones (Fig. 2.10). They are found along the margin of Exuma Sound in areas of high current energy and are usually associated with migrating sand dunes (Dravis 1982; Dill et al. 1986). They form a range of large-scale morphologies including domes, columns, and mounds up to 2 m in height. Growth rates of stromatolite heads in hypersaline ponds in San Salvador were measured at 0.16 mm/year (Paull et al. 1992). Stromatolites along the east coast of Stocking Island occur in the back-reef and reef-flat zone of an algal fringing reef and reach a thickness of up to 1 m (Macintyre et al. 1996).

Bahamian stromatolites are built primarily by prokaryotic communities that alternate between filamentous cyanobacteria, a bacterial biofilm community, and endolithic cyanobacteria (Reid et al. 2000). Stromatolite surfaces that are not actively growing may be colonized by a variety of eukaryotic organisms, such as macroalgae, diatoms, and sponges. Table 2.15 summarizes the types of stromatolites found in the Bahamas and the environments in which they form.

2.4.8   Bioerosion

Bioerosion is the most important factor in reef destruction and rocky shore erosion on the Bahamas and is a major source of loose sediment in intertidal and shallow subtidal environments (Hoskin et al. 1986). Bioerosion rates in carbonate environments are on the same order as the rate of reefal growth (Hein and Risk 1975; compare Hassan 1998).

In general, the most important boring organisms in today's reefs are sponges, specifically those of the family Clionidae (see Section 2.4.5). For example in the Florida reef tract, Hein and Risk (1975) attribute the bulk of destruction on coral heads to boring sponges. Other volumetrically important bioeroders are parrotfish (Scaridae). Bioerosion rates by Scaridae are highest for dead coral substrate inhabited by endolithic algae and for Bonaire reach rates of up to 12,000 g/m^2/year for the shallow reef and 1,900 g/m^2/year for the reef slope (Bruggemann et al. 1996).

Important bioeroders in the Bahamas include sponges, echinoderms, bivalves, grazing fish, gastropods, polychaete worms, barnacles, and microorganisms such as bacteria, fungi and algae (Hutchings 1986). In the shallow subtidal zone of LBB, Hoskin et al. (1986) measured bioerosion rates by rock infauna excluding sea urchins of 1,830 g/m^2/year (see Table 2.16 for bioerosion rates).

Microbioerosion is about one order of magnitude lower than macrobioerosion. Experimental data of Hoskin et al. (1986) show that microbioerosion is most

Table 2.16 Example of erosion rates in Black Rock, Little Bahama Bank (From Hoskin et al. 1986)

Environment	Process	Average erosion rate (g/m²/year)
Intertidal	Bioerosion by microborers (unprotected limestone cubes)	259
	Bioerosion by the chiton *Acanthopleura granulata* (calculated from fecal pellet production)	104
	Physical abrasion (tethered, unprotected limestone plates)	55
	Chemical solution (filter-protected limestone cubes)	13
Shallow subtidal	Bioerosion by *Echinometra lucunter* (sea urchin)	6,670
	Bioerosion by all other rock infauna	1,830

Table 2.17 Bioerosion rates of microborers in various environments near Lee Stocking Island, Bahamas (From Vogel et al. 2000)

Environment	Bioerosion rates (g/m²/year)	Microborers
Lagoonal patch reef	80–300	Red algae, green algae, cyanobacteria, heterotrophic bacteria, hyellids
Sea grass meadow	100–420	Green algae, cyanobacteria, heterotrophic bacteria, hyellids
Stromatolite reef	120–510	Green algae, cyanobacteria, heterotrophic bacteria, hyellids
A. palmata reef	200–280	Red algae, green algae, cyanobacteria, heterotrophic bacteria, hyellids
Shelf edge reef	90–160	Red algae, green algae, cyanobacteria, heterotrophic bacteria, hyellids
Slope, 30–130 m depth	10–20	Green algae, cyanobacteria, heterotrophic bacteria

intense in the intertidal zone (259 g/m²/y) and decreases with increasing depth. Vogel et al. (2000) found that rates of microbioerosion are lower in shelf edge and slope environments compared to other shallow water environments (Table 2.17).

Chemical and physical erosion play a secondary role in eroding the submarine substrate (Table 2.16). Chemical dissolution rates reported for the Bahamas are fairly constant with depth and are about 13 g/m²/year (Hoskin et al. 1986). Physical abrasion depends on water motion and for an experiment above the intertidal zone averages at 55 g/m²/year (Hoskin et al. 1986).

2.5 Geometries of Facies Belts

Relatively simple facies relationships, so called "level-bottom" or "facies prosaic" geometries, are typical for the Bahamas (Fig. 2.12; Enos 1974, 1983; Enos and Perkins 1977). This is in contrast to the compartmentalized "facies mosaics" of, e.g., Florida Bay or the Belize Lagoon. Bahamian sedimentary facies distribution reflects the control by antecedent topography and energy flux. For example, facies

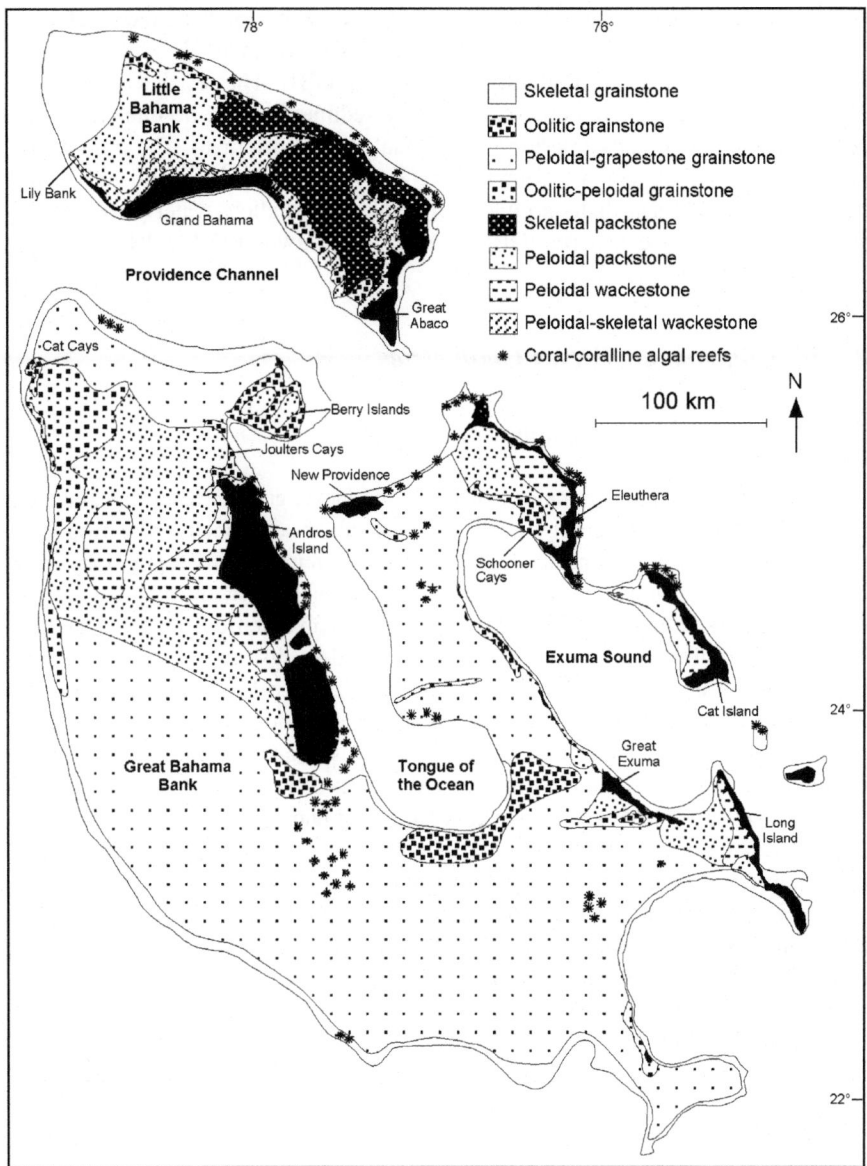

Fig. 2.12 Sedimentary facies distribution on GBB and LBB, and locations mentioned in text (Modified after Enos 1974; with data from Purdy 1963b; Reijmers et al. 2008)

distribution is distinct on the windward and leeward sides of the margins (Hine et al. 1981b). More extensive grainstone belts are present on the higher energy windward side, whereas packstone and wackestone belts dominate the lee side, especially where Pleistocene bedrock islands are present.

The following dominant depositional environments are described below in detail: Sand bodies are the most variable facies and dominate along the platform margins, but also blanket the platform interior on GBB. The restricted interior of northwestern GBB mostly consists of muddy sediment and is dominated by two distinct facies, the muddy interior and the tidal-flat environment. The high-energy systems of coralgal reef are found along bank margins. Slope and deep-water deposits occur as linear belts around each bank. The shallow-water facies dimensions of several transects across GBB and LBB are shown in Table 2.18.

2.5.1 Controls on Facies Geometries

2.5.1.1 Antecedent Topography

Antecedent topography is a major control on the facies geometries of the Bahamas by providing large-scale energy barriers that result in a distinctive windward and leeward facies configuration. In addition, variations in underlying topography influence localized deposition. Antecedent topography of the present-day depositional system of the Bahamas is composed of Pleistocene eolianite ridge and beach deposits, partially lithified Holocene eolian and beach deposits, and modern sediments (Carew and Mylroie 1995). The Pleistocene deposits provide a topographic base for Holocene deposition of eolian dunes, intertidal beach and tidal-flat sediments, and reefs (Fig. 2.13). On a large scale, the islands and associated subtidal ridges are barriers to the energy flux onto the Bahamian platforms and create a depositional windward and leeward side (Fig. 2.11). In general, higher energy deposits such as skeletal grainstones and oolitic grainstones are found on the windward sides of islands and elsewhere along the platform edges. For example, the Joulters ooid shoal of northern Andros Island is a mobile sand fringe that follows the windward margin of GBB, along a highly agitated environment (Harris 1979). Lower energy deposits including pelletal wackestones and packstones are found in the lee of islands (Fig. 2.11).

Variations in Pleistocene topography also control sedimentation on a localized level. In the Joulters Cays, areas of thick deposits of pelletal wackestones are found in small depressions, whereas areas of higher relief are associated with high-energy ooid grainstones (Harris 1979). In addition, preexisting topography controls sediment thickness in the Joulters Cays area. Accumulated sediment is thickest in areas of low preexisting topography and thinnest where topography was higher. Any paleo-high, such as ridges and relict reefs, can act as an energy barrier and a sediment trap (Table 2.19; Hine et al. 1981b). Hardgrounds may provide a substrate for modern patch reef growth (Palmer 1979). Localized influence on deposition is least in the lee of islands where underlying topography has been buried by high sedimentation rates of the low energy area such as in the tidal flat environment (Table 2.19; Hine et al. 1981b).

During the earliest part of the Holocene as sea-level began to rise, variations in the underlying Pleistocene surface may have had a larger control on facies development

Table 2.18 Lateral extension of facies elements in sections across GBB and LBB, measured on the map by Enos (1974) (for location of the sections see Fig. 2.12). Section A–A' is an example for the influence of an island on the deposition of a leeward muddy area. Section B–B' is an example for an unrimmed part of the platform and high-energy tidal bar belts at the TOTO. Section C–C' is an example for the windward GBB east of the TOTO. Section D–D' is a section through LBB where the island is located at the leeward side, on Providence Channel

Facies (in order; windward to leeward for GBB sections; S–N for LBB section.)	GBB A–A' (km)	GBB B–B' (km)	GBB C–C' (km)	LBB D–D' (km)
Marine sand belt (skeletal, oolitic, and oolitic-pelletal grainstone)	11	2	–	14.5
Interior sand blanket (pelletal grainstone)	13	104	63	–
Muddy interior (pelletal and skeletal wackestone or packstone)	94	–	–	68.5
Island	25.5	–	–	9.5
Tidal bar belt (oolitic grainstone)	–	17	5	–
Marine sand belt (skeletal, oolitic, and oolitic-pelletal grainstone)	7	1.5	2	1

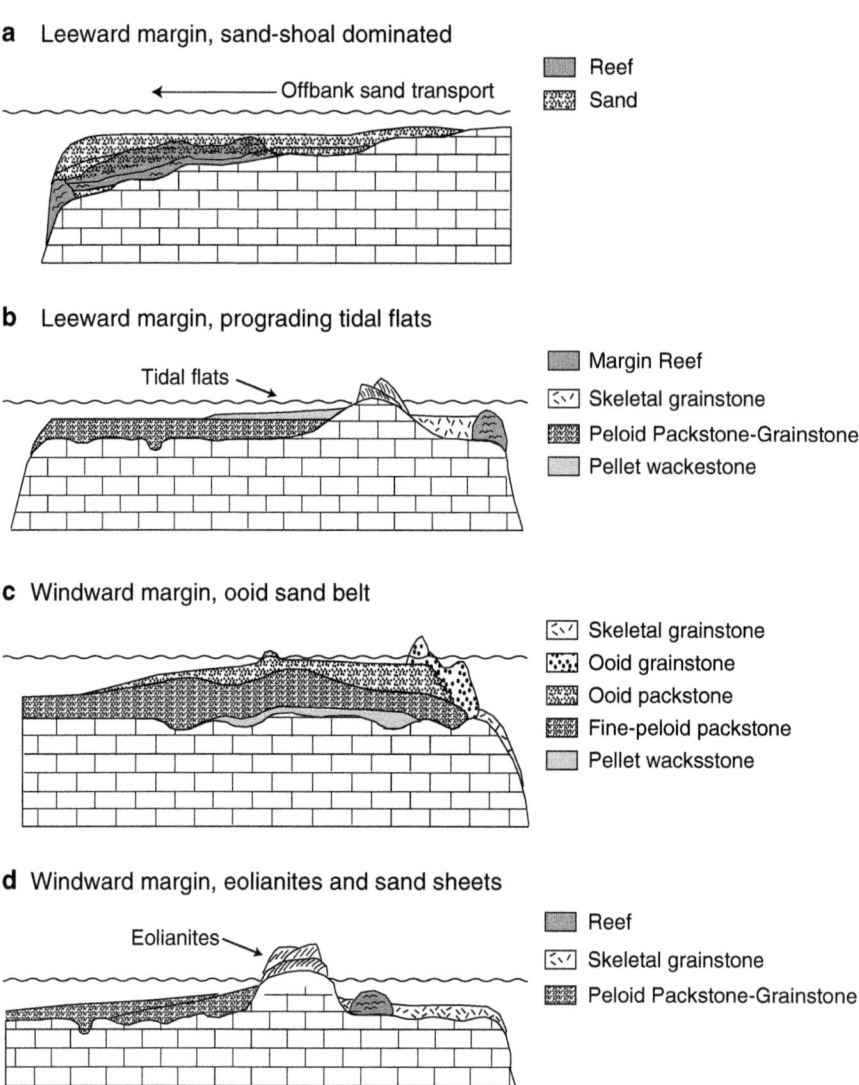

a Leeward margin, sand-shoal dominated

←————— Offbank sand transport

▨ Reef
▨ Sand

b Leeward margin, prograding tidal flats

Tidal flats ↘

▨ Margin Reef
▨ Skeletal grainstone
▨ Peloid Packstone-Grainstone
▨ Pellet wackestone

c Windward margin, ooid sand belt

▨ Skeletal grainstone
▨ Ooid grainstone
▨ Ooid packstone
▨ Fine-peloid packstone
▨ Pellet wacksstone

d Windward margin, eolianites and sand sheets

Eolianites ↘

▨ Reef
▨ Skeletal grainstone
▨ Peloid Packstone-Grainstone

Fig. 2.13 Growth history of carbonate margin settings. (**a**) Leeward margin, LBB. With sea-level rise, margin reefs develop then are buried by sand transported offbank as sea-level continues to rise (From Hine and Neumann, 1977). (**b**) Leeward margin, GBB. (**c**) Windward margin, Joulters Cay, Northern Andros. As the margin is initially flooded, the antecedent topography controls the initial accumulation of pelletal mud and sand. With increasing sea-level, ooid formation begins, a marine sand belt develops along the margin edge and creates its own hydrodynamic regime that controls further shoal development. The Marine sand belt eventually grows bankward and marginal sands grow up to sea-level forming islands that restrict energy moving across the platform (From Harris 1979). (**d**) Windward margin, Exuma Sound

Table 2.19 Influence of antecedent topography on localized facies (From Hine et al. 1981b)

Bank-margin type	Influence of antecedent topography
Windward, open	Accumulation of sands in low-energy zones, which form behind relict reef ridges
Windward, protected	Pleistocene rock ridges provide a substrate for shallow, linear reefs and control their morphology
Leeward, open	None, underlying topography buried by off-bank transport of sand
Leeward, protected	Significant, inter-island gaps locally accelerate tidal currents forming small ooid bodies; in regions of submarine ridges, ooid shoals are more laterally extensive; energy barriers prevent off-bank sand transport allowing reefs to grow
Tidal dominated	Very little along open margins, water flows unrestricted, subaerial highs accumulate sediment

than at the present. Local Holocene features, however, had a larger influence on modern localized deposition. For example, along the high-energy platform edge of the Tongue of the Ocean, a 70 km long sand belt produces an obstacle to wave-energy, behind which lower energy deposits accumulate (Palmer 1979). Rapid fresh-water diagenesis is important in creating Holocene islands that influence sediment deposition (Strasser and Davaud 1986). In addition, modern reefs locally act as sediment dams where sand bodies accumulate.

2.5.1.2 Energy Flux

Sedimentation on the leeward western margin and slope differs in several respects from sedimentation at the windward eastern side. A reduction in energy from east to west is the combined product of damping of waves, shorter fetch, blocking of the ocean swell by windward barriers, and reduction of tidal resonance. On the modern GBB, the orientation of energy flux controls the growth position of reefs and the sediment accumulation rates and composition on the adjacent slope. Wind direction is another major parameter controlling sediment distribution. Wind velocity and direction remain virtually unchanged across the platforms (Hine et al. 1981b). "Leeward" does not necessarily imply that a wind shadow exists, but that a net off-bank energy flux is observed as a result of offshore-directed wind. Net on-bank (windward) versus net off-bank (leeward) energy flux is determined by wind direction in combination with the exposure at any given locality (Table 2.20).

Reefs grow preferentially on the windward sides of the platforms, where on-bank energy flux provides low nutrient, clear water and prevents sedimentation on the reef. In addition, the windward sides of the banks contain more grainstones than the leeward sides (Table 2.20). Large sand belts also develop preferentially on the wind-ward margins. On GBB, the mud content in the sediment increases by approximately 20–30% towards the western (leeward) side. Another effect of this net leeward transport on the bank top is the burial of early Holocene reefs that formed during the initial

Table 2.20 Influence of setting with respect to wave energy flux (W/m or kg·m/s³) on bank margins of the Bahamas (From Hine et al. 1981b)

	Windward open	Windward protected	Leeward open	Leeward protected	Tide dominated
Wave energy flux	High; 5–22*10² W/m	High; 9–22 *10² W/m	Low; 2–4*10² W/m	Lowest: no wave energy into lagoon; 2–4*10² W/m	Variable; 2–22*10² W/m
Energy barrier	Wave energy slowly absorbed over gently rising bank	Islands and small submerged rock ridges	Rapid shallowing of marginal escarpment; no deep-water waves into interior	Islands and slightly submerged rock ridges	Mostly open margin with some rock ridges
Reefs	Mostly patch reefs; few shallow (0–5 m); deep platform-edge reefs (15–20 m)	Shallow: rock-ridge-controlled, deeper: shelf-edge reefs; corals not prolific; algal sands	No active reefs due to burial by off-bank transport of sands	Deeper shelf-edge reefs; coral growth not prolific	Deeper shelf-edge reefs; coral growth not prolific; spur and groove
Sediment transport direction	Onto bank towards south for N-LBB; SW: large relict bed forms. GBB: mostly W or SW-ward	Movement seaward to the deep by gravity processes or by storm return flows	Off bank to the wes and to deep flanks of margin	Offbank to W along eeper margin; bankwards to E at shallow margin between inter-island gaps; N at NW Providence channel	Sand transport onto bank but some transport of fines off bank
Sand bodies	In low-energy topographic lows behind deep reef-rock structures; 1–3 m thick, thinning bankward	Well-developed beaches and nearshore bars in discontinuous pockets; large (up to 10 m thick), sand bodies behind marginal escarpment	Sand carried to bank edge by dominant off bank transport; up to 20 m thick, wedge shaped, thinning bankward; large seaward oriented sand waves	Poorly developed to none; thin sandy talus debris seaward of reefs, bare rock surfaces, small connected and isolated ooid shoals opposite inter-island gaps along shallow margin	Large bankward spillover lobes and active sand waves; accumulation of sand in lees of rocks and islands; little accumulation on deep reef or near marginal escarpment

transgression on the bank (Hine et al. 1981a). Consequently, windward reef margins are vertically aggrading while leeward reefs are intercalating with platform-derived sediment and are prograding basinward (Eberli et al. 2001).

Sediment produced on the platform top is continuously exported to adjacent deep-water areas. Neumann and Land (1975) estimated that green-algal production was 1.5–3 times the mass of aragonite mud and *Halimeda* sand present in the Bight of Abaco – the excess material being subject to export. Fine-grained, bank-derived carbonate sediment is transported as far as 120 km into the adjacent basin, creating a halo of periplatform sediments (i.e. mixtures of bank derived and pelagic sediments) around the isolated banks (Kier and Pilkey 1971; Boardman and Neumann 1984). Along western GBB, a 5–90 m thick Holocene mud package has accumulated (Wilber et al. 1990). This thick package documents the high volume of fine-grained sediment that is exported to the deeper water regions. Wilber et al. (1990) calculated that the amount of Holocene sediment transported and deposited offbank is volumetrically 40–80% as large as the sediment presently found on the bank. Sedimentation rates on the windward slopes are lower than on the leeward slopes. Grammer et al. (1993) observed using submersible observations and high-resolution seismic data around the Tongue of the Ocean that the Holocene sediment thickness is up to 80 m on the leeward slopes and only 20 m on windward slopes, indicating a fourfold higher sedimentation rate on the lee side of Tongue of the Ocean. The leeward side sediment wedge onlaps the slope from 240 m to over 360 m water depth.

Storms and density currents are the two of the possible mechanisms for off-bank transport. Storms, generally coming from the northwest, are capable of transporting sediment off bank and hurricanes can transport sediment in variable directions depending on the direction of movement of the hurricane.

Density currents flowing off the bank-top is another mechanism responsible for carrying sediment or even eroding sediment (Wilson and Roberts 1992; Wilber et al. 1993). Density currents can form during winter cold fronts that lower temperatures of bank-top water below that of surface waters of adjacent deeper water areas. Density currents also form in the summer when evaporation is high and bank-top waters become more saline thus raising water density (Wilson and Roberts 1992; Smith 2001). These density plumes entrain ambient water as they flow off the bank and descend to just below the mixed layer at which point they spread laterally along isopycnals (Hickey et al. 2000). Density plumes in Exuma Sound flow off the shelf via tidal channels and reach water depths of around 45–75 m (Hickey et al. 2000).

Tidal currents affect sediment transport, offbank shedding, island erosion and accretion, and the shape and orientation of sand bodies in the Bahamas. The threshold of motion for moderately sorted carbonate sands with a mean grain diameter of 0.35 mm is 16 cm/s (similar for leeward sand bodies of Hine et al. 1981b). Where peak tidal currents exceed 100 cm/s, parallel-to-flow sand bodies (tidal-bar belts) will form, and where peak tidal currents exceed 50 cm/s but are less than 100 cm/s, normal-to-flow sand bodies, i. e. parallel to bank edge (marine sand belt), will form (Halley et al. 1983).

Increased tidal range at the heads of embayments results in an increase in the size of the sand belts and the individual sand bars, and in an increase in grain size

(Table 2.21). On the Bahamas, the fastest currents produced by the tides (200 cm/s) are found in the Schooner Cays just north of Exuma Sound and at the southern end of TOTO (Table 2.21). In these areas, both sand belts and individual sand bars reach their largest dimensions (Table 2.21). The ooid sand belt at TOTO is characterized by longitudinal dunes that run parallel to the tidal flow and roughly perpendicular to the platform margin. There is clearly a direct relationship between tidal strength and size of sand bodies. In addition, the orientation of the belts is related to the nature of the margin. At the end of narrow straits, e.g., the southern end of the TOTO, the tidal bulge will produce a radial arrangement of tidal bars, while along open margins, sand belts are linear features running parallel to the margin with the internal bars perpendicular to the margin.

2.5.1.3   Holocene Rise in Sea-level

The modern sediment facies geometries are partially controlled by the Holocene sea-level rise and the rate of the rise. As sea-level began to rise in the Holocene, the amount of accommodation space available for carbonate sediment production increased, and as the platforms of the Bahamas flooded the energy regime became highly variable across the platform. Consequently, the Holocene succession of sediment varies widely in different locations on the bank. This variability is illustrated in Fig. 2.13. Figure 2.13a illustrates a leeward open margin at the northern margin of LBB, where reefs formed only to be flooded by offbank transported sediments. Seismic and drilling data document that Holocene sands up to 25 m thick lie on top of early Holocene fringing reefs (Hine et al. 1981a). Figure 2.13b is a cross section of GBB at Andros, showing persistant reef-development at the windward margin and the muddy tidal-flat deposits in the lee of an aeolianite island, consisting of bioturbated, unlayered, pelleted mud and silt (Shinn et al. 1965; Hardie 1977). Only a thin cap of layered sediments exists over the bioturbated sediment in the channel belt.

Figure 2.13c shows the cross-section of the Holocene sandshoal at Joulter's Cays. The ooid shoal grew in three stages: (1) an early bank-flooding stage in which muddy sands of fine peloids and pellets accumulated in protected Pleistocene lows, (2) a shoal forming stage during which ooid production began on bedrock highs where bottom-agitation was focused, (3) the shoal development to the present size and physiography. During this stage ooid sands were transported further bankwards as the bank of active bars broadened. Eventually the exchange of water between the seaward and bankward side of the shoal was increasingly restricted by three mechanisms: (1) widespread sediment build-up approaching sea-level, (2) restriction of tidal channel flow, (3) island formation along the shoal's ocean-facing margin. As a result, the series of bars and channels became an intertidal sandflat where sediments are a mixture of ooid and peloid sands (Harris 1979). Figure 2.13d is the situation at the Exuma Cays margin, with some windward reef development and sandy tidal flats in the lee of the aeolianite islands. The platform interior in this place shows a progressive coarsening-upwards succession from mudstone at 6,700 year bp, grapestone/packstone at 4,460 year bp, grainstone at 4,170 year bp. Since then, lithified grainstone and skeletal debris cover most of the platform interior (Taft et al. 1968).

Table 2.21 Six basic types of Pleistocene Bahamian islands based on platform and oceanic setting as described by Kindler and Hearty (1997)

Island type	Morphology	Energy	Composition	Examples
Windward along narrow, steep shelf	Long, narrow, high	High	Vertically stacked interglacial deposits	Eleuthera, South Abaco, Cat Island, Long Island
Windward along broad, shallow shelf	Long, broad, with offshore cays	High	Numerous accretion ridges	North and Central Abaco, north Eleuthera, Crooked Island, Acklins Island, Turks and Caicos Islands
Semi-protected along a deep-water margin	High topography on windward, low topography on leeward	Moderate	High deposits along windward and beach ridges along leeward	Bimini and Berry Islands, Exuma Islands, Ragged Island, New Providence Island
Leeward protected, away from margin or protected by another island	Large, broad, and low	Low	Accreting beach ridges and leeward mud deposits	Andros Island, Grand Bahama Island
Isolated on small platform	High topography on windward, low topography on leeward	High	Vertical buildup on windward, accretion on leeward	San Salvador Island, Mayaguana and Inagua Islands, Samana, Rum and Plana Cays
Isolated mid-bank remnant	Small, isolated and vertically sided	Low	Remnants of Pleistocene islands being eroded away	Moore's Island, LBB, Rocky Dundas

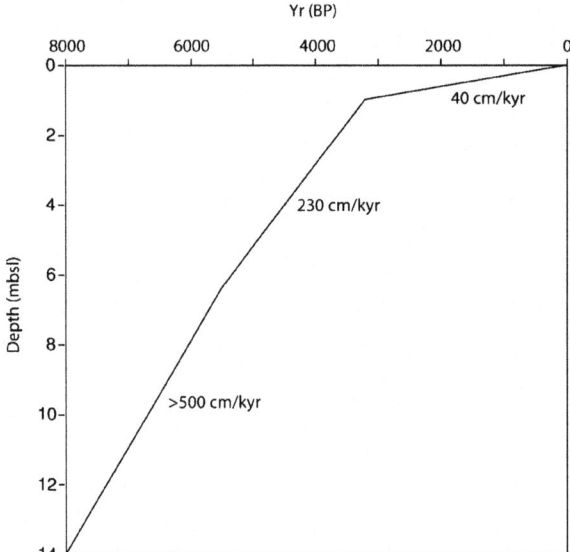

Fig. 2.14 Holocene sea-level curve and rate of rise for the past 8,000 year (BP) compiled by Kindler (1995) from various regional eustatic curves of the Bahamas

2.5.2 Facies Geometries

2.5.2.1 Islands

Islands cover about 11,400 km², which is about 3% of the total area of the Bahamas or about 6% of the shallow-water areas (Meyerhoff and Hatten 1974). Islands are found predominantly on the windward, eastern side of the platforms and are oriented parallel to the margins (Table 2.21). They are primarily composed of Pleistocene and Holocene eolianite ridges, beach and intertidal deposits, partially lithified Holocene eolian and beach deposits, and modern sediments including supratidal and intertidal deposits, eolian dunes, and piles of shallow marine reef talus (e.g., Doran 1955; Milliman 1967; Garrett and Gould 1984; Strasser and Davaud 1986; Carew and Mylroie 1995; Kindler and Hearty 1995; Hearty and Kindler 1997). Island deposits can be divided into eight stratigraphic units, two of which are Holocene in age (Kindler and Hearty 1996). The Holocene units include an older partially lithified, oolitic-peloidal limestone, eolian in origin and formed when sea-level was lower. The second unit is Late Holocene in age, partially lithified skeletal grainstone, marine to eolian in origin.

The eolianite ridges that formed during the Pleistocene reach an elevation as high as 60 m, but most are 10–20 m high (Fig. 2.13; Carew and Mylroie 1995, 1997; Wilson et al. 1995). Holocene eolianites often cap Pleistocene eolianites and reach elevations over 40 m such as on Lee Stocking Island (Kindler 1995). Parabolic dunes on the NW Bahamas indicates that the formation of these islands during the Pleistocene (stage 5e) occurred under dryer conditions than at present, but most dunes are not parabolic and pre-date stage 5e (Kindler and Strasser 2000). Bahamian island morphology and associated deposits are controlled by their platform and oceanic setting which ultimately controls the energy regime of modern depositional facies (Kindler and Hearty 1997; Table 2.22). For example, Andros Island, the largest island of the Bahamian archipelago, is situated on the eastern margin of GBB along the TOTO. It is 153 km long and 29–64 km wide and is predominantly composed of Pleistocene oolites that form low hills on the windward side of the island (Vaughan 1914; Smith 1940). The leeward side, in contrast, is dominated by modern low-energy facies, tidal flats and mangrove swamps (Smith 1940; Shinn et al. 1969; Gebelein 1974).

The islands are also a controlling factor on the distribution of modern sand deposits (Section 2.5.1). For example, the Exuma Islands are a discontinuous chain of Pleistocene dune ridges formed at the shelf edge of Exuma Sound. With the Holocene rise of sea-level, saddles between the crests of the dune ridges have been flooded and tidal currents pass through these breaks. The bottleneck effect causes swift currents and agitation sufficient for oolite deposition in the form of spits and tidal deltas extending onto the bank between islands (Reid et al. 1995; Gonzalez and Eberli 1997).

The interior of the islands in some cases is occupied by marine and hypersaline lakes. For example, on San Salvador Island, the low areas between dune ridges are occupied by lakes and ponds that cover an area of 26% of the island, whereby surface streams are absent (Wilson et al. 1995). These lakes and ponds are less than 1–3 m deep and in some cases are connected to caves that form passages to the ocean. On San Salvador, the salinity of these lakes varies strongly with the seasons and between the lakes and can reach up to 200 ppt (Wilson et al. 1995). On Lee Stocking Island, a 1,500-year old pond preserves coastal and lacustrine sedimentation (Dix et al. 1999). This pond started as a coastal embayment that was probably closed off by a change in alongshore deposition about 700 year bp. It is likely that several ponds are remnants of earlier embayments. In contrast to the numerous sinkholes that are located on larger islands, these ponds are shallow and usually have muddy shorelines and evaporative sediments.

2.5.2.2 Sand Bodies

The prerequisites for the formation of sand bodies on carbonate platforms are an autochthonous sediment source and a mechanism to remove fine material (sorting). Several other factors influence the development of sand bodies: the antecedent topography, oceanic setting (physical processes that include wind, tidal, and wave-generated currents), sea-level fluctuations, and early diagenesis. The antecedent topography and oceanic setting are the most important factors in determining the geometry, internal structure, composition, and texture of a sand body (Table 2.22). Sand bodies are subdivided into five types as described by Ball (1967) and Halley et al. (1983): marine sand belts, tidal bar belts, eolian ridges, platform-interior sand blankets, and tidal deltas (Table 2.23).

Table 2.22 Characteristics of tide-influenced deposits (*Sources*: Ball 1967; Ginsburg and James 1974; Harris 1979; Gonzalez and Eberli 1997; United States Naval Oceanographic Office 1978)

	Cat Cay	Joulters Cays	Lee Stocking Island	Schooner Cays	South TOTO
Depositional facies	Marine sand belt	Tidal bar belt	Tidal inlet	Tidal bar belt	Tidal bar belt
Maximum tidal range (cm)	119	129	110	124	≥129
Depth (m)	0–2	0–6	4–8	0–4.6	0–6
Current velocity (cm/s)	0.5–150	–	60–100	≤200	≤200
Sand belt dimensions (km × km)	42 × 1.7	25 × 2	4 × 0.5	8 × 0.8	100 × 20
Sand bar dimensions					
– Width (km)	0.5 1.0	0.2–0.6	0.2–0.5	0.3–0.7	05.–1.5
– Length (km)	–	5.0	–	8.0	12.0–20.0
Thickness (m)	0.5–4.0	–	–	–	3.0–9.0
Sediment type	Ooids	Ooids	Ooids and mixed	Ooids	Ooids
Grain size (μm)	200–400	300–500	250–500	250–500	Up to 1 mm
Exposure	Leeward open platform edge	Windward protected platform edge	Windward protected platform edge	Leeward protected cul de sac of deep embayment	Windward and leeward protected cul de sac of deep embayment
Current flow direction	Parallel to platform edge	Normal to platform edge	Restricted to channel	Normal to platform edge	Normal to platform edge

Table 2.23 Subdivision of modern sand bodies (After Ball (1967))

	Marine sand belt	Tidal bar belt	Eolian ridge	Platform interior sand blanket
Setting	Leeward slope break; also windward	High-energy, tidal-dominated bank margin, at open embay-ments without island rims	Seaward ridges of islands; adjacent to marine sands	Platform interior where not sheltered by islands
Orientation to platform margin	Belt parallel to slope break	Belt; parallel; individual bars perpendicular	Ridge parallel to slope breaks	Blanket in platform interior
Internal structure	Cross-beds dip perpendicular to belt's long axis	Cross-beds dip perpendicular to long axes of bars, in channels; parallel to channel axis	Large spillover or parallel cross-bed sets dip perpen-dicular to ridge long axis towards platform interior	Burrows
Composition	Dominantly oolitic (>40%), skeletal, peloidal	Dominantly oolitic, skeletal, peloidal	Skeletal, peloidal, oolitic	Peloidal (dominant), skeletal, oolitic
Belt dimensions	1–4 km wide 25–75 km long	Southern end TOTO: 100 km wide and 20 km long	–	Mackie Shoal: 1.5 km wide, 30 km long
Individual bars	–	0.5–1.5 km wide 12–20 km long	> 30 m high	–
Water depth	0–5 m, crests can be exposed during low tide	0–10 m	Above sea-level	5–15 m, typically 7–10 m
Grain sizes (ooids and peloids)	0.25–1.0 mm, moderately to poorly sorted	Up to 1 mm	–	About 0.5 mm
Transport of grains	Off-bank direction by storms only	–	Wind-transport onto bank top	Wind-generated currents

Marine Sand Belts

Marine sand belts are generally found along open, leeward bank margins but they also occur on windward margins (Hine et al. 1981b). They are oriented roughly parallel to the bank edge and have dimensions of 1–4 km in width and 25–75 km in length but vary widely in size and geometry (Fig. 2.15; Halley et al. 1983). Channels cut the sand belt normal to the sand body axis (Fig. 2.16; Ball 1967). Tidal deltas develop at both ends of the channels and are controlled by tidal flow. As Ball (1967) pointed out, geometry and internal structure of bars reflect both ebb and flood currents, but flood currents are by far dominant. Channels are usually well-developed features with little fine-grained sediment but with abundant coarse-grained biocastic and cemented cobbles (Ball 1967). Sediment transport in channels may be most active during storms that generate off-bank currents (Hine et al. 1981a).

Along windward margins, particularly those fronting large islands sand waves can form in water depths to 30–50 m. The sand bodies are composed of sand waves 10–100 m in spacing and heights of 0.5–4 m with smaller-scale ripples superimposed on the sand waves (Fig. 2.17; Halley et al. 1983; see also Boothroyd and Hubbard 1975). Even though significant areas of sand-wave crests emerge at low tide, the sand waves and ripples do not necessarily fill up to sea-level even during low tide.

Lily Bank is an active tide-dominated ooid shoal system on the northeastern side of LBB, 27 km long and 2–4 km wide, with its long axis generally parllel to the shelf

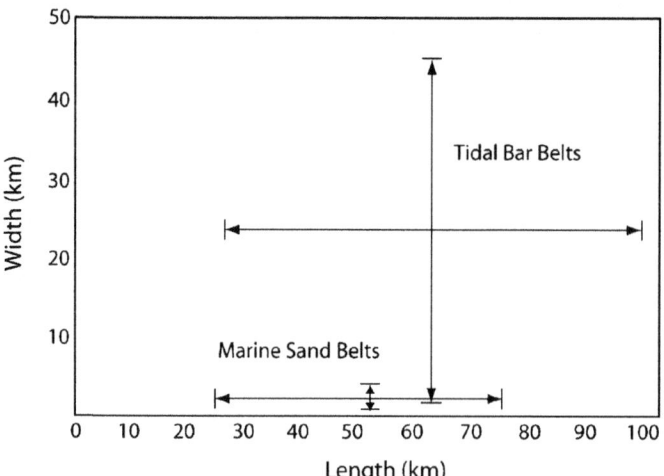

Fig. 2.15 Comparison of dimensions of marine sand belts and tidal bar belts of the Bahamas. Tidal bar belts have a larger range in size (includes data from TOTO, the Exumas, Joulters Cays and Schooner's Cay) while the range in width of marine sand belts is small (includes data from the Cat Cays, Lily Bank and the Berry Islands) (Sources of quantitative data: Ball 1967; Hine 1977; Harris 1979, 1983; Halley et al. 1983)

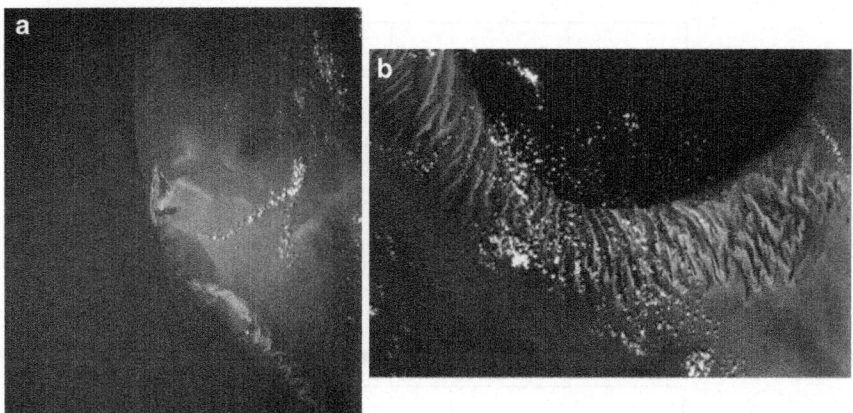

Fig. 2.16 Satellite data of Bahamian sand bodies. (**a**) The Cat Cay ooid belt is a marine sand belt along the leeward margin of GBB. The axis of the sand body is parallel to the margin and is cut by channels perpendicular to the margin. (**b**) The ooid belt at the southern end of Tongue of the Ocean is a tidal bar belt containing individual sand bars perpendicular to the margin (Courtesy of NASA ISS Earthkam

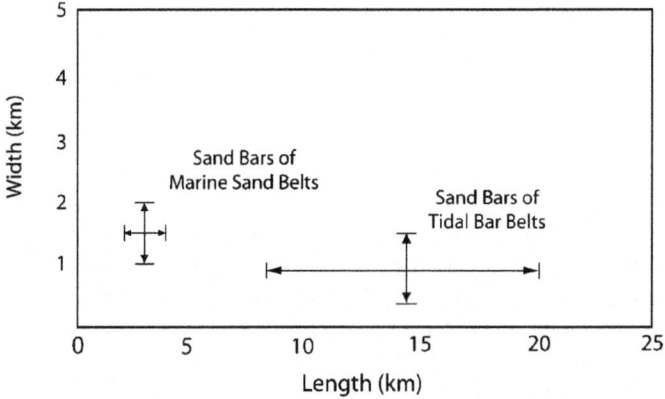

Fig. 2.17 Comparison of dimensions of individual bars in marine sand belts and tidal bar belts. Bars in tidal bar belts have a much larger variety in length (includes data from TOTO, the Exumas, and Schooner's Cay) in comparison to marine bar belts (includes data from the Cat Cays and Lily Bank). The sources of quantitative data are from Ball 1967; Hine 1977; Halley et al. 1983, and the data are listed in the text

break (Hine 1977; Rankey et al. 2006). Tidal range is about 1 m and tidal flow velocities in shallow areas can exceed 80 cm/s (Rankey et al. 2006). Flood-oriented tidal deltas dominate over ebb-oriented tidal deltas reflecting the high energy of the windward margin. Ooidal sediment is about 4 m thick and lies on top of 1–2 m of mud (Hine 1977). A typical bar measures about 4 km by 2 km. Migrating sand waves (spacing >6 m) covered with dunes are abundant on the tidal deltas and zones

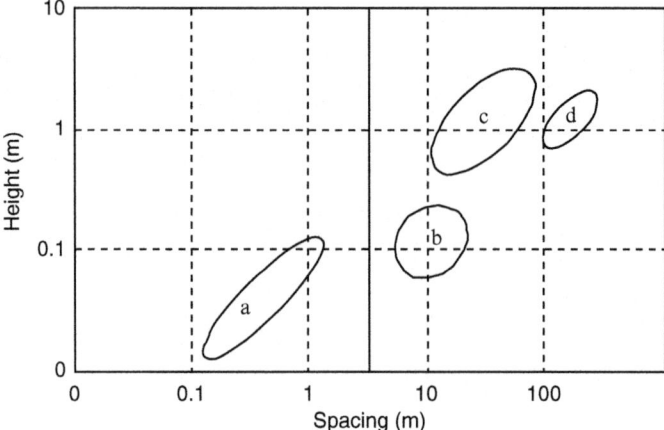

Fig. 2.18 Height and spacing relationships of bedforms. (**a**) Subtidal ripples on Lily Bank and subtidal bedforms on Walker's Cay; (**b**) intertidal sand waves on Lily Bank; (**c**) subtidal active sand waves on Lily Bank; (**d**) relict sand waves on Lily Bank (After Hine 1977)

between the lobes and shoals (Hine 1977). Figure 2.18 shows the various bedforms found on Lily Bank and their height/space relationship.

Tidal Bar Belts

Belts of tidal sand bars are found along high-energy, tide-dominated bank margins and at open embayments without island rims. Tidal energy is the dominant control-ling factor on geometry, so sand bars are oriented perpendicular to the bank margin (Fig. 2.16). Compared to marine sand belts, tidal bar belts show a wider range in both, width and length (Fig. 2.15), while individual bars in tidal bar belts have a wider range in length (Fig. 2.17). Tidal bar belts reach up to 100 km in length and 45 km in width (Halley et al. 1983). Individual tidal bars range from 0.5–1.5 km in width, 12–20 km in length and around 3–9 m in thickness. They usually are covered by sand waves oriented obliquely to bars (Halley et al. 1983; Fig. 2.17). Channels between sand bars, through which tidal energy travels to form lobes, are between 1–3 km wide and 2–7 m deep (Halley et al. 1983).

While well-sorted ooids dominate the composition of tidal bar belts, skeletal, peloidal, and aggregate grains increase toward the bank interior. The grain sizes vary according to the flow velocities acting on the sand belts. For example, on the wind-ward side of GBB at Joulters Cays, the ooids are considerably larger (300–500 μm) than the ooids on the leeward side of GBB at Cat Cay (200–400 μm; Table 2.22). At the southern end of the Tongue of the Ocean, where flow velocities are highest in the Bahamas, the ooids are the largest (up to 1 mm).

Tidal bar belts are found in the areas of Joulters Cays, Schooners Cay, the southern end of TOTO, and Frazers Hog Cay (southern Berry Islands). The most prominent tidal bar belt in the Bahamas is located at the southern end of TOTO and extends for nearly 150 km along the bank margin and is 13–20 km wide (Newell 1955; Newell and Rigby 1957; Ball 1967; Dravis 1977; Palmer 1979; Harris and Kowalik 1994). Within this area, broad sand ridges parallel to tidal flow generally radiate from around the curving shelf margin (Fig. 2.16).

The Joulters Cays area, immediately north of Andros Island on GBB, displays a variety of environments in which ooid sands accumulate. The Joulters Cays shoal is a 400 km² sand flat cut by numerous tidal channels and fringed by mobile sands along the margin edge (Harris 1979, 1983). The mobile sand belt is 0.5–2 km wide and 2–3 m thick and extends 25 km, terminating abruptly at the shelf margin (Harris 1979, Major et al. 1996). Well-sorted ooid sands occur in the center of the sand complex (Major et al. 1996). Platform-ward the tidal bar belt grades into a stabilized sand flat (Harris 1979). The Joulters Cays are three Holocene islands that lie within the active area of the shoal.

In the Schooners Cays area, at the northern end of Exuma Sound, the tidal bar belt is 56 km long and 20 km wide (Budd and Land 1989). The individual bars composing the belt are 8 km long, 400–800 m wide and are asymmetrical and convex-shaped reaching 7 m thick (Ball 1967; Harris and Kowalik 1994). Bar crests are covered by medium scale ripples subparallel to the trends of the bars (Dravis 1979). Channels between sand bars reach depths of 5–8 m and contain little sediment except in spillover lobes (Dravis 1979; Harris and Kowalik 1994). Sediment accumulation rates are around 0.73 mm/year (Dravis 1979).

Eolian Ridges

Eolian ridges compose the highlands of most Bahamian islands and extend parallel to the bank margin. For eolian dunes to develop, a sediment source and appropriate wind energy is needed to carry carbonate sand onto the bank (McKee and Ward 1983). Sediment sources for the carbonate eolian dunes in the Bahamas are marine sands inform tidal bars and beaches. Holocene dune height is highly variable but can reach up to 40 m (McKee and Ward 1983; Kindler 1995). Carbonate eolian dune ridges develop almost exclusively as coastal dune ridges at a relatively short distance (1–3 km) from the shelfbreak (Abegg et al. 2001). However, a dune ridge might be composed of several generations of Pleistocene and eolian deposits. The internal structure of eolian ridges is dominated by large-scale foreset cross-beds dipping toward the bank interiors, indicating that the material within the dune was/is generated in the narrow marine belt in front of the eaolian dune ridge. Finer-scale sedimentary structures include distinctive, millimeter scale, inversely-graded laminations couplets. The carbonate eolianites typically also contain a diverse assemblage of ichnocoensis and animal trace fossils (Curran and White 2001). Eolian dune composition depends on the source of the marine sands and includes well-sorted skeletal grains, ooids, or pellets. Kindler and Hearty (1996) have argued that a distinctive allochem composition exists for each Pleistocene sea-level highstand while Carew and Mylroie (1995, 1997, 2001) question such a clear relationship because eolianites may be deposited during all phases of a highstand.

Lee Stocking Island is composed of two types of Holocene eolian dune ridges. The first type are oolitic-peloidal dunes 6–8 m high; the second type reach up to 43 m in height and contain predominantly skeletal grains, but also reworked ooids (Kindler 1995). Modern eolian sedimentation on Lee Stocking Island is limited due to the combination of high sea-level and reefal growth that cuts off sediment supply (Kindler 1995). Lithified Holocene eolianites on San Salvador are elongate transverse dunes oriented perpendicular to the wind and composed of coalesced dunes (White and Curran 1988). They contain typical eolian features including paleosols, subaerial crusts, terrestrial fossils, and vadose cements (Caputo 1993).

Platform-Interior Sand Blankets

Platform-interior sand blankets cover large areas in the platform interior and are found between high-energy slope-break deposits and low-energy island-shadow deposits. Because water depths on bank tops are consistently shallow (about 2–10 m), and energy is low, sand blankets show a relatively constant thickness of about 6 m, thinning out above highs of antecedent topography. They extend over long distances, and in many cases are limited in size only by the size of the carbonate platform. The largest sand blanket covers most of GBB. Sand blanket composition grades from muddy sand (grainstone, higher energy) to sandy mud (packstone, lower energy) and is dominated by peloidal and aggregate grains (grapestone) (Ball 1967; Hine et al. 1981b). Interior sands of GBB average at ~27% peloids, 32% grapestones, 15% ooids, 5% pellets and 7% mud (Purdy 1963b). The interior sands near Ragged Island and southwest of New Providence Island contain over 50% aggregate grains and 26% peloids (Illing 1954). The remaining percentages are a mix of skeletal grains and pellets (Illing 1954). Burrowing is widespread in this facies. In particular, the burrowing of the shrimp *Callianassa* is abundant and is expressed by mounds up to half a meter in height that can be densely crowded or spaced several meters apart (Bathurst 1975). Algal and sea grass cover (*Thalassia* being the most common) varies from sparse to dense. Some active bank-interior shoals form bathymetric highs and are created by wind-generated currents.

Tidal Deltas

Tidal deltas are lobate or fan-shaped sand bodies formed by tidal currents flowing between barriers such as islands (Halley et al. 1983; Reeder and Rankey 2009). The constriction of the currents funnels sand into a lobe-shaped feature, which forms where the current slows down (Ball 1967). A variety of bedforms is found within tidal deltas including sand waves and ripples. Bedform migration is controlled by tidal range and strength (Gonzalez and Eberli 1997). Gonzalez and Eberli (1997) observed sand-wave migration on the order of 4 m during monthly tidal cycle (Table 2.24). The inter-island channels in some examples are scoured free of sediment (Illing 1954). Transport of sediment is assumed to be highest

Table 2.24 Characteristics of a tidal-inlet sand body; Lee Stocking Island (From Gonzalez and Eberli 1997)

Length	4.0 km
Width	0.25 km
Volume of sand body	$5.5–6.0 \times 10^5$ m^3
Distance from margin	0.5 km
Water depth	2.0–10.0 m
Tidal range (neap-spring)	0.6–1.1 m
Current speed	0.6–1 m/s, 1.5 m/s maximum in shallow areas
Height of sand waves (maximum)	≤4.0 m
Sand wave migration in monthly tidal cycle (maximum)	4 m/cycle
Maximum grain size	0.5 mm
Minimum grain size	0.25 mm
Sediment type	Ooid sand (<90% ooids) and mixed ooids with bioclasts (up to 45% bioclasts)
Wind velocity	0.8–7.2 m/s
Exposure	Leeward protected shelf edge at deep-water embayment (exuma sound)

during storm events (Halley et al. 1983). Grain composition consists of skeletal grains, ooids and rare aggregate grains. A tidal delta located in Gun Point Channel in the southeast of GBB, the skeletal content drops from nearly 100% on the marginal shelf to less than 25% on the tidal delta located opposite the channel, and the remaining sediment is composed of ooids and composite ooids (grapestones) (Illing 1954).

2.5.2.3 Muddy Environments

Despite the large size of the Bahama platforms muddy sediments are found only at some sheltered locations on the platform where energy is lowest. Depth in the muddy interior is slightly less than that of the platform interior sand blankets, averaging at 2 m, and vegetation is sparse. The areas on GBB of mud-dominated sediments are for example in the lee of Andros and Eleuthera Islands. On LBB, muddy sediments are found in the interior of the platform as well as in leeward positions of Great Abaco and Grand Bahama Islands. The grain composition consists of moderately to poorly sorted fecal pellets, peloids, skeletal grains, and aragonite needles. In the lee of Andros Island, within the pelletal wackestone facies, the sediment is composed of 66% mud, 17% skeletal grains, 12.5% pellets, 2% each of peloids and skeletal grains (Purdy 1963b). The pelletal packstone facies is 48% mud, 33% pellets, 11% skeletal grains, 6% ooids and 2% peloids (Purdy 1963b).

The muddy environment is commonly referred to as the "carbonate mud factory", and the origin of the mud is widely debated. The large quantities of mud produced there may originate from the breakdown of calcareous algae and direct precipitation on resuspended sediment (see Section 2.3.2 for more detailed discussion).

Many authors have attempted to estimate how much mud is being produced on GBB through various means. Neumann and Land (1975) estimate that the breakdown of calcareous algae produces 1.5–3 times the amount of mud found on the bank today. Shinn et al. (1989) estimate mud production in whitings at three times that of calcareous algae. Whatever the source of mud (see Section 2.3.2), the numbers illustrate the large quantities being produced. Most of this mud is being exported to the deep water areas surrounding the banks (Wilber et al. 1990; Wilson and Roberts 1992). The pelletization of mud may be a primary control in the stabilization of the muddy facies, reducing the transport of fine-grained mud to other locations during high-energy events.

2.5.2.4   Tidal Flats

Tidal flats are located predominantly on the western, leeward side of islands and form belts oriented parallel to the islands. Tidal flats are composed of three geomorphic zones: (1) a high algal (or microbial) marsh zone that is entirely supratidal and brackish, (2) a channel belt which includes subtidal (channels, ponds), intertidal (algal mat zone), and supratidal depositional facies (levees and beach ridges), and (3) a marginal marine zone (Fig. 2.19; Shinn et al. 1969; Hardie 1977; Hardie and Shinn 1986). Aragonitic mud, pellets and some skeletal grains, especially miliolid foraminifers, compose sediment in both zones. The aragonitic mud is produced offshore and is transported by storms and spring tides to the interior of the tidal flats. The high algal marsh is exposed for prolonged periods and

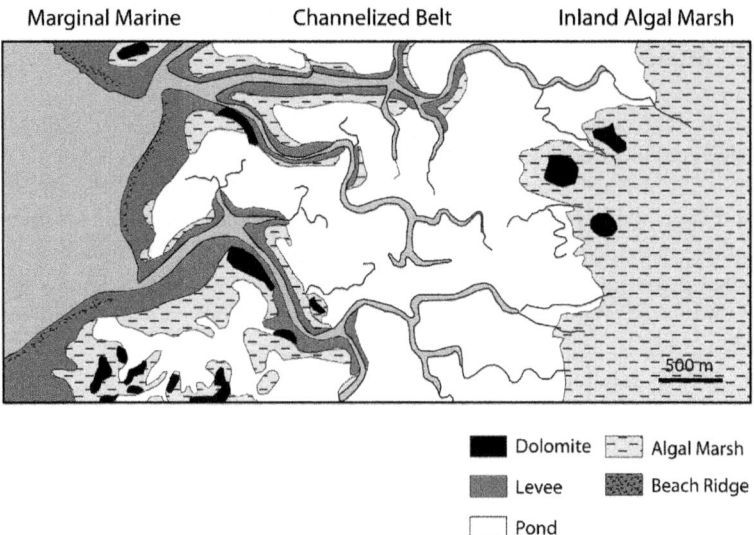

Fig. 2.19 Facies and geomorphic features of the tidal flats environment of Andros Island (After Hardie 1977)

covered by water only during storm events. Organisms dominating these supratidal settings are cyanobacteria, primarily *Scytonema* and *Schizothrix*. Sedimentary structures in the supratidal zone include desiccation cracks, storm laminae, algal structures, and fenestrae (Shinn 1983).

Environments in the channel belt zone include tidal channels, levees, ponds, and microbial mats. The portion of the channel belt is exposed two times a day during low tide. This zone contains a network of meander-shaped channels up to 3 m deep and active and stabilized channel bars (Shinn 1983; Rankey and Morgan 2002). The meanders do not migrate laterally but exhibit headward erosion and landward progradation toward the pond facies at around 2 m/year (Rankey and Morgan 2002). Channel levees are laminated due to the intermittent sedimentation and the subsequent colonization of cyaonbacteria (*Schizothrix*). Sedimentary structures in the channel belt zone include mangrove roots, halophyte grass roots, fiddler crab burrows, and worm burrows (Shinn 1983). Gastropods that graze on microbial mats are common. Mangroves are concentrated in margins and shallower areas of ponds. Beach ridges develop where tidal flats meet the subtidal. They generally consist of coarse-grained skeletal material.

The tidal flats on the leeside of Andros Island are cited as classical tidal flat example in a humid environment (Shinn 1983). The supratidal marsh has a width of 2 km and is several tens of kilometers long. Total length of the tidal flats is over 70 km and width reaches 5 km (Rankey 2002). Quantitative analysis of remote sensing data covering the Three Creeks area along northwest Andros Island shows a quantitative breakdown into the following morphological subfacies within the channelized belt: high and low algal marsh 54%; open pond 23%; mangrove ponds 20%; levee 1–2%; beach ridge 1%. Although the lateral facies distribution in these tidal flats are highly complex, they are not random (Rankey 2002). Crusts of aragonite, high-Mg calcite, and dolomite form in the supratidal adjacent to the hammocks and ponds (Shinn et al. 1965; Gebelein et al. 1980). The fractal nature of northwest Andros Island suggest a self-organizing system that is the result of the absence of a template or external forcing (Rankey 2002). The absence of external forcing is further illustrated by the minimum ipact of extreme events such hurricanes (Rankey et al. 2004).

Tidal flats on the southern flanks of North Caicos, Middle Caicos, and East Caicos extend 60 km in length and 10 km in width. Sediment thickness usually reaches 3–4 m, although some areas even reach 5–6 m (Gebelein et al. 1980; Wanless and Dravis 1989). Caicos tidal flats receive significantly less rainfall than Andros Island tidal flats; 50–75 cm/year versus 165 cm/year, respectively, and only half as many cold fronts pass over Caicos in contrast to Andros (Wanless and Dravis 1989). Due to the drier climate on the Caicos tidal flats, more evaporation takes place, and gypsum-cemented crusts form (Wanless and Dravis 1989).

2.5.2.5   Coralgal Reefs

Reefs are a subordinate feature of the present-day Bahamas. Coral reefs cover 324 km^2 of LBB, and 1,832 km^2 (2.2%) of GBB (Linton et al. 2002).

On GBB, discontinuous shelf-edge reefs are found predominantly on the windward sides of the platforms. Where the shelf-edge reefs form an energy barrier patch reefs develop in the quieter lagoon. Patch reefs are also found scattered in the bank interior and are most abundant west of the southern end of TOTO. Other examples include the Yellow Banks southeast of New Providence Island. Channels reefs occur within the tidal channels of the Exuma Cays (Table 2.11). Few reefs occur at the leeward margin of GBB, e.g., along the Cuba Channel. Leeward reefs are more common in the Turks and Caicos Islands.

The Andros reef system is a barrier reef of 217 km length, making it one of the longest reef systems in the Western Atlantic. It is situated 1–5 km from the Andros shoreline and separated by a shallow 2–4 m deep lagoon (Harris and Kowalik 1994). The Andros Barrier Reef has a fore-reef and backreef zone and sediments are composed of clean coralgal skeletal sand (Kramer et al. 2003). The post-mortem breakdown of reefal organisms is the source for the skeletal sand of the reef facies. Organisms boring or grazing on the reef produce fecal pellets and other sand to mud-sized grains within the reef system (James 1983). The reef exhibits spur-and-groove structures. The spurs that are colonized by active coral reefs are around 60 m wide and are cut by grooves measuring 3–6 m wide and 3 m deep (Kramer et al. 2003). In the grooves, sediment movement inhibits coral growth. Toward the open ocean, the reef is characterized by a vertical wall at 40–50 m water depth that is characterized by irregular features such as caves and notches (Eberli et al. 1998).

Other barrier reefs occur along the northern edge of the Caicos platform (130 km in length; Harris and Kowalik 1994) and along southern Eleuthera, the latter being the most luxuriant coral reef in the Bahamas (70 km in length). Small and infrequent fringing reefs occur in the Exumas; they are at best hundreds of meters long and about 40 m wide (Chiappone et al. 1997a).

On northeastern LBB, a deep shelf-edge reef complex in 30 m water depth and a shallow reef-veneered rock ridge measure up to 175 km long (Storr 1964; Macintyre 1972; Hine and Neumann 1977). This deep reef is a Holocene reef that did not keep up with sea-level rise and subsequently drowned (Hine and Neumann 1977). Fore-reef talus is virtually absent and sand accumulation takes place in the back-reef and inter-reef zones preventing the deep reef from being covered by sediment (Hine and Neumann 1977).

In the Exuma Islands, patch reefs are circular with a relief of typically 1–3 m (Table 2.11; Chiappone et al. 1997a). On the windward side of the island chain individual patch reefs reach up to 1,000 m² in area while leeward patch reefs reach only areas of up to 700 m² (Chiappone et al. 1997a).

2.5.2.6 Adjacent Slopes

Generally, the slopes of the present-day Bahamas are characterized by high angles that decrease with depth. The slopes encompass the area of transition between the shallow-water platform and the deep-water environment and can be divided into an upper slope, lower slope and toe-of slope. The upper slope, as described from the

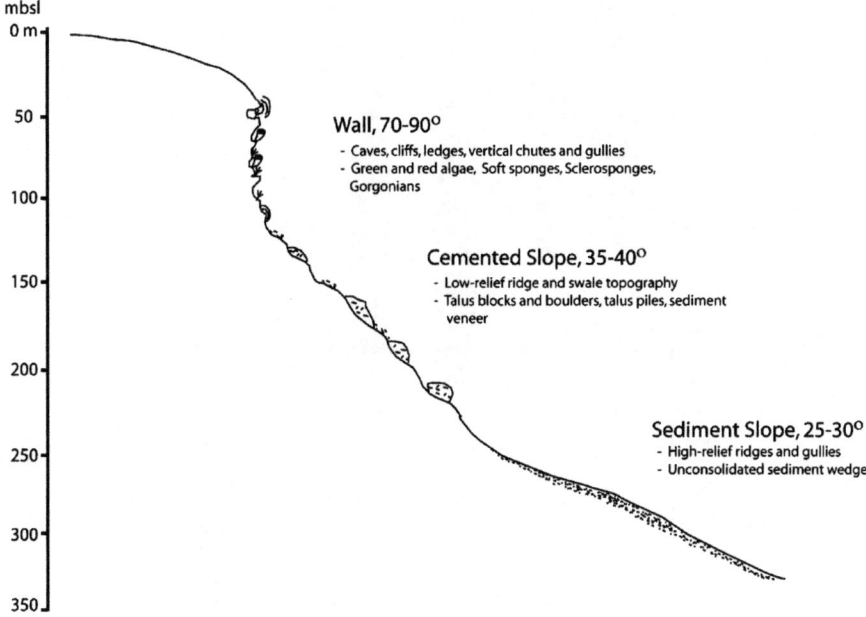

Fig. 2.20 Windward and leeward upper slopes of TOTO are be divided into three regions, the wall, cemented slope, and soft sediment slope (Modified from Grammer et al. 1993)

windward and leeward slopes of TOTO, is composed of three parts: (1) the wall, (2) the cemented slope, and (3) the soft-sediment slope (Fig. 2.20; Grammer and Ginsburg 1992). The uppermost slope or wall has angles of 70–90°, begins at depths around 40–60 m and extends down to 130–140 m (Grammer et al. 1993). The wall contains caves, ledges, and overhangs, as well as encrusted vertical surfaces. The encrusting biota consist of platy growth forms of zooxanthellate corals, coralline algae, *Halimeda*, and sponges (Figs. 2.20 and 2.21; Grammer et al. 1993). The cemented slopes have angles of 35–34° and extend to depths of about 365 m (Grammer et al. 1993). They have a thin veneer of sediment, ridges and swales, often described as a gullied slope, and localized slope-failure scarps (Mullins and Neumann 1979). Talus debris in the lowermost portion of the cemented slope contains large blocks and rubble (Grammer et al. 1993). The soft-sediment slope has angles of 25–28° and consists of a wedge of sand and mud derived from both the slope wall and the platform (Grammer et al. 1993). Sedimentation processes along the upper slope include slumps and turbidity currents, the latter being responsible for the development of incised channels or gullies (Betzler et al. 1999). Average slope angles in the Exuma Sound on the upper slope are 30–35° and channels range from 1–3 m in width and 80–100 m in depth (Crevello and Schlager 1980).

The lower slope begins at around 400 m and exhibits gentler angles than the upper slope, averaging at 3.5° on the leeward side of GBB, 3–5° in the Exuma Sound, and 6–9° in the TOTO (Schlager and Chermak 1979; Crevello and Schlager 1980;

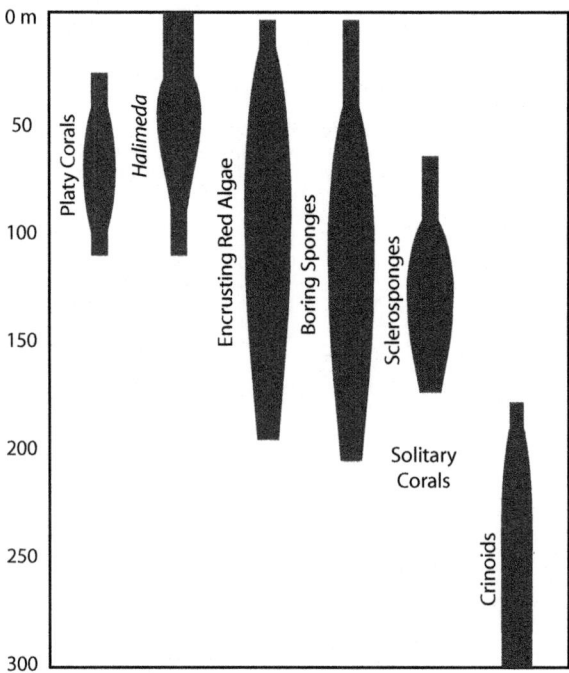

Fig. 2.21 Depth ranges of organisms on the slopes of the TOTO (After Ginsburg et al. 1991 and Grammer et al. 1993)

Betzler et al. 1999). Sediments consist of prograding turbidite sequences and slump deposits similar to the upper slope, and exhibit turbidite channels and gullies (Schlager and Chermak 1979; Betzler et al. 1999). The gullies in TOTO measure tens to hundreds meters wide and 20–100 m deep and contain erosional cliffs (Schlager and Chermak 1979). The gully floors contain coarse sand and boulders and are often lined with erosional limestone outcrops (Schlager and Chermak 1979).

The toe-of-slope, or continental rise, is the area of transition between the slope and basin (Crevello and Schlager 1980). The toe-of-slope exhibits angles of 0.5–2.5° in Exuma Sound, 2.4° in the lee of GBB, and 0.5° in the southern TOTO (Schlager and Chermak 1979; Crevello and Schlager 1980). In Exuma Sound, the toe-of-slope is 1.8–4.6 km in width and is cut by some gullies with a relief of 10–220 m that extend down from the lower slope (Crevello and Schlager 1980). In the southern TOTO, the toe-of-slope begins around 1,150 m, and sediments are composed of carbonate ooze, turbidites, and debris flow and slump deposits that originate on the upper slope (Schlager and Chermak 1979).

The slopes are loci of voluminous deposition of carbonate material from the shallow-water carbonate factory. The excess of fine-grained material that is produced on the platform interior is swept off the banks by waves and currents, especially during storm events (Neumann and Land 1975). The off-bank transport and deposition of shallow-marine material on the slopes is the main process responsible for the rapid lateral progradation of the Bahama banks (Eberli and Ginsburg 1987, 1989). Downslope processes include turbidity currents and mass wasting which are strong enough to produce channels on the upper and lower slope ("gullied slope"; Mullins and Neumann 1979; Anselmetti et al. 2000), but also density cascading of fine-grained material (Wilson and Roberts 1995). Lateral growth on the leeward side of GBB led to the progradation of more than 25 km westward into the present-day Straits of Florida since the Miocene; thus, progradation is one of the most striking features governing the development of this carbonate platform (Fig. 2.2; (Eberli and Ginsburg 1987, 1989); Eberli 2000; McNeill et al. 2001).

The slopes of the Bahamas are characterized by a strong asymmetry between windward and leeward slopes with respect to sediment accumulation rates and geometries (Mullins and Neumann 1979; Eberli and Ginsburg 1989). On the windward side of the Bahamas Escarpment, the slope has a very narrow width of about 5 km; in contrast, the slope northwest of LBB is about 100 km wide (Mullins and Neumann 1979). The dominant direction of winds, currents, and consequent transport of sediment is towards the west; therefore, accumulation and progradation rates are much higher on the western, leeward sides of the platforms. Accumulation rates on the leeward slope of western GBB amounts to 80–110 m/ka and the lateral progradation to 11–15 m/ka (Wilber et al. 1990). Grammer and Ginsburg (1993) observed a Holocene sediment wedge 80 m thick along leeward slopes of TOTO, whereas the wedge on the windward slopes, measures 20 m in thickness.

Deposition on slopes adjacent to the platform is a combination of pelagic and platform-derived sediments. Fine-grained periplatform ooze is characterized by a mixture of fine-grained shallow-water constituents such as aragonite needles and small benthic foraminifers and pelagic grains such as Globigerinids and coccoliths, whereas coarser-grained turbidites and other gravity-displaced deposits contain platform-derived grains such as *Halimeda* plates and typically contain only sparse pelagic components (e.g. Westphal 1998; Westphal et al. 1999; Rendle 2000; Kenter et al. 2002; Betzler et al. 1999). Slope sediment at 500 m depth in the Northwest Providence Channel is composed of 80% bank-derived and 20% planktic-derived material (Pilskaln et al. 1989).

While most Bahamian slopes are depositional in style, some are dominated by erosional processes and exhibit limited amounts of progradation. For example, the windward slope along the eastern margin of LBB (the Bahama Escarpment) is one of the steepest margins in the world with slope angles over 40° that drop down to over 4,000 m into the Atlantic Ocean (Emiliani 1965). The margin exhibits high amounts of mass wasting along the upper and lower slope with talus sediments that accumulate along the toe-of-slope (Mullins and Neumann 1979). Erosional slopes are also found along the northeast margin of GBB (seaward of the Berry Islands) and southeast margin of LBB. The slopes measure over 20° and are less than 5 km wide (Mullins and Neumann 1979). Slopes observed off the Blake-Bahama Escarpment near Cat Island have an average angle of 25° and extend below 2,000 m water depth (Freeman-Lynde et al. 1981).

2.5.2.7   Deep-Water Environment

Deep-water sediments around the Bahamian platforms consist of a mixture of pelagic sediments and material derived from the bank top, the so-called periplatform ooze (Schlager and James 1978). Sediments are transported to the deep-water environment episodically via turbidity currents, debris flows, and density driven flows (Crevello and Schlager 1980; Wilson and Roberts 1995), but also as a continuous rain of fine-grained sediment through the water column.

The deep-water environment surrounding the Bahamian platforms include the deep-water reentrants of TOTO and Exuma Sound, open seaways including the Straits of Florida, Santaren Channel and Providence Channel, and the open Atlantic Ocean. The deep reentrants are flat-floored, u-shaped basins (Schlager and Chermak 1979). The TOTO is around 1,300–2,000 m deep whereas Exuma Sound reaches 800 m to over 900 m water depth (Schlager and Chermak 1979; Crevello and Schlager 1980). In the Exuma Sound, 25% of the sediment is composed of gravity-displaced sediment and 75% of periplatform ooze with open-marine components such as pteropod shells (Crevello and Schlager 1980). The gravity-displaced sediment consists of 50–70% components derived from shallow water and 30–50% derived from slope and basin. Turbidites, debris flows and other mass-transport deposits thin towards the basin axis and have areal extents up to 400 km^2 (Crevello and Schlager 1980). Gravity-flow frequency in Exuma Sound is estimated at one in 10,000–13,000 year, while for the TOTO a range of one in 500–10,000 year is assumed (Rusnak and Nesteroff 1964; Crevello and Schlager 1980).

Sedimentation is further influenced by offshore currents that form drift deposits and other features such as sediment waves and erosional channels on the basin floor. The convergence of currents northwest of LBB and GBB has formed two streamlined mounds of sediment extending northward from the northwest extremities of LBB and GBB (Fig. 2.22; Mullins et al. 1980). The Great Bahama Drift covers an area of about 85 km in length and 60 km in width; and the Little Bahama Drift covers an area of about 100 km in length and 60 km in width (Mullins et al. 1980). The surface currents flowing through the Santaren Channel off the west slope of GBB have formed the Santaren Drift, a symmetrical mound that reaches a thickness of 1,000 m (Fig. 2.22; Anselmetti et al. 2000). Submersible observations in the Straits of Florida have revealed other modern current features including sand waves up to 1–2 m high between 538 and 222 m water depth off the coast of Bimini and north-south oriented sand ridges up to 12–15 m high at around 719 m depth off the Miami Terrace Escarpment (Neumann and Ball 1970).

2.5.3   Sea-Level Fluctuations

Fluctuating sea-level is one of the major controlling factors for sediment production, distribution and accumulation on the modern Bahama platforms. Sediment production on the flooded platform is very high; in fact more sediment is produced

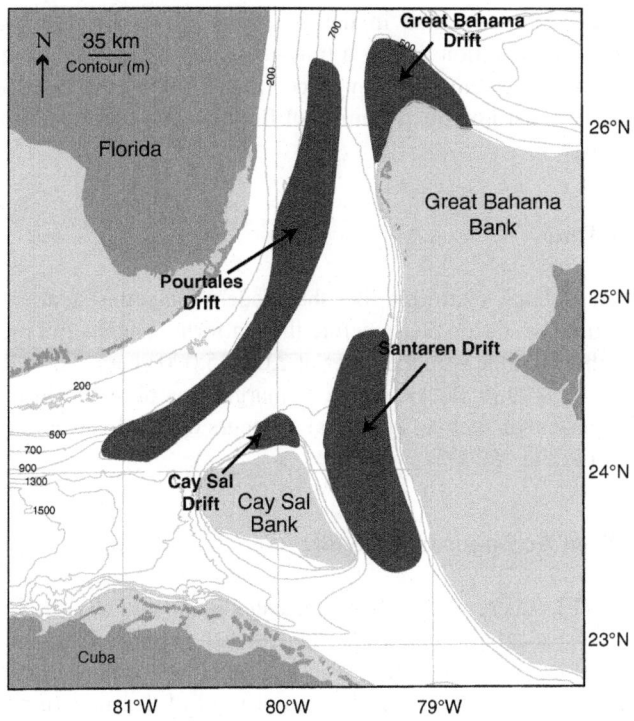

Fig. 2.22 Depositional facies map showing the location of drift deposits and other current features in the Florida Straits. The Pourtales and Santaren drifts are formed by surface currents associated with The Gulf Stream while the Great Bahama Drift forms due to converging currents along the northwest corner of GBB (From Bergman 2005)

than can be accumulated on the platform top (Schlager 2005). On the present-day Bahamas, the area available for carbonate production during highstands is two orders of magnitude larger than the area that is productive during lowstands (Schlager et al. 1994). A sea-level drop of 10 m is sufficient to expose most of the present-day GBB, restricting production to a narrow rim on the slope (Traverse and Ginsburg 1966; Burchette and Wright 1992). Consequently, sedimentation rates of interglacial periods exceed that of the glacials by a factor of 4–6. This increased sediment export during sea-level highstand was termed highstand shedding (Droxler and Schlager 1985). This highstand shedding puts the shallow-water carbonates 180° out of phase with the siliciclastics environment where most of the sediment export into the deepwater areas occurs during sea-level lowstands. During sea-level highstand, fine-grained carbonate mud that is produced in the vast shallow-water areas are exported to form fine-grained, aragonite-rich periplatform sedimentation. In contrast, lowstand sediment is coarser-grained and shows a lower percentage of aragonite mud (Droxler et al. 1983; Droxler and Schlager 1985; Reijmer et al. 1988, Haak and Schlager 1989; Westphal 1998; Westphal et al. 1999).

The Holocene sea-level rise in the Bahamas allows for the evaluation of the process of sedimentation during a transgression and sea-level highstand, i.e. the process of filling accommodation space created during the last sea-level rise. Important aspects regarding the dynamics of flooding and sediment production are discussed below.

2.5.3.1 Lag Time

On flat-topped isolated platforms like the modern Bahamas, sea-level rise may result in a lag time or startup phase before the platform resumes full production. In the Holocene this lag time is in the order of 2,000–5,000 years (Fig. 2.23; Schlager 1981; Harris et al. 1993). Backstepping of margins, as observed in many ancient examples, suggests that such a lag time also occurs during long-term (third-order) sea-level rises (Schlager 2005).

2.5.3.2 Unfilled Accommodation Space

In the Bahamas/Florida region, carbonate production should be sufficient to fill the accommodation space that is created by the Holocene sea-level rise (Neumann and Land 1975; Nelsen and Ginsburg 1986; Bosence 1989). Nevertheless, there is strong indication that accommodation space of most of the modern platforms will never be filled. The amount of filling of accommodation space is controlled by facies and location. Reefs, for example, easily fill the available accommodation space. Other facies, such as ooids accumulating in high-energy areas fill accommodation space or even overfill it to create islands. In protected areas where sediment accumulates in muddy tidal flats, e.g. behind the Andros Island paleo-high, Holocene sediment is also above sea-level. Most areas of Great Bahama Bank, however, are not filled with modern sediments and probably never will, because base level is lower than the current mean sea-level. Currently, water depth is 7 m in the platform interior whereas the Holocene sediment thickness is about 2.5 m.

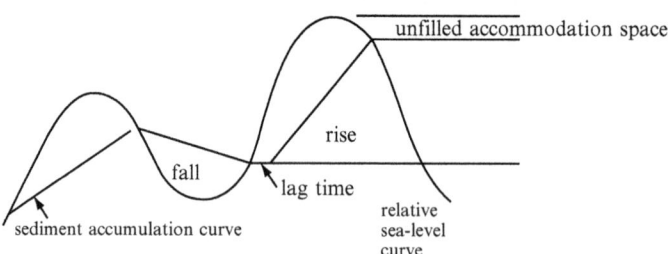

Fig. 2.23 Sedimentation during a sea-level rise. Sedimentation lags behind the flooding of the platform by 2–5 ky. Accommodation space created by the rise is filled differently on different locations on the platform. In addition, most of the platform is never filled completely

Biogenic production during the last 6,000 years, when the platform was flooded, would have been enough to fill this space.

Unfilled accommodation space is not a function of sediment production but rather of sediment accumulation. The energy of waves, tides and winds across the platform prevents filling of accommodation space in many places. Only in protected areas, such as behind topographic highs or in high accumulation areas, is complete filling or overfilling of accommodation space possible. Consequently, sediment accumulation is probably not occurring in the top 5 m on a carbonate platform. If the present is a key to the past, 2/3 of the accommodation space on flat-topped isolated platforms remain unfilled in each sea-level cycle.

Boss and Rasmussen (1995) concluded, based on the Holocene sediment thickness, that there is no correlation between accommodation space and sediment thickness. Similarly, accommodation space remained unfilled during the last interglacial when sea-level is estimated to have been about 6 m higher than at present. If accommodation space would have been filled completely and uniformly across GBB during the last interglacial, the bank still would be above modern sea-level.

Unfilled accommodation space is not a function of sediment production but rather of prevented sediment accumulation. The energy across the platform prevents filling of accommodation space in many places. Thus, on modern GBB, the amount of filling of accommodation space is controlled by facies and location. Reefs, for example, have a growth potential of approximately 10 mm/year (Bosscher and Schlager 1992), which easily fills the available accommodation space. In addition, reefs are preserved even in high-energy environments. This is in contrast to other facies where sedimentation rate is generally less than production rate. Other facies, such as ooids accumulating in high-energy areas, fill accommodation space and locally overfill it to create strom beach ridges. Overfill also occurs in protected areas such as in the muddy tidal flats behind the Andros Island paleo-high.

2.5.3.3 Stratigraphic Evolution

Changing amplitudes and durations of sea-level changes result in variations in the rate of progradation and the architecture of platforms (Eberli and Ginsburg 1987, 1989; Pomar 1993). For the late Tertiary and the Quaternary, seismic sequences as well as morphology modifications can be related to sea-level changes (see various chapters in Ginsburg 2001; Eberli et al. 1997b; Swart et al. 2000).

Cores from seven sites drilled during ODP Leg 166 along western Great Bahama Bank retrieved the sedimentary record and the timing of high and low-frequency sea-level changes throughout the Neogene (Eberli et al. 1997b). Facies successions within the cores contain indications of sea-level changes on two different scales. First, there are high-frequency alternations between meter thick layers with platform-derived material and thin layers with more pelagic sediments. Carbonate-rich intervals are interpreted to reflect periods of high sea-level while the thin intervals correspond to times of increased pelagic and siliciclastic input

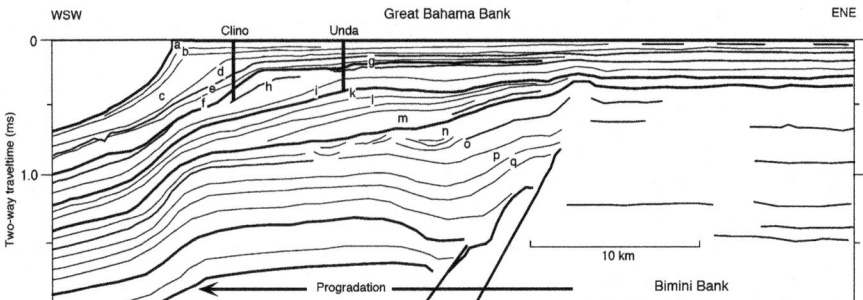

Fig. 2.24 Detail of the Western Line showing the leeward clinoforms that prograde westward from the Bimini Bank (From Eberli et al. 1997b)

during sea-level lowstands. The duration of these alternations (20–40 ky.) correlates to orbitally induced high-frequency climate and sea-level changes (Bernet et al. 1998; Kievman 1998).

Longer-term sea-level changes with duration of 0.5–2 Ma are recorded by alternating high (up to 20 cm/ky) and low sedimentation rates (<2 cm/ky) that document a long-term pattern of bank flooding with concomitant shedding to the slope, and periods of bank exposure with reduced shallow-water carbonate production and largely pelagic sedimentation in the basin (Eberli 2000). The longer-term changes coincide with progradation pulses that are imaged on the seismic data as depositional sequences (Fig. 2.24). These carbonate depositional sequences display five major elements. On the platform top, the sediments are arranged in shallow-water packages separated by exposure horizons (Beach 1982; Kievman 1998). On the upper slope, the prograding pulses are characterized by fine-grained platform-derived material. The middle to lower slope has a variable facies assemblage consisting of periplatform, pelagic and redeposited carbonates. Small-scale channeling and lobes of turbidites produce irregular depositional surfaces. At the toe-of-slope, redeposited carbonates accumulate during both sea-level highstand and sea-level lowstands. These carbonate turbidite series are arranged in mounded lobes with feeder channels. The distal portion of the sequences is dominated by cyclic marl/limestone alternations with few turbidites. Highstand shedding is recognized in the thickness of the entire sedimentary package rather than in the number of turbidites (Bernet et al. 2000).

The morphological evolution of GBB is, however, not solely determined by fluctuating sea-level. For example, the upper Pliocene record shows prograding geometries and a steepening of the flanks of the platform. In Pliocene-Pleistocene platform-top sediments, Beach and Ginsburg (1980) and Beach (1982) observed a change in composition from a dominance of skeletal to non-skeletal grains. This change is interpreted to reflect the morphologic evolution from an open bank with gentle slopes into a flat-topped bank that resembled the modern morphology (see also Schlager and Ginsburg 1981; McNeill et al. 1988; Reijmer et al. 1992; Westphal 1998; Westphal et al. 2000). The upper, dominantly non-skeletal, formation represents a major change in sediment deposition due to changed circulation and shallowing of the bank (Beach and Ginsburg 1980). This change in depositional style from skeletal to non-skeletal observed in the platform interior is also reflected in the composition of adjacent toe-of-slope turbidites from ODP Leg 101 (Reijmer et al. 1992).

The Pliocene-Pleistocene limestone is bounded above by an upper Pleistocene unconformity that forms the present-day surface of many islands along the margin of the banks, and underlies unlithified Holocene sediments in the interior of GBB. The onset of glaciation on the northern hemisphere in the Pliocene resulted in high-frequency sea-level fluctuations with rises that only slightly overstepped the platform top (flooding during isotope stages 13, 9, 7, 5e).

References

Abegg FE, Loope DB, Harris PM (2001) Carbonate eolianites – depositional models and diagenesis. In: Abegg FE, Harris PM, Loope DB (eds) Modern and ancient carbonate eolianites. SEPM Spec Publ 71:17–30

Abel CE, Tracy BA, Vincent CL, Jensen RE (1989) Hurricane hindcast methodology and wave statistics for Atlantic and Gulf Hurricanes from 1956–1975. WIS Report 19. Coastal Engineering Research Center. Vicksburg, Mississippi, p 85

Acker KL, Risk MJ (1985) Substrate destruction and sediment production by the boring sponge *Cliona caribbaea* on Grand Cayman Island. J Sed Petrol 55:705–711

Adey WH, Macintyre IG (1973) Crustose coralline algae: a re-evaluation in the geological sciences. GSA Bull 84:883–904

Agassiz A (1894) A reconnaissance of the Bahamas and of the elevated reefs of Cuba in the team yacht "Wild Duck", January to April, 1893. Bull Mus Comp Zool Harvard Coll 17:1–281

Andrews JE, Shepard FP, Hurley RJ (1970) Great Bahama Canyon. GSA Bull 81:1061–1078

Anselmetti FS, Eberli GP, Ding Z (2000) From the Great Bahama Bank into the Straits of Florida: a margin architecture controlled by sea-level fluctuations and ocean currents. GSA Bull 112:829–844

Atkinson LP, Berger T, Hamilton P, Waddell E, Leaman K, Lee TN (1995) Current meter observations in the Old Bahama Channel. J Geophys Res 100:8555–8560

Ball MM (1967) Carbonate sand bodies of Florida and the Bahamas. J Sed Petrol 37:556–591

Ball MM (1972) Exploration methods for stratigraphic traps in carbonate rocks. In: King RE (ed) Stratigraphic oil and gas fields – classification, exploration methods and case histories. AAPG Mem 16:64–81

Ball MM, Harrison CGA, Hurley RJ, Leist CE (1969) Bathymetry in the vicinity of the northeastern scarp of the Great Bahama Bank and Exuma Sound. Bull Mar Sci 19:243–252

Bathurst RGC (1975) Carbonate sediments and their diagenesis. Developments in Sedimentology 12, Elsevier, p 658

Beach DK (1982) Depositional and diagenetic history of Pliocene-Pleistocene carbonates of Northwestern Great Bahama Bank; Evolution of a carbonate platform. Ph.D. thesis, University of Miami, p 447

Beach DK, Ginsburg RN (1980) Facies succession of Pliocene-Pleistocene carbonates, northwestern Great Bahama Bank. AAPG Bull 64:1634–1642

Bergman KL (2004) Seismic analysis of paleocurrent features in the Florida Straits: insights into the paleocurrent, upstream tectonics, and the Atlantic-Caribbean connection. Ph.D. thesis, University of Miami, p 190

Bernet KH, Eberli GP, Anselmetti FS (1998) The role of orbital precession in creating marl/limestone alternations, Neogene, Santaren Channel, Bahamas. 15th international sedimentological congress, Alicante, Abstract: 191–192

Bernet K, Eberli GP, Gilli A (2000) Turbidite frequency and composition in the distal part of the Bahamas. In: Swart PK, Eberli GP, Malone M, Sarg JF (eds) Proceedings of ODP, Sci. Results, 166: College Station, TX (Ocean Drilling Program), pp 45–60

Betzler C, Reijmer JJG, Bernet K, Eberli GP, Frank T, Anselmetti FS (1999) Sedimentary patterns and geometries of the Bahamian outer carbonate ramp (Miocene and Lower Pliocene, Great Bahama Bank). Sedimentology 46:1127–1144

Black M (1933) The precipitation of calcium carbonate on the Great Bahama Bank. Geol Mag 832:455–466

Blair SM, Norris JN (1988) The deep-water species of *Halimeda lamouroux* (Halimedaceae, Chlorophyta) from San-Salvador Island, Bahamas – species composition, distribution and depth records. Coral Reefs 6:227–236

Boardman MR, Neumann CA (1984) Sources of periplatform carbonates: Northwest Providence Channel, Bahamas. J Sed Petrol 54:1110–1123

Boothroyd JC, Hubbard DK (1975) Genesis of bedforms in mesotidal estuaries. In: Cronin LE (ed) Estuarine research, Geology and Engineering, 2. Academic, New York, pp 217–234

Bosart LF, Schwartz BE (1979) Autumnal rainfall climatology of the Bahamas. Mon Weather Rev 107:1663–1672

Bosence D (1989) Aspects of carbonate deposition in the Caribbean. Proceedings Cumberland Geological Society 5(2):235–236

Boss SK, Neumann AC (1993) Impact of Hurricane Andrew on carbonate platform environments, northern Great Bahama Bank. Geology 21:897–900

Boss SK, Rasmussen KA (1995) Misuse of Fischer plots as sea-level curves. Geology 23:221–224

Bosscher H, Schlager W (1992) Computer-simulation of reef growth. Sedimentology 39:503–512

Bourrouilh-Le Jan FG (1980) Hydrologie des nappes d'eau superficielles de l'ile Andros, Bahama; dolomitisation et diagenese de plaine d'estran en climat tropical humide. Bull des Centres de Recherches Exploration-Production Elf-Aquitaine 4(2):661–707

Broecker WS, Sanyal A, Takahashi T (2000) The origin of Bahamian whitings revisited. Geophys Res Lett 27:3759–3760

Broecker WS, Takahashi T (1966) Calcium carbonate precipitation on the Bahamas Banks. J Geophys Res 71:1575–1602

Bruggemann JH, van Kessel AM, van Rooij JM, Breeman AM (1996) Bioerosion and sediment ingestion by the Caribbean parrotfisch *Scarus vetula* and *Sparisoma viride*: implications of fish size, feeding mode and habitat use. Mar Ecol-Prog Ser 134:59–71

Budd DA, Land LS (1989) Geochemical imprint of meteoric diagenesis in Holocene ooid sands, Schooner Cays, Bahamas; correlation of calcite cement geochemistry with extant groundwaters. J Sed Petrol 60:361–378

Bullard EG, Everett JE, Smith AG (1965) The fit of the continents around the Atlantic. Philos Trans R Soc Lond Ser A 1088:41–51

Burchette TP, Wright VP (1992) Carbonate ramp depositional systems. Sed Geol 79:3–57

Caputo MV (1993) Eolian structures and textures in oolitic-skeletal calcarenites from the Quaternary of San Salvador Island, Bahamas; a new perspective on eolian limestones. In: Keith BD, Zuppann CW (eds) Mississippian oolites and modern analogs. AAPG Stud Geo 35:243–259

Carew JL, Mylroie JE (1995) Depositional model and stratigraphy for the Quaternary geology of the Bahama Islands. In: Curran HA, White B (eds) Terrestrial and shallow marine geology of the Bahamas and Bermuda. GSA Special Paper 300:5–32

Carew JL, Mylroie JE (1997) Geology of the Bahamas. In: Vacher HL, Quinn TM (eds) Geology and Hydrogeology of carbonate islands. Developments in Sedimentology 54, Elsevier, pp 91–140

Carew JL, Mylroie JE (2001) Quaternary carbonate eolianites of the Bahamas: useful analogues for the interpretation of ancient rocks? In: Abegg FE, Harris PM, Loope DB (eds) Modern and ancient carbonate eolianites. SEPM Spec Publ 71:33–46

Chalker BE, Barnes DJ, Dunlap WC, Jokiel PL (1988) Light and reef-building coral. Interdiscipl Sci Rev 13:22–237

Chiappone M, Sullivan KM, Sluka R (1997a) Reef invertebrates of the Exuma Cays, Bahamas: Part 1 – corals. Bahamas J Sci 4:30–36

Chiappone M, Sullivan KM, Sluka R (1997b) Reef invertebrates of the Exuma Cays, Bahamas: Part 2 – octocorals, Part 1 – corals, continued. Bahamas J Sci 4:28–30

Chiappone M, Sullivan KM, Lott C (1996) Hermatypic Scleractinian corals of the southeastern Bahamas: a comparison to western Atlantic reef systems. Carib J Sci 32:1–13

Cloud PE Jr (1962) Environment of Calcium Carbonate deposition west of Andros Island, Bahamas. USGS Prof Paper 35:494

Craton M (1986) A history of the Bahamas, 3rd edn. San Salvador Press, Waterloo Ontario, p 332

Crevello PD, Schlager W (1980) Carbonate debris sheets and turbidites, Exuma Sound, Bahamas. J Sed Petrol 50:1121–1148

Crutcher HL, Quayle RG (1974) Mariners world-wide climatic guide to tropical storms at sea. Naval Weather Service Command, U.S. Government Printing Office, Washington DC, p 246

Cry GW (1965) Tropical cyclones of the North Atlantic Ocean: tracks and frequencies of hurricanes and tropical storms, 1871–1963. U. S. Weather Bureau Technical Paper 55:148

Curran HA, White B (2001) The inchology of Holocene carbonate eolianites of the Bahamas. In: Abegg FE, Harris PM, Loope DB (eds) Modern and ancient carbonate eolianites. SEPM Spec Publ 71:47–56

Dietz RS, Holden JC (1973) Geotectonic evolution and subsidence of Bahama platform, reply. GSA Bull 84:3477–3482

Dietz RS, Holden JC, Sproll WP (1970) Geotectonic evolution and subsidence of Bahama platform. GSA Bull 81:1915–1927

Dill RF, Shinn EA, Jones AT, Kelly K, Steinen RP (1986) Giant stromatolites forming in normal salinity water. Nature 324:55–58

Dix GR, Patterson RT, Park LE (1999) Marine saline ponds as sedimentary archives of late Holocene climate and sea-level variation along a carbonate platform margin; Lee Stocking Island, Bahamas. Palaeogeogr Palaeoclim Palaeoecol 150:223–246

Doran E (1955) Land forms of the southeastern Bahamas. University of Texas Publications: 5509, p 38

Dravis JJ (1982) Hardened subtidal stromatolites, Bahamas. Science 219:385–386

Dravis JJ (1979) Rapid and widespread generation of Recent oolitic hardgrounds on a high energy Bahamian platform, Eleuthra Bank, Bahamas. J Sed Petrol 49:195–208

Dravis JJ (1977) Holocene sedimentary depositional environments on Eleuthera Bank, Bahamas. M.S. thesis, University of Miami, p 386

Droxler AW, Schlager W (1985) Glacial versus interglacial sedimentation rates and turbidite frequency in the Bahamas. Geology 13:799–802

Droxler AW, Schlager W, Whallon CC (1983) Quaternary aragonite cycles and oxygen-isotope record in Bahamian Carbonate ooze. Geology 11:235–239

Dunham RJ (1962) Classification of carbonate rocks according to depositional texture. In: Ham WE (ed) Classification of carbonate rocks. AAPG Mem 1:108–121

Eberli GP (2000) The record of Neogene sea-level changes in the prograding carbonates along the Bahamas Transect – Leg 166 synthesis. Proceedings of ODP, Sci. Results, 166: College Station, TX (Ocean Drilling Program), pp 167–177

Eberli GP, Ginsburg RN (1987) Segmentation and coalescence of platforms, Tertiary, NW Great Bahama Bank. Geology 15:75–79

Eberli GP, Ginsburg RN (1989) Cenozoic progradation of NW Great Bahama Bank – A record of lateral platform growth and sea-level fluctuations. In: Crevello PD, Wilson JL, Sarg JF, Read JF (eds) Controls on carbonate platform and basin evolution. SEPM Spec Pub 44:339–351

Eberli GP, Grammer GM, Harris PM (1998) Sequence stratigraphy and reservoir distribution in a modern carbonate platform. AAPG core workshop and field seminar, Guidebook, p 670

Eberli GP, Kendall CGStC, Moore P, Whittle GL, Cannon R (1994) Testing a seismic interpretation of Great Bahama Bank with a computer simulation. AAPG Bull 78:981–1004

Eberli GP, Swart PK, Malone M (1997a) Scientific Party. Proceedings of ODP, Init Repts, 166: College Station, TX (Ocean Drilling Program), p 850

Eberli GP, Swart PK, McNeill DF, Kenter JAM, Anselmetti FS, Melim LA, Ginsburg RN (1997b) A synopsis of the Bahamas Drilling Project: results from two deep core borings drilled in the Great Bahama Bank. In: Eberli GP, Swart PK, Malone MJ et al (eds) Proceedings of ODP, Init Repts, 166: College Station, TX (Ocean Drilling Program), pp 23–41

Eberli GP, Anselmetti FS, Kenter JAM, McNeill DF, Ginsburg RN, Swart PK, Melim LA (2001) Calibration of seismic sequence stratigraphy with cores and logs. In: Ginsburg RN (ed) Subsurface geology of a prograding carbonate platform margin, Great Bahama Bank: results of the Bahamas drilling project. SEPM Spec Publ 70:241–266

Emiliani C (1965) Precipitous continental slopes and considerations on the transitional crust. Science 147:145–148

Enos P (1974) Map of surface sediment facies of the Florida-Bahama Plateau. GSA Map Series MC-5, Boulder, CO

Enos P (1983) Shelf environment. In: Scholle PA, Bebout DG, Moore CH (ed) Carbonate depositional environments. AAPG Mem 33:267–295

Enos P (1991) Sedimentary parameters for computer modeling. In: Franseen EK, Watney WL, Kendall CSCG, Ross W (eds) Sedimentary modeling: computer simulation and methods for improved parameters definition. Bulletin of the Kansas Geological Survey 233:63–99

Enos P, Perkins RD (1977) Quaternary sedimentation in South Florida. GSA Mem 147:198

Ericsson DB, Ewing M, Heezen B (1952) Turbidity currents and sediments in the North Atlantic. AAPG Bull 36:489–511

Fabricius FH (1977) Origin of marine ooids and grapestones. Contributions in Sedimentology 7, Schweitzerbart, Stuttgart, p 113

Freeman-Lynde RP, Cita MA, Jadoul F, Miller EL, Ryan WVF (1981) Marine geology of the Bahama escarpment. Mar Geol 44:119–156

Freile D, Milliman JD, Hillis L (1995) Leeward bank margin *Halimeda* meadows and draperies and their sedimentary importance on the western Great Bahama Bank slope. Coral Reefs 14:27–33

Fütterer DK (1974) Significance of the boring sponge *Cliona* for the origin of fine grained material of carbonate sediments. J Sed Petrol 44:79–84

Gabb WM (1873) Topography and geology of Santo Domingo. Trans Am Phil Soc, n ser 15:49–259

Garrett P, Gould SJ (1984) Geology of New Providence Island, Bahamas. GSA Bull 95:209–220

Gebelein CD (1974) Guidebook for the modern bahamian platform environments. GSA Annual Meeting Fieldtrip Guide, p 93

Gebelein CD, Steinen RP, Garrett P, Hoffman EJ, Queen JM, Plummer LN (1980) Subsurface dolomitization beneath the tidal flats of central west Andros Island, Bahamas. In: Zenger DH, Dunham JB, Ethington RL (eds) Concepts and models of dolomitization. SEPM Spec Publ 28:31–49

Gibson TG, Schlee J (1967) Sediments and fossiliferous rocks from the eastern side of the Tongue of the Ocean, Bahamas. Deep-Sea Research 14:691–702

Ginsburg RN, Shinn EA (1964) Distribution of the reef building community in Florida and the Bahamas. AAPG Bull 48:527

Ginsburg RN (1976) Sedimentary record of paleoclimate in carbonate tidal flats. AAPG Bull 60:874–875

Ginsburg RN (ed) (2001) Subsurface geology of a prograding carbonate platform margin. SEPM Spec Publ 70:207

Ginsburg RN, James NP (1974) Holocene carbonate sediments of continental shelves. In: Burk CA, Drake CL (eds) Continental margins. Springer, New York, pp 137–155

Ginsburg RN, Harris PM, Eberli GP, Swart PK (1991) The growth potential of a bypass margin, Great Bahama Bank. J Sed Petrol 61:976–987

Ginsburg RN, Shinn EA (1994) Preferential distribution of reefs in the Florida reef tract: the past is the key to the present. In: Ginsburg RN (ed) Proceedings colloquium on global aspects of coral reefs: health, hazards, and history. RSMAS Univ Miami, p 21–27

Gonzalez R, Eberli GP (1997) Sediment transport and sedimentary structures in a carbonate tidal inlet; Lee Stocking Island, Exumas Islands, Bahamas. Sedimentology 44:1015–1030

Glockhoff C (1973) Geotectonic evolution and subsidence of Bahama platform; discussion. GSA Bull 84:3473–3476

Grammer GM, Ginsburg RN (1992) Highstand versus lowstand deposition on carbonate platform margins – insight from quaternary foreslopes in the Bahamas. Mar Geol 103:125–136

Grammer GM, Ginsburg RN, Harris PM (1993) Timing of deposition, diagenesis, and failure of steep carbonate slopes in response to a high-amplitude/high-frequency fluctuation in sea level, Tongue of the Ocean, Bahamas. In: Loucks RG, Sarg JF (eds) Carbonate sequence stratigraphy. AAPG Mem 57:107–131

Graus RR, Macintyre IG, Herchenroder BE (1984) Computer simulation of the reef zonation at Discovery Bay, Jamaica: hurricane disruption and long term physical oceanographic control. Coral Reefs 3:59–68

Greenstein GJ (1993) Is the fossil record of regular echinoids really so poor? A comparison of living and subfossil assemblages. Palaios 8:587–601

Greenstein BJ, Meyer DL (1990) Mass mortality of *Diadema antillarum* adjacent to Andros Island Bahamas In: Myloroie J, Gerace D (eds) Fourth symposium on geology of the Bahamas. Bahamian field station, San Salvador, pp 159–168

Haak AB, Schlager W (1989) Compositional variations in calciturbidites due to sea-level fluctuations, late Qaternary, Bahamas. Geol Rundsch 78:477–486

Halley RB, Harris PM, Hine AC (1983) Bank margin. In: Scholle PA, Bebout DG, Moore CH (eds) Carbonate depositional environments. AAPG Mem 33:463–506

Hallock P, Cottey TL, Forward LB, Halas J (1986) Population biology and sediment production of Archias Angulatus (foraminifera) in Largo Sound, Florida. J Foraminifer Res 16:1–8

Hardie LA (1977) Sedimentation on the modern carbonate tidal flats of northwest Andros Island, Bahamas. John Hopkins Univ Stud Geol 22:202

Hardie LA, Shinn EA (1986) Carbonate depositional environments, modern and ancient; Part 3: tidal flats. Colorado School Mines Quart 81(1):1–74

Harris PM (1979) Facies anatomy and diagenesis of a Bahamian ooid shoal. Sedimenta 7, University of Miami, FL, p 163

Harris PM (1983) The Joulters ooid shoal, Great Bahama Bank. In: Peryt TM (ed) Coated grains. Springer, New York, pp 132–141

Harris PM, Kowalik WS (1994) Satellite images of carbonate depositional settings – examples of reservoir- and exploration-scale geologic facies variation. AAPG, Methods in exploration 11:147

Harris PM, Kerans C, Bebout DG (1993) Ancient outcrop and modern examples of platform carbonate cycles – implications for subsurface correlation and understanding reservoir heterogeneity. In: Loucks RG, Sarg JF (eds) Carbonate sequence stratigraphy. AAPG Mem 57:475–492

Hassan M (1998) Modification of carbonate substrata by bioerosion and bioaccretion on coral reefs of the Red Sea. Shaker Verlag, Aachen, p 124

Hearty PJ, Kindler P (1997) The stratigraphy and surficial geology of New Providence and surrounding Islands, Bahamas. J Coastal Res 13:798–812

Hein FJ, Risk MJ (1975) Bioerosion of coral heads: Inner patch reefs, Florida Reef Tract. Bull Mar Sci 25:133–138

Herrera A de (1601–1615) Historia general de los hechos de los castellanos en las Islas y Tierra Firme del mar Océano que llaman Indias Occidentales

Hess HH (1933) Submerged river valleys of the Bahamas. AGU transactions (14th anniversary meeting), pp 168–170

Hess HH (1960) The origin of the Tongue of the Ocean and other great valleys of the Bahama Banks. 2nd Caribbean geological conference, Mayaguez, Puerto Rico, pp 160–161

Hickey BM, MacCready P, Elliott E, Kackel NB (2000) Dense saline plumes in Exuma Sound, Bahamas. J Geophy Res 105:11471–11488

Hidore JJ, Oliver JE (1993) Climatology: an atmospheric science. MacMillan, New York, p 423

Hilgard EW (1871) On the geological history of the gulf of Mexico. Am J Sci 102:391–404

Hilgard EW (1881) The later tertiary of the gulf of Mexico. Am J Sci 122:58–65

Hillis H (1997) Coralgal reefs from a calcareous green alga perspective and a first carbonate budget. Proceedings of the 8th international coral reef symposium. Panama 1:761–766

Hine AC (1977) Lily Bank, Bahamas; history of an active oolite sand shoal. J Sed Petr 47:1554–1581

Hine AC, Neumann AC (1977) Shallow carbonate-bank-margin growth and structure, Little Bahama Bank, Bahamas. AAPG Bull 61:376–406

Hine AC, Wilber RJ, Bane JM, Neumann AC, Lorenson KR (1981a) Offbank transport of carbonate sands along open, leeward bank margins: northern Bahamas. Mar Geol 42:327–348

Hine AC, Wilber RJ, Neumann AC (1981b) Carbonate sand bodies along contrasting shallow bank margins facing open seaways in northern Bahamas. AAPG Bull 65:261–290

Hoskin CM, Reed JK (1985) Carbonate sediment production by the rock-boring urchin, *Echinometra lucunter* and associated endolithic infauna at Black Rock, Little Bahama Bank. In: Reaka ML (ed) The ecology of coral reefs. Symposia series for undersea research 3(1):151–162

Hoskin CM, Reed JK, Mook DH (1986) Production and off-bank transport of carbonate sediment, Black Rock, southwest Little Bahama Bank. Mar Geol 73:125–144

Hutchings PA (1986) Biological destruction of coral reefs: a review. Coral Reefs 4:239–252

Illing MA (1952) Distribution of certain foraminifera within the littoral zone on the Bahama Banks. Ann Mag Nat Hist Ser 12(5):275–285

Illing LV (1954) Bahamian calcareous sands. AAPG Bull 38:1–95

Isemer HJ, Hasse L (1985) The Bunker climate atlas of the North Atlantic Ocean. Springer Verlag, New York, p 342

James NP (1983) Reef. In: Scholle PA, Bebout DG, Moore CH (ed) Carbonate depositional environments. AAPG Mem 33:345–440

Jeans CV, Rawson PF (1980) Andros island, chalk and oceanic oozes – unpublished work of Maurice Black, 5th edn. Yorkshire Geol Soc Occasional Publications, p 100

Johns E, Wilson WD, Molinary RL (1999) Direct observations of velocity and transport in the passages between the Intra-Americas Sea and the Atlantic Ocean, 1984–1996. J Geophys Res 104:25805–25820

Jokiel PL, Coles SL (1977) Effect of temperature on the mortality and growth of Hawaiian reef corals. Mar Biol 43:201–208

Kenter JAM, Anselmetti FS, Kramer P, Westphal H, Vandamme MGM (2002) Acoustic properties of "young" carbonate rocks, Ocean Drilling Program Leg 166 and Holes Clino and Unda, Western Great Bahama Bank. J Sed Res 72:129–137

Kier JS, Pilkey OH (1971) The influence of sea-level changes on sediment carbonate mineralogy, Tongue of the Ocean, Bahamas. Mar Geol 11:189–200

Kievman CM (1998) Match between late Pleistocene Great Bahama Bank and deep-sea oxygen isotope records of sea level. Geology 26:635–638

Kindler P (1995) New data on the Holocene stratigraphy of Lee Stocking island (Bahamas) and its relation to sea-level history. In: Curran HA, White B (eds) Terrestrial and shallow marine geology of the Bahamas and Bermuda. GSA Spec Paper 300:105–116

Kindler P, Hearty PJ (1995) Pre-Sangamonian eolianites in the Bahamas? New evidence from Eleuthera Island. Mar Geol 127:73–86

Kindler P, Hearty PJ (1996) Carbonate petrography as an indicator of climate and sea-level changes: new data from Bahamian Quaternary units. Sedimentology 43:381–399

Kindler P, Hearty PJ (1997) Geology of the Bahamas: architecture of Bahamian Islands. In: Vacher HL, Quinn TM (eds) Geology and hydrogeology of carbonate islands. Develoments in Sedimentology 54, Elsevier, pp 141–160

Kindler P, Strasser A (2000) Palaeoclimatic significance of co-occurring wind- and water-induced sedimentary structures in the last-interglacial coastal deposits from Bermuda and the Bahamas. Sediment Geol 131:1–7

Kramer PA (2003) Synthesis of coral reef health indicators for the western Atlanitic: results of the AGRRA program 1997–2000. Atoll Res Bull 496:1–58

Kramer PA, Kramer PR, Ginsburg RN (2003) Assessment of the Andros island reef system, Bahamas (Part1: stony corals and algae). Atoll Res Bull 496:77–100

Lang JC (1974) Biological zonation at the base of a reef. Am Scientist 62(3):272–281

Lang JC, Wicklund RI, Dill RF (1988) Depth- and habitat-related bleaching of zooxanthellate reef organisms near Lee Stocking Island, Exuma Cays, Bahamas. Proceedings of the 6th international coral reef symposium, Townsville, Australia, pp 269–274

Leaman KD, Vertes PS, Atkinson LP, Lee TN, Hamilton P, Waddell E (1995) Transport potential vorticity, and current/temperature structure across Northwest Providence and Santaren Channels and the Florida Current off Cay Sal Bank. J Geophys Res 100:8561–8569

Leg 101 Scientific Party (1988) Leg 101 – an overview. In: Austin JA Jr, Schlager W et al (ed) Proceedings of ODP, Sci Results, 101: College Station, TX (Ocean Drilling Program), pp 455–472

Le Pichon X, Fox PJ (1971) Marginal offsets, fracture zones, and the early opening of the north Atlantic. J Geophys Res 76:6294–6308

Liddell WD, Avery WE, Ohlhorst SL (1997) Pattern of benthic community structure, 10–250 m, the Bahamas. Proceedings of the 8th International Coral Reef Symposium. Panama 1:437–442

Linton D, Smith R, Alcolado P, Hanson C, Edwards P, Estrada R, Fisher T, Fernandez RG, Geraldes F, McCoy C, Vaughan D, Voegeli V, Warner G, Wiener J (2002) Status of coral reefs in the northern Caribbean and Atlantic Node of the GCRMN. In: Wilkinson C (ed) Status of coral reefs of the world. Australian Institute of Marine Science, Townsville, pp 277–302

Littler MM, Littler DS (1984) A relative dominance model for biotic reefs. Advances in reef sciences. Proceedings of the joint meeting of the Atlantic Reef committee and international society of Reef studies, Miami, Florida, Abstract: 73–74

Littler MM, Littler DS, Hanisak MD (1991) Deep-water rhodolith distribution, productivity and growth history at sites of formation and subsequent degradation. J Exp Biol Ecol 150:163–182

Logan BW (1961) Cryptozoon and associate stromatolites from the Recent, Shark Bay, Western Australia. J Geo 69:517–533

Lugo-Fernandez A (1989) Wave height changes and mass transport on Tague Reef, North Coast of St. Croix, U.S. Virgin Islands. Unpub. Ph.D. disseration, The Louisiana State University and Agricultural and Mechanical College, p 205

Lynts GW (1970) Conceptual model of the Bahamian platform for the last 135 million years. Nature 225:1226–1228

Macintyre IG (1972) Submerged reefs of eastern Caribbean. AAPG Bull 56:720–738

Macintyre IG, Reid PR (1992) Comment on the origin of aragonite needle mud; a picture is worth a thousand words. J Sed Petrol 62:1095–1097

Macintyre IG, Burke RB, Stuckenrath R (1977) Thickest recorded Holocene reef section, Isla Pérez core hole, Alacran Reef, Mexico. Geology 5:749–754

Macintyre IG, Prufert-Bebout L, Reid RP (2000) The role of endolithic cyanobacteria in the formation of lithified laminae in Bahamian stromatolites. Sedimentology 47:915–921

Macintyre IG, Reid RP, Steneck RS (1996) Growth history of stromatolites in a Holocene fringing reef, Stocking Island, Bahamas. J Sed Res 66:231–242

Major RP, Bebout DG, Harris PM (1996) Facies heterogeneity in a modern ooid sand shoal – an analog for hydrocarbon reservoirs. Geological Circular 96-1, Bureau of Economic Geology, Univ Texas, Austin, p 30

Maldonado M, Young CM (1996) Bathymetric patterns of sponge distribution on the Bahamian slope. Deep-Sea Res 43:897–915

Masaferro JL, Eberli GP (1999) Jurassic-Cenozoic structural evolution of the southern Great Bahama Bank. In: Mann P (ed) Caribbean basins: sedimentary basins of the world, Elsevier, pp 167–193

Masaferro JL, Poblet J, Bulnes M, Eberli GP, Dixon TH, McClay K (1999) Palaeogene-Neogene/ present day (?) growth folding in the Bahamian foreland of the Cuban fold and thrust belt. J Geol Soc London 156:617–631

McKee ED, Ward WC (1983) Eolian Environment. In: Scholle PA, Bebout DG, Moore CH (eds) Carbonate depositional environments. AAPG Mem 33:131–170

McNeill DF, Ginsburg RN, Chang SR, Kirschvink JL (1988) Magnetostratigraphic dating of shallow-water carbonates from San Salvador, Bahamas. Geology 16:8–12

McNeill DF, Eberli GP, Lidz BH, Swart PK, Kenter JAM (2001) Chronostratigraphy of prograding carbonate platform margins: A record of dynamic slope sedimentation, Western Great Bahama Bank. In: Ginsburg RN (ed) Subsurface geology of a prograding carbonate platform margin, Great Bahama Bank. SEPM Spec Publ 70:101–134

Meyerhoff AA, Hatten CW (1974) Bahamas salient of North America, Tectonic framework, stratigraphy and petroleum potential. AAPG Bull 58:1201–1239

Milliman JD (1967) The geomorphology and history of Hogsty Reef, a Bahamian atoll. Bull Mar Sci 17:519–543

Milliman JD, Freile D, Steinen RP, Wilber RJ (1993) Great Bahama Bank aragonitic muds: mostly inorganically precipitated, mostly exported. J Sed Petrol 63:589–595

Monty CLV (1976) The origin and development of cryptalgal fabrics. In: Walter MR (ed) Stromatolites, developments in Sedimentology 20, Elsevier, pp 193–250

Moore HB (1972) An estimate of carbonate production by macro-benthos in some tropical, soft-bottom communities. Mar Biol 17:145–148

Morse JW, He S (1993) Influences of T, S, and PCO_2 on the pseudo-homogenous precipitation of $CaCO_3$ from seawater: implications for whiting formation. Mar Chem 41:291–297

Morse JW, Mackenzie FJ (1990) The geochemistry of sedimentary carbonates. Elsevier, Amsterdam, p 707

Morse JW, Millero FJ, Thurmond V, Brown E, Ostlund HG (1984) The carbonate chemistry of Grand Bahama Bank waters: after 18 years another look. J Geophys Res 89:3604–3614

Morse JW, Gledhill DK, Millero FJ (2003) $CaCO_3$ precipitation kinetics in waters from the Great Bahama Bank: implications for the relationship between bank hydrochemistry and whitings. Geochim Cosmochim Acta 67:2819–2826

Mullins HT (1975) Stratigraphy and structure of Northeast Providence Channel, Bahamas and origin of the northwestern Bahama platform. M.S. thesis, Duke University, Durham, NC, p 203

Mullins HT, Lynts GW (1977) Origin of the northwestern Bahama platform: review and reinter-pretation. GSA Bull 88:1447–1461

Mullins HT, Neumann AC (1979) Deep carbonate bank margin structure and sedimentation in the northern Bahamas. In: Doyle LJ, Pilkey OH (eds) Geology of continental slopes. SEPM Spec Publ 28:165–192

Mullins HT, Neumann AC, Wilber RJ, Hine AC, Chinburg SJ (1980) Carbonate sediment drifts in northern Straits of Florida. AAPG Bull 64:1701–1717

Mullins HT, Heath KC, Van Buren HM, Newton CR (1984) Anatomy of a modern open-ocean carbonate slope: northern Little Bahama Bank. Sedimentology 31:141–168

National Buoy Data Center (1973) Environmental conditions within specified geographic regions: offshore East and West coast of the United States and in the gulf of Mexico. U.S. Department of Commerce, p. 7/150–153

Nelsen JE, Ginsburg RN (1986) Calcium-carbonate production by epibionts on *Thalassia* in Florida Bay. J Sed Petrol 56:622–628

Nelson RJ (1853) On the geology of the Bahamas and on coral formations generally. Quaterly J Geol Soc London 9(35):200–215

Neumann AC, Ball MM (1970) Submersible observations in the Straits of Florida: geology and bottom currents. GSA Bull 81:2861–2874

Neumann AC, Land LS (1975) Lime mud deposition and calcareous algae in the Bight of Abaco, Bahamas: a budget. J Sed Petrol 45:763–786

Neumann AC, Gebelein CD, Scoffin TP (1970) The composition, structure and erodability of subtidal mats, Abaco, Bahamas. J Sed Petrol 40:274–297

Newell ND (1955) Bahamian platforms. In: Poldervaart A (ed) Crust of the Earth. Boulder, CO, pp 303–316

Newell ND, Imbrie J (1955) Biogeological reconnaissance in the Bimini area, Great Bahama Bank. Trans N Y Acad Sci 18:3–14

Newell ND, Rigby JK (1957) Geological studies in the Great Bahama Bank. In: Le Blanc RJ, Breeding JG (eds) Regional aspects of carbonate sedimentation. SEPM Spec Publ 5:15–79

Newell ND, Imbrie J, Purdy EG, Thurber DL (1959) Organism communities and bottom facies, Great Bahama Bank. Bull Am Mus Nat Hist 117:117–228

Newell ND, Purdy EG, Imbrie J (1960) Bahamian oolitic sand. J Geo 68:481–497

Palmer MS (1979) Holocene facies geometry of the Leeward Bank Margin, Tongue of the Ocean, Bahamas. M.S. thesis, University of Miami, p 200

Paull CK, Neumann AC, Bebout B, Zabielski V, Showers W (1992) Growth rate and stable isotopic character of modern stromatolites from San Salvador, Bahamas. Palaeogeogr Palaeoclim Palaeoecol 95:335–344

Paulus FJ (1972) The Geology of Site 98 and the Bahama Platform. In: Hollister CD, Ewing JI et al (eds) Proceedings of ODP, Init Repts, 11: College Station, TX (Ocean Drilling Program), pp 877–897

Payri CE (1997) *Hydrolithon reinboldii* distribution, growth and carbon production of a French Polynesian reef. 8th international coral reef symposium. Panama 1:755–760

Perkins RD, Enos P (1968) Hurricane Betsy in the Florida-Bahamas area; geologic effects and comparison with Hurricane Donna. J Geo 76:710–717

Perkins RD, Dwyer GS, Rosoff DB, Fuller J, Baker PA, Lloyd RM (1994) Salina sedimentation and diagenesis; West Caicos Island, British West Indies. In: Purser B, Tucker M, Zenger D (eds) Dolomites; a volume in honour of Dolomieu. IAS Spec Publ 21:37–54

Pilskaln CH, Neumann CA, Bane JH (1989) Periplatform carbonate flux in the northern Bahamas. Deep-Sea Res 36:1391–1406

Pomar L (1993) High-resolution sequence stratigraphy in prograding Miocene carbonates; application to seismic interpretation. In: Loucks RG, Sarg JF (eds) Carbonate sequence stratigraphy; recent developments and applications. AAPG Mem 57:389–407

Purdy EG (1963a) Recent calcium carbonate facies of the Great Bahama Bank. 1. Petrography and reaction groups. J Geo 71:334–355

Purdy EG (1963b) Recent calcium carbonate facies of the Great Bahama Bank. 2. Sedimentary facies. J Geo 71:472–497

Queen JM (1978) Carbonate sedimentology and ecology of some pelleted muds west of Andros Island, Great Bahama Bank. Ph.D. thesis, State University of New York, Stony Brook, p 401

Rankey EC (2002) Spatial patterns of sediment accumulation on a Holocene carbonate tidal flat northwest Andros Island, Bahamas. J Sed Res 72:591–601

Rankey EC, Morgan J (2002) Quantified rates of geomorphic change on a modern carbonate tidal flat, Bahamas. Geology 30:583–586

Rankey EC, Enos P, Steffen K, Druke D (2004) Lack of impact of hurricane Michelle on tidal flats, Andros island, Bahamas: integrated remote sensing and field observations. J Sed Res 74:654–661

Rankey EC, Riegl B, Steffen, K (2006) Form, function, and in a tidally dominated ooid shoal, Bahamas. Sedimentology 53:1191–1210

Reaka-Kulda ML, O'Connell DO, Regan JD, Wicklund RI (1994) Effect of temperature and UV-B on different components of coral reef communities from the Bahamas. Proceedings colloquium on global aspects of coral reefs: health, hazards, and history. RSMAS Univ Miami, 126–131.

Reeder SL, Rankey EC (2009) Controls on morphology and sedimentology of carbonate tidal deltas, Abacos, Bahamas. Marine Geology 267:141–155

Reid RP, Macintyre IG, Browne KM, Steneck RS, Miller T (1995) Modern marine stromatolites in the Exuma Cays, Bahamas: uncommonly common. Facies 33:1–17

Reid RP, Visscher PT, Decho AW, Stolz JF, Bebout BM, Dupraz C, Macintyre IG, Paerl HW, Pinckney JL, Prufert-Bebout L, Steppe TF, DesMarais DJ (2000) The role of microbes in accretion, lamination and early lithification of modern marine stromatolites. Nature 406:989–992

Reijmer JJG, Schlager W, Bosscher H, Beets CJ, McNeill DF (1992) Pliocene/Pleistocene platform facies transition recorded in calciturbidites (Exuma Sound, Bahamas). Sediment Geol 78:171–179

Reijmer JJG, Schlager W, Droxler AW (1988) Site 632: Pliocene-Pleistocene sedimentation cycles in a Bahamian basin. In: Austin JA Jr, Schlager W et al (eds) Proceedings of ODP, Sci Results 101: College Station, TX (Ocean Drilling Program): 213–220

Rendle, RH (2000) Quaternary slope development and sedimentology of the Western, Leeward Margin of Great Bahama Bank (ODP Leg 166). Ph.D. thesis, University of Kiel, Germany, p 199

Richardson WS, Schmitz WJ Jr, Niiler PP (1969) The velocity structure of the Florida Current from the Straits of Florida to Cape Fear. Deep-Sea Res 16:225–231

Riegl B, Piller WE (2003) Possible refugia for reefs in times of environmental stress. Int J Earth Sci 92:520–531

Riegl B, Manfrino C, Hermoyian C, Brandt M, Hoshino K (2003) Assessment of the coral reefs of the Turks and Caicos Islands (Part 1: stony corals and algae). Atoll Res Bull 496:461–480

Robbins LL, Blackwelder PL (1992) Biochemical and ultrastructural evidence for the origin of whitings: a biologically induced calcium carbonate precipitation mechanism. Geology 20:464–468

Robbins LL, Tao Y, Evans CA (1997) Temporal and spatial distribution of whitings on Great Bahama Bank and a new lime mud budget. Geology 25:947–950

Roberts HH (1979) Reef-crest wave and current interactions and sediment transport. AAPG Bull 63:517

Roberts HH, Rouse LJ Jr, Walker ND, Hudson JH (1982) Cold-water stress in Florida Bay and northern Bahamas; a product of winter cold-air outbreaks. J Sed Petrol 52:145–155

Rose PR, Lidz B (1977) Diagnostic foraminiferal assemblages of shallow-water modern environments: South Florida and the Bahamas. Sedimenta 6, University of Miami, FL, p 55

Rusnak GA, Nesteroff WD (1964) Modern turbidites: Terrigenous abyssal plain versus bioclastic basin. In: Miller RL (ed) Papers in marine geology. Macmillan, New York, pp 488–507

Schlager W (1981) The paradox of drowned reefs and carbonate platform. GSA Bull 92:197–211

Schlager W (2005) Carbonate sedimentology and sequence stratigraphy: SEPM concepts in Sedimentology and Paleontology 8, p 200

Schlager W, James NP (1978) Low-magnesian calcite limestone forming at the deep-sea floor, Tongue of the Ocean, Bahamas. Sedimentology 25:675–702

Schlager W, Chermak A (1979) Sediment facies of platform-basin transition, Tongue of the Ocean, Bahamas. In: Doyle LJ, Pilkey OH (eds) Geology of continental slopes. SEPM Spec Pub 27:193–208

Schlager W, Ginsburg RN (1981) Bahama carbonate platforms – the deep and the past. Mar Geol 44:1–24

Schlager W, Austin JA et al (1985) Ocean drilling program; rise and fall of carbonate platforms in the Bahamas. Nature 315:632–633

Schlager W, Bourgeois F, Mackenzie G, Smit, J (1988) Boreholes at Great Isaac and Site 626 and the history of the Florida Straits. In: Austin JA Jr, Schlager W et al (eds) Proceedings of ODP, Sci Results, 101: College Station, TX (Ocean Drilling Program), 425–437

Schlager W, Reijmer JJG, Droxler AW (1994) Highstand shedding of carbonate platforms. J Sed Petrol 64:270–281

Schuchert C (1935) Historical geology of the Antillean-Caribbean region or the lands bordering the gulf of Mexico and the Caribbean Sea. Wiley, New York, p 811

Sealey NE (1994) Bahamian Landscapes: an introduction to the physical geography of the Bahamas. Media Publishing, Nassau, Bahamas, p 128

Sheridan RE (1971) Geotectonic evolution and subsidence of Bahama platform; discussion. GSA Bull 82:807–809

Sheridan RE (1974) Atlantic continental margin of North America. In: Burk CA, Drake CL (eds) The geology of continental margins. Springer, New York, pp 391–407

Sheridan RE (1976) Sedimentary basins of the Atlantic margin of North America. Tectonophysics 36:113–132

Sheridan RE, Crosby JT, Bryan GM, Stoffa PL (1981) Stratigraphy and structure of Southern Blake Plateau, Northern Florida Straits and Northern Bahama platform from multichannel seismic reflection data. AAPG Bull 65:2571–2593

Shinn EA (1983) Tidal flat. In: Scholle PA, Bebout DG, Moore CH (eds) Carbonate depositional environments. AAPG Mem 33:345–440

Shinn EA, Ginsburg RN, Lloyd RM (1965) Recent supratidal dolomite from Andros Island, Bahamas. In: Pray LC, Murray RC (eds) SEPM Spec Publ 13:112–123

Shinn EA, Lloyd RM, Ginsburg RN (1969) Anatomy of a modern carbonate tidal flat. J Sed Petrol 53:1202–1228

Shinn EA, Steinen RP, Dill RF, Major RP (1993) Lime-mud layers in high energy tidal channels: a record of hurricane deposition. Geology 21:603–606

Shinn EA, Steinen RP, Lidz BH, Swart PK (1989) Whitings, a sedimentologic dilemma. J Sed Petrol 59:147–161

Shore and Beach (1972) Surface water temperature and density – Atlantic Coast, North and South America. Shore Beach 40:37–43

Smith CL (1940) The Great Bahama Bank. J Mar Res 3:147–189

Smith FGW (1948) Atlantic reef corals; a handbook of the common reef and shallow-water corals of Bermuda, the Bahamas, Florida, the West Indies, and Brazil. University of Miami Press, p 112

Smith NP (2001) Weather and hydrographic conditions associated with coral bleaching: Lee Stocking Island, Bahamas. Coral Reefs 20:415–422

Smith JD, Hopkins TS (1972) Sediment transport on the continental shelf off of Washington and Oregon in light of recent current measurements. In: Swift DJP, Duane DB (eds) Shelf sediment transport; process and pattern. Dowden, Hutchinson and Ross, Stroudsburg, PA, pp 143–180

Stafford-Smith MG (1992) Mortality of the hard coral *Leptoria phrygia* under persistent sediment influx. Proceedings of the 7th International Coral Reef Symposium. Guam 1:289–299

Steneck RS, Testa V (1997) Are calcareous algae important to reefs today or in the past? Symposium summary. Proceedings of the 8th international coral reef symposium. Panama 1:685–698

Stevens J (1726) The general history of the vast continent and islands of America

Stockman KW, Ginsburg RN, Shinn EA (1967) The production of lime mud by algae in south Florida. J Sediment Geol 37:633–648

Storr JF (1964) Ecology and oceanography of the coral-reef tract, Abaco Island, Bahamas. GSA Spec Paper 79:98

Strasser A, Davaud E (1986) Formation of Holocene limestone sequences by progradation, cementation, and erosion; two examples from the Bahamas. J Sed Res 56:422–428

Suess E (1885–1909) Das Antlitz der Erde. 3rd edn. F Tempsky, Prague, 1989 p.

Swart PK, Eberli GP, Malone MJ, Sarg JK (eds) (2000) Proceedings of the ODP, Sci. Results, 166: College Station, TX, p 195

Taft WH, Arrington F, Haimovitz A, MacDonald C, Woolheater C (1968) Lithification of modern marine carbonate sediments at Yellow Banks, Bahamas. Bull Mar Sci 18:762–828

Talwani M, Worzel JL, Ewing M (1960) Gravity anomalies and structure of the Bahamas. Transactions of the 2nd Caribbean geological conference, University of Puerto Rico, pp 156–160

Thompson JB (2000) Microbial whitings. In: Riding RE, Awramik SM (eds) Microbial sediments. Springer, Berlin, pp 250–260

Traverse A, Ginsburg RN (1966) Palynology of the surface sediments of Great Bahama Bank, as related to water movement and sedimentation. Mar Geol 4:417–459

Tucker ME, Wright VP (1990) Carbonate sedimentology. Blackwell Science, Oxford, p 482

Uchupi E, Milliman JD, Luyendyk BP, Bowin CO, Emery KO (1971) Structure and origin of southeastern Bahamas. AAPG Bull 55:687–704

United States Naval Oceanographic Office (1973) Surface currents. Naval Oceanographic Office Special Publication 1400-NA 1, Naval Oceanographic Office NSTL Station MS

Vaughan TW (1914) Preliminary remarks on the Geology of the Bahamas, with special reference to the origin of the Bahaman and Floridian Oölites. Carnegie Institution publication no. 182:47–54

Vogel K, Gektidis M, Golubic S, Kiene WE, Radtke G (2000) Experimental studies on microbial bioerosion at Lee Stocking Island, Bahamas and One Tree Island, Great Barrier Reef, Australia: implication for paleoecological reconstructions. Lethaia 33:190–204

Wang J, Mooers CNK (1997) Three-dimensional perspectives of the Florida Current: transport, potential vorticity, and related dynamical properties. Dyn Atmos Oceans 27:135–149

Wanless HR, Dravis JJ (1989) Carbonate environments and sequences of Caicos platform. Field Trip Guidebook T374, 28th International Geological Congress, p 75

Westphal H (1998) Carbonate platform slopes – a record of changing conditions. The Pliocene of the Bahamas. Lecturer Notes in Earth Sciences 75, Springer, Heidelberg, p 197

Westphal H, Reijmer JJG, Head MJ (1999) Input and diagenesis on a carbonate slope (Bahamas): response to morphology evolution and sea-level fluctuations. In: Harris PM, Saller AH, Simo JA, Handford CR (eds) Advances in carbonate sequence stratigraphy – application to reservoirs, outcrops and models. SEPM Spec Pub 63:247–274

Westphal H, Head MJ, Munnecke A (2000) Differential diagenesis of rhythmic limestone alternations supported by palynological evidence. J Sedim Research 70:715–725

White B, Curran HA (1988) Mesoscale physical sedimentary structures and trace fossils in Holocene eolianites from San Salvador, Bahamas. Sediment Geol 55:163–184

Wilber RJ, Milliman JD, Halley RB (1990) Accumulation of Holocene banktop sediment on the western margin of Great Bahama Bank: rapid progradation of a carbonate megabank. Geology 18:970–974

Wilber RJ, Whitehead JA, Halley RB, Milliman JD (1993) Carbonate-periplatform sedimentation by density flows; a mechanism for rapid off-bank and vertical transport of shallow-water fines: comment. Geology 21:667–668

Wilkinson CR, Buddemeier RW (1994) Global climate change and coral reefs: implication for people and reefs. In: UNEP-IOC-ASPEI-IUCN global task team on the implication of climate on coral reefs, p 124

Wilson JL (1974) Characteristics of carbonate platform margins. AAPG Bull 58:810–824

Wilson PA, Roberts HH (1992) Carbonate-periplatform sedimentation by density flows: a mechanism for rapid off-bank and vertical transport of shallow water fines. Geology 20:713–716

Wilson PA, Roberts HH (1995) Density cascading: off-shelf transport, evidence and implications, Bahama Banks. J Sed Res 65:45–56

Wilson WL, Mylroie JE, Carew J (1995) Caves as a geologic hazard; a quantitative analysis from San Salvador Island, Bahamas. In: Beck B (ed) Karst Geohazards; engineering and environmental problems in karst terranes. Proceedings – multidisciplinary conference on sinkholes and the Engineering and Environmental impacts of Karst 5:487–495

Woelkerling WJ (1976) South Florida benthic marine algae; keys and comments. Sedimenta 5, University of Miami, FL, p 145

Wood R (1999) Reef evolution. Oxford University Press, Oxford, p 414

Young IR, Holland G (1996) Atlas of the oceans – Wind and wave climate. Pergamon Press, Oxford, p 414

Chapter 3
Belize: A Modern Example of a Mixed Carbonate-Siliciclastic Shelf

Donald F. McNeill, Xavier Janson, Kelly L. Bergman, and Gregor P. Eberli

3.1 Introduction

The Belize shelf is located on the eastern side of the Yucatan Peninsula (Fig. 3.1). It extends along approximately 300 km in a north-south direction and 10–40 km in an east-west direction. The Belize shelf lagoon is a mixed carbonate-siliciclastic rimmed platform developed under a humid tropical climate. The southern shelf lagoon depositional system is attached to a mountainous mainland of the Yucatan peninsula. The main relief features are the Maya Mountains that culminate at a height approximating 1,000 m above sea level.

The Belize shelf is composed of a more or less continuous reefal barrier that separates a lagoon from the deep open-ocean to the east. The shelf can be subdivided into five main regions based on water depth and morphology (Fig 3.1). The five subdivisions include: (1) Chetumal Bay, a shallow (2 m or less), restricted embayment at the northern end of the Belize Lagoon and southern Mexico; (2) Northern Shelf Lagoon, the northern lagoon ranging from Belize City to Ambergris Cay with water depths up to 6 m; (3) Central Shelf Lagoon, from the latitude of Belize City to that around the Settee River with water depths up to 22 m; (4) Southern Shelf Lagoon, from the latitude of the Settee River and the start of the abundant lagoon reefs, southward to the end of the barrier reef at Sapodilla Cays, water depths up to 40 m; and (5) Gulf of Honduras, the area comprising the southernmost part of the Belize lagoon and ranging into open water (>50 m) where no shallow reefs occur. Seaward of the Belize shelf three atoll platforms lie offshore (Turneffe, Lighthouse, and Glovers) (Fig. 3.1).

D.F. McNeill (✉) and G.P. Eberli
Comparative Sedimentology Laboratory, Rosenstiel School of Marine and Atmospheric Science, University of Miami, Miami, Florida, USA
e-mail: geberli@rsmas.miami.edu

X. Janson
Bureau of Economic Geology, University of Texas, Austin, Texas, USA

K.L. Bergman
Chevron Energy Technology Company, San Ramon, California, USA

H. Westphal et al. (eds.), *Carbonate Depositional Systems: Assessing Dimensions and Controlling Parameters*, DOI 10.1007/978-90-481-9364-6_3,
© Springer Science+Business Media B.V. 2010

The non-carbonate mountain relief associated with the humid climate is responsible for the drainage of siliciclastic into the near-shore shallow-water area. A well-established long-shore current has confined this bedload siliciclastic deposit to the coastal area. Suspended siliciclastic mud are transported further offshore, mostly during the wet season (June–October). These mud plumes often reach the central part of the lagoon, and sometimes as far as the barrier reef. Mud plumes that originate from drainage around The Gulf of Honduras have also been observed to carry mud-laden water and cover much of the lagoon southward from the latitude of Placencia.

Pure carbonate sedimentation occurs at several different settings on the shelf lagoon. The carbonate deposits consist of reefal build-ups and skeletal sand near the barrier reef and the inner lagoon build-ups; carbonate marl in the deeper lagoon and sandy mud and carbonate mud in the more restricted lagoon area. The facies organization in Belize is the result of the combined influence of the structural background of the shelf and the antecedent morphology, the geographical setting and the climate, the oceanographic condition and the type of sediment produced. In the following section, each of these controlling parameters will be shortly described and the resulting facies organization of the shelf will be presented in detail. The aim is to illustrate the numerous physical and biological processes associated with feedback mechanism that control the carbonate deposition in such a setting.

3.2 Geography, Climate, and Oceanographic Setting

3.2.1 Climate

The Belize shelf lies within the tropical climate belt and the marine environment is tropical to subtropical (Portig 1976). The climate has been summarized by Wright et al. (1959) and is characterized by a pronounced wet and dry season. The rainy season lasts from June to October with yearly rainfall averages ranging from ~125 cm in northern Belize (at Chetumal) to 450 cm in the south (near the Guatemala border) (Fig. 3.1). Humidity averages 78% between March and June with a range of 58–96% (Rützler and Macintyre 1982a). Figure 3.2 shows monthly rainfall averages at Carrie Bow Cay and Table 3.1 shows hurricanes that have passed within a 50 km radius of Carrie Bow Cay between the years of 1960 and 1980.

Stoddart (1963, 1969) showed that the Belize reefs were affected by a major hurricane on the average of once per every 6-years based on storm frequency between 1931 and 1961.

3.2.2 Air and Water Temperature

Air temperatures in Belize average 27°C in summer and 24°C in winter (Wright et al. 1959). The range of air temperatures is from 10°C to 36°C; the average range in Belize City is from 23°C to 33°C (Rützler and Macintyre 1982a).

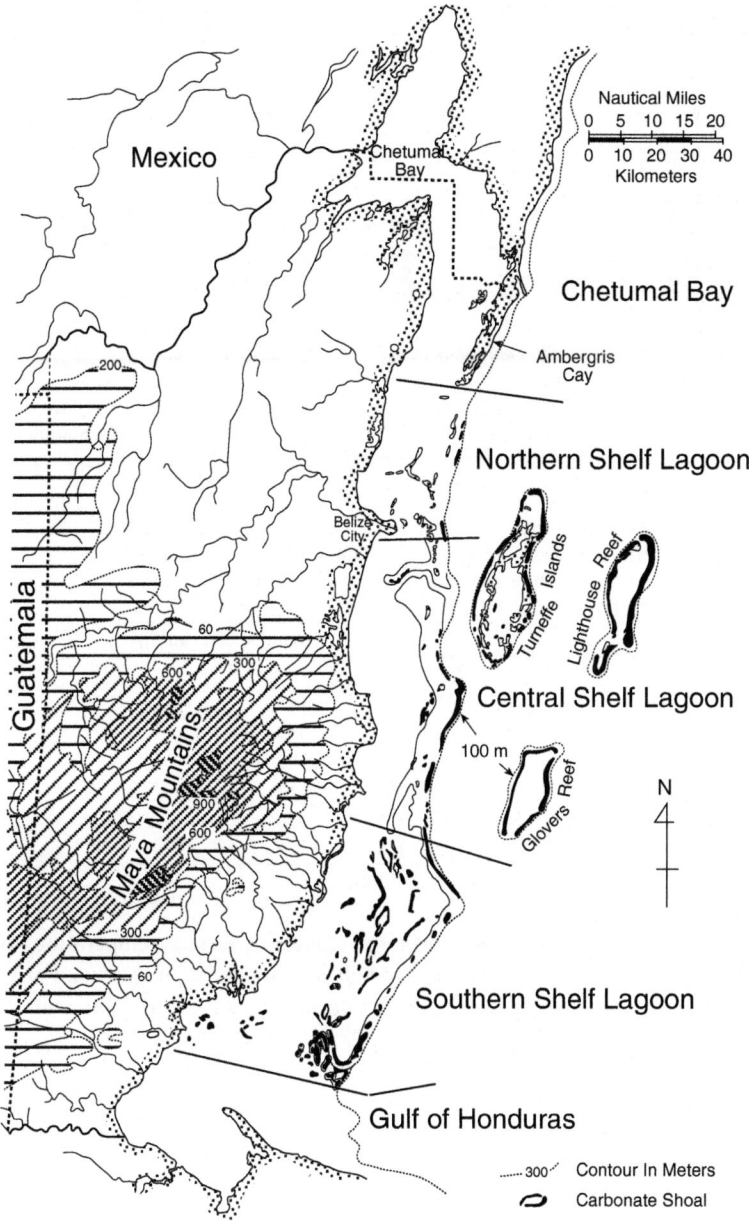

Fig. 3.1 Location of the five main regions that comprise the Belize shelf lagoon

In June of 1975, mean day and night water temperatures recorded at Carrie Bow Cay were 28.6°C and 28.0°C, respectively (Kjerfve 1978). A comparison of water and air temperatures at Carrie Bow Cay and the Belize mainland is shown in

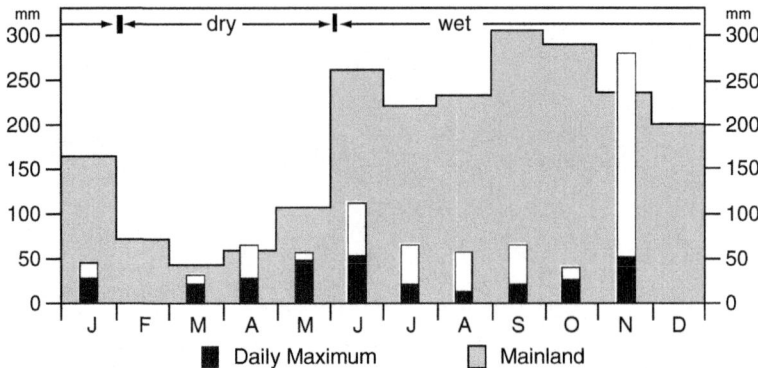

Fig. 3.2 Monthly average (*white bars*) and daily maximum rainfall (*black bars*) at Carrie Bow Cay (1976–1980), as well as mainland monthly rainfall (*gray*) at Melinda Forest Station averaged over a 71-year period (1906–1977) (Figure 58 of Rützler and Ferraris 1982)

Table 3.1 Hurricanes that passed within a 50 km radius of Carrie Bow Cay (1960–1980), including maximum sustained wind speed while center was within 50-km radius (Data is from Table 8 of Rützler and Ferraris (1982))

Name	Month/year	Wind speed (km/h)
Abby	July 1960	128
Anna	July 1961	148
Hattie	October 1961	259
Francelia	August 1969	182
Laura[a]	November 1971	111
Fifi	September 1974	176
Greta	September 1978	176

[a]Officially declared a tropical storm

Table 3.2. Fuglister (1947) reported that average sea surface temperatures off the northern Belize shelf vary from 25.5°C in February to 28.5°C in August. Purdy et al. (1975) report central lagoon water temperatures in summer of 1961 and 1962 to range from 26.4°C to 30.9°C and average 28.9°C (n = 215).

In January 1972 they measured water temperatures in the southern shelf to range between 24.6°C and 27.0°C with a mean of 26.2°C (n = 64). Temperatures of approximately 29°C were recorded by Pusey (1964) near the reef on the northern shelf and 30°C in the shelf lagoon and bay with daily warming of 2°C. The wind effect due to "northers" on water temperatures over the reef is insignificant (Stoddard 1962), but its effect on the more restricted, shallow lagoon is unknown (Ebanks 1967).

3.2.3 Wind

Although Belize is located near the tropical doldrums belt, it is strongly affected by easterly tradewinds (Fig. 3.3) (Ebanks 1967, 1975). The tradewinds blow steadily

Table 3.2 Means and ranges of water temperatures in °C at Carrie Bow Cay from 1976–1980. Mainland data is monthly mean from the 10-year period, 1965–1975 (Data from Figure 55 of Rützler and Ferraris 1982)

	Air	Sand	Reef flat	Lagoon	Mainland
January	24.5	25.5	25.0	25.5	–
	21.0–31.0	21.0–35.0	22.0–28.0	23.0–29.0	22.0–29.0
February	24.5	25.5	25.0	25.5	–
	21.0–29.0	19.0–29.0	22.0–28.0	23.0–29.0	21.0–30.0
March	27.0	30.5	28.0	28.0	–
	21.0–32.0	25.0–38.0	25.5–33.0	25.0–31.0	23.0–31.0
April	27.5	28.5	28.0	27.5	–
	21.5–34.0	21.5–37.0	23.0–34.5	24.0–32.0	24.0–32.0
May	30.0	31.0	29.5	29.0	–
	25.0–36.0	25.0–40.0	25.5–36.0	25.0–33.0	25.0–33.0
June	29.0	30.0	29.0	29.0	–
	25.0–35.5	24.0–37.5	26.5–35.5	26.0–32.0	25.0–32.0
July	–	–	–	–	–
					25.0–32.0
August	–	–	29.0	30.0	–
			28.0–31.0	28.0–34.0	25.0–33.0
September	28.0	30.0	29.5	28.5	–
	25.0–32.0	24.0–37.0	24.5–33.0	27.0–32.0	25.0–32.0
October	28.0	30.0	29.5	28.5	–
	25.0–32.0	24.0–37.0	24.5–33.0	27.0–32.0	24.0–31.0
November	26.5	30.0	27.5	25.5	–
	24.0–33.0	23.0–40.0	21.0–34.0	20.0–30.0	23.0–30.0
December	–	–	–	–	–
					23.0–29.0

from the east and northeast at velocities of 12–15 knots (~6–7.5 m/s) (U.S. Naval Oceanographic Office 1963), but afternoon onshore breezes can increase these tradewinds to 15–20 knots (~7.5–10 m/s). Offshore breezes in the mornings can be felt at the mainland shoreline, but rarely affect the cays along the barrier reef tract. The normal easterly tradewinds are disrupted in the winter by cold air masses that move southward off continental North America. These "northers" are frequently accompanied by 30-knot (~15 m/s) winds and rainfall (Ebanks 1967). These cold air masses can bring temperatures as low as 10°C (Stoddart 1962) and often produce heavy rainfall. The strong winds (and waves) likely impact sediment movement in the shallow lagoon and reef crests. Wind data collected from the barrier reef at Carrie Bow Cay document the seasonal direction and strength of the winds over the Belize lagoon (Table 3.3).

Hurricanes (wind speeds >118 km/h) made landfall in Belize during late summer and fall months, on an average of every 6 years within the 30-year period from 1931 to 1961 (Stoddard 1963). Tropical storms (wind speeds >88 km/h) occur every 2.5 years (Gentry 1971). Although the mainland, shallow reefs, and exposed

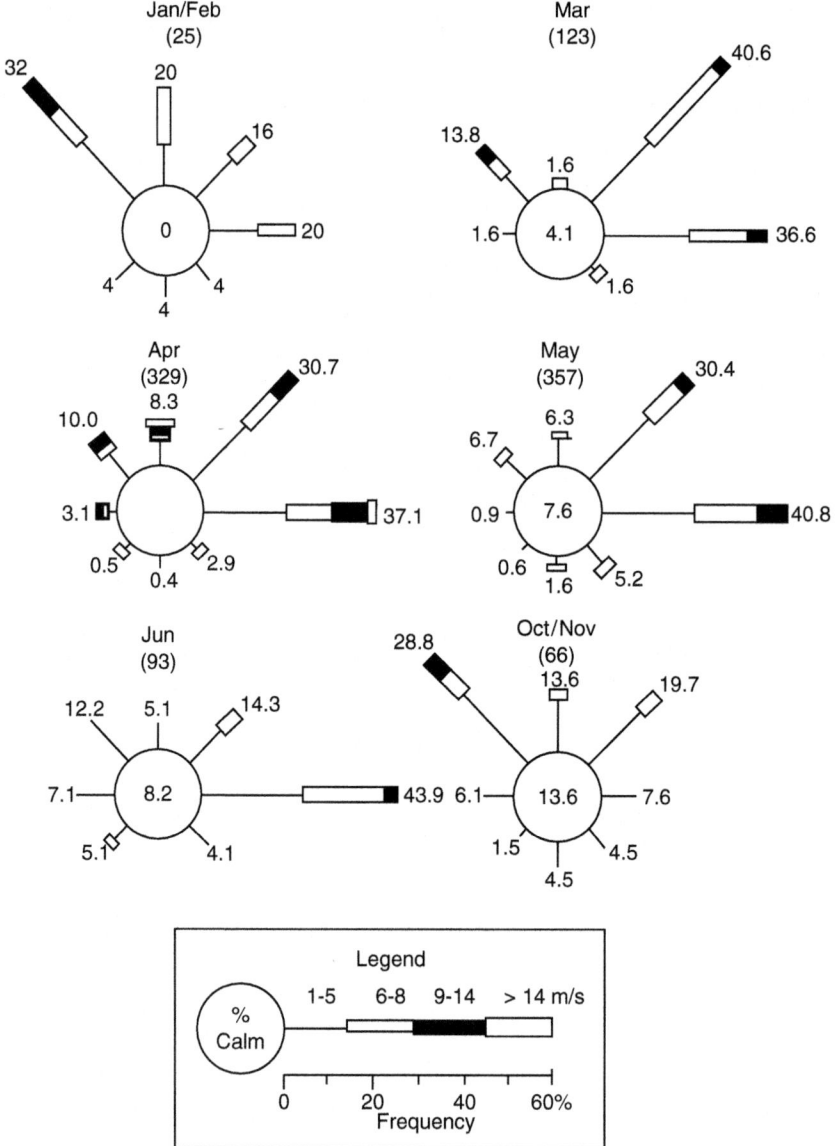

Fig. 3.3 Wind observations at Carrie Bow Cay: northeasterly trade winds prevail at velocities of 4–5 m/s during about 70% of the year; wind roses indicate direction, speed, and frequency. Numbers in parentheses are number of observations during 1976–1980 (Figure redrawn from Figure 56 of Rützler and Ferraris 1982)

cays can be drastically affected by hurricanes, the deeper submarine depositional environments are not significantly altered (Stoddard 1962; Vermeer 1963; Ebanks 1967). Because wind events associated with storms are closely related to short term

Table 3.3 Summary of monthly wind speed frequencies at Carrie Bow Cay during the period 1976–1980 (From Rützler and Ferraris 1982)

Month	Dominant direction	Frequency (%)	Secondary direction	Frequency %	Speed (m/s)	Frequency (%)
January/	NW	32.0	NE	20	Calm	0
February				20	1–5	60
					6–8	32
					9–14	8
March	NE	40.6	E	36.6	Calm	4
					1–5	48
					6–8	40
					9–14	8
April	E	37.1	NE	30.7	Calm	7
					1–5	52
					6–8	25
					9–14	14
					>14	2
May	E	40.8	NE	30.4	Calm	8
					1–5	58
					6–8	26
					9–14	8
June	E	43.9	NE	14.3	Calm	8
					1–5	65
					6–8	22
					9–14	5
October/	NW	28.8	NE	19.7	Calm	14
November					1–5	71
					6–8	12
					9–14	3

draining and flooding (surge) events in the lagoon, sediment transport likely to be most affected by wind-induced currents and waves during the passage of storms. Recently, with the passing of Hurricane Iris (2001), large coral heads and palmate coral boulders were transported to the reef flat along some of the coastal reefs south of Placencia (D.F. McNeill, unpublished observation, 2001).

3.2.4 Wave Energy

Most of the wave energy arriving in Belize dissipates along the extensive barrier reef complex that shields the Belize lagoon from the open Caribbean Sea. Wave energy is also dissipated along the windward, eastern portions of the three isolated offshore platforms of Belize (Turneffe, Lighthouse, Glovers). The leeward flanks of these offshore atoll platforms, the immediate leeward side of the barrier platform, and the siliciclastic mainland shoreline of the Belize lagoon are exposed to far less wave power.

The mean wave propagation direction is 74.5° (ENE), which lies at the center of a 45° sector that comprises 87% of wave frequency (Fig. 3.4).

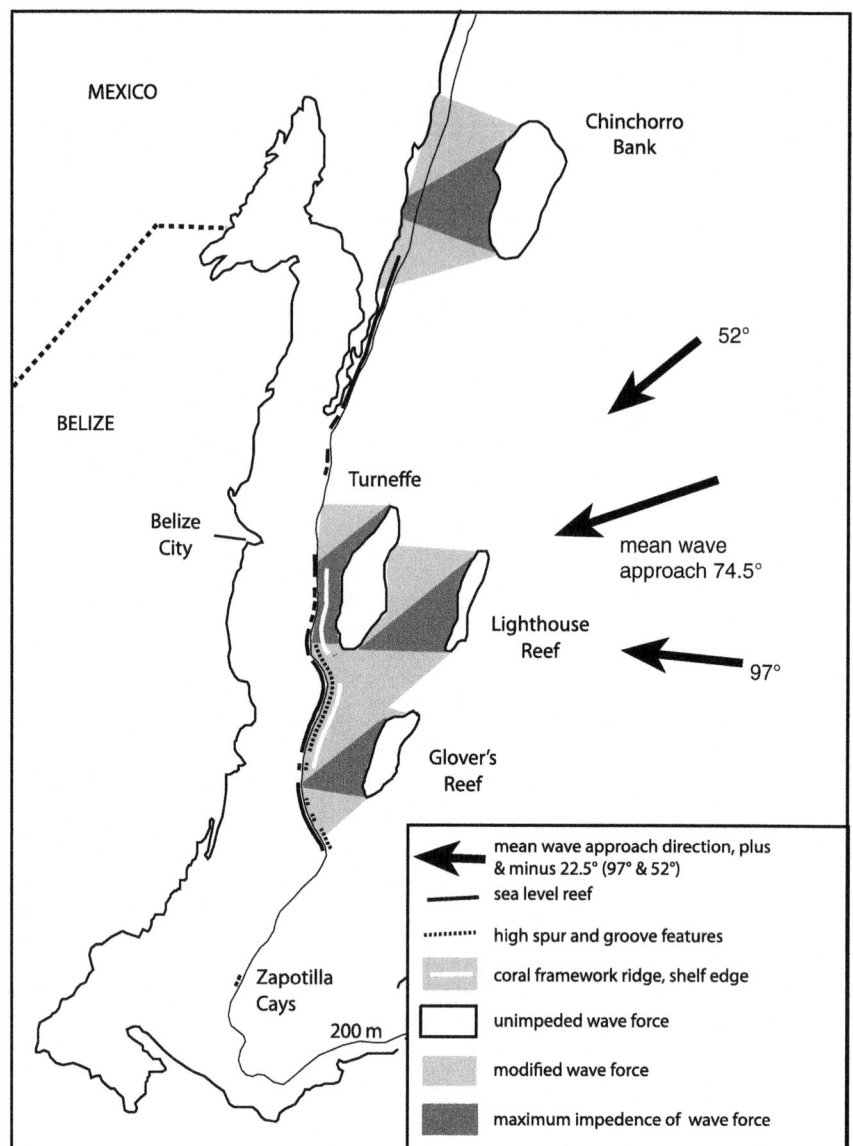

Fig. 3.4 Diagram of the relation between the magnitude of wave energy and sea level reefs, spur and groove features, and shelf-edge ridges along the Belize barrier reef. Wave force impact on the reefs is controlled by wind speed and direction and by the impedance and modifying effect of the offshore carbonate platforms (Figure redrawn from Burke 1982)

This direction influences the location of fore-reef structures on the seaward side of the barrier reef, in particular the high spur and groove features and coral framework ridges (Burke 1982). It is interesting to note that where the unimpeded wave force hits the Belize barrier, the reefs do not reach sea level, and they are often discontinuous laterally (e.g., the barrier interval between the wave shadow of Chinchorro Bank and Turneffe, and the area south of the Glovers reef platform).

The maximum wave height reported for this region is 10 m with a period of 12.7 s, which occurred during hurricane Greta in 1978 (Macintyre et al. 1987). Hurricane conditions are likely needed to move large branching fronds of the coral *Acropora palmata* (Macintyre et al. 1987). Tropical storm force conditions are interpreted to be capable of moving sand-size sediment across most of the barrier apron, while normal trade winds do not normally move sediment across much of the barrier reef sediment apron (Macintyre et al. 1987).

3.2.5   Tides and Currents

Tides in the Belize lagoon (as measured on the barrier reef) are microtidal and of the mixed semidiurnal type (Form number F between 0.25 and 1.50) (Kjerfve et al. 1982). At Carrie Bow Cay (F = 1.13) the mean tidal range is 15 cm (Kjerfve et al. 1982). The mean predicted ranges at Belize City and Punta Gorda is 16 cm with diurnal mean ranges of 18.5 and 21 cm, respectively (International Marine Tide Tables 1998). At Ambergris Cay maximum tidal ranges are of 33–40 cm on exposed seaward shores (Ebanks 1967). Lesser ranges exist on exposed leeward shores and often no tides at all are observed in restricted intra-island lagoons. Predicted maximum monthly tidal ranges are listed in Table 3.4 for Belize City and Punta Gorda. Although having less of an influence upon sediment transport than do wind induced currents, the persistence of the tidal currents set up by these weak tides is reflected in channels through mud shoals on the shelf and in lineations of seagrass beds near the mouth of Chetumal Bay (Ebanks 1967). Figures 3.5 and 3.6 illustrates the general circulation pattern of currents off and on the Belize shelf, respectively.

Greer and Kjerfve (1982) measured current velocities on the barrier reef platform at South Water Cut and Carrie Bow Cut. The currents in these channels clearly exhibit tide-dominated periodicities with peak velocities of approximately 40 cm/s. Flood tide current velocities exceed ebb tide current velocities. This can be attributed to the fact that during flood tide the water moves with the wind and against it during the ebb tide. On the lee side of Carrie Bow Cay, weak currents (2–6 cm/s) are set up by refracted waves that penetrate nearby cuts. Current measurements in the back-reef lagoon at Carrie Bow Cay indicate a maximum velocity of 33 cm/s, but show very little tidal influence and are therefore attributed to the persistent northeasterly tradewinds (4–6.5 m/s) observed at the site. Mean and maximum tidal ranges at Carrie Bow Cay are 15 and 40 cm, respectively (Greer and Kjerfve 1982). Changing wind directions result in alternately draining and flooding of the shallow Belize shelf (Stoddard 1963; Ebanks 1967, 1975). Water level is reported to commonly fall by about 70 cm during "northers" which illustrates the relative importance of wind-induced currents in contrast to tide-induced currents. As an example, during Hurricane Keith in 2000, the "northers" associate with the depression nearly emptied the Chetumal Bay for several hours (U.S. National Hurricane Center web archive, url: www.nhc.noaa.gov/2000keith.html).

Table 3.4 Maximum monthly tidal ranges (cm) (From International Marine Tide Tables 1998)

	January	February	March	April	March	June	July	August	September	October	November	December
BC	34	32	34	38	39	38	34	32	31	35	39	39
PG	34	32	34	38	39	38	34	32	31	35	39	39

BC = Belize City, PG = Punta Gorda

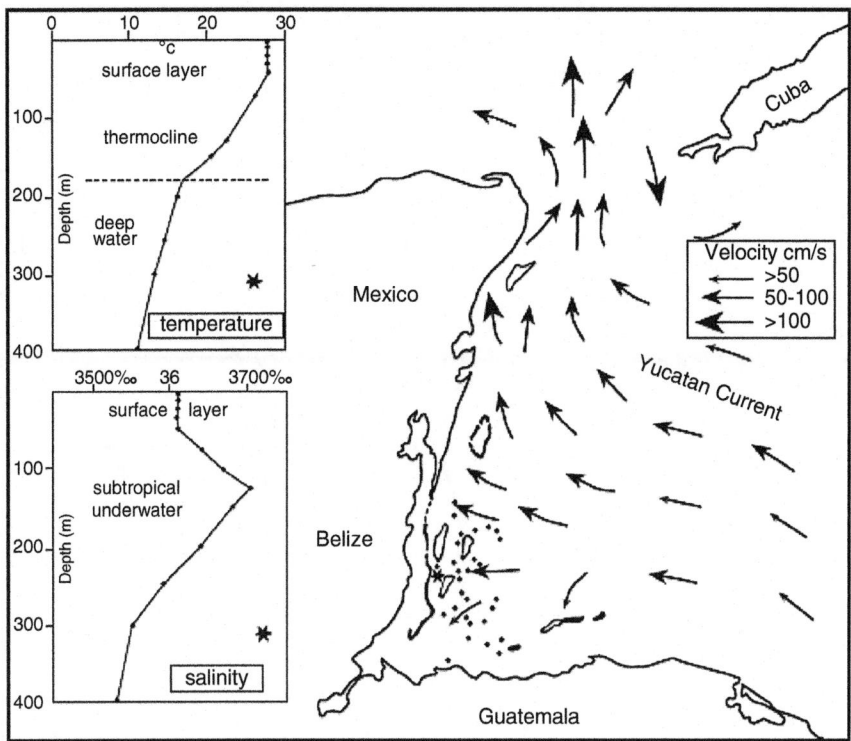

Fig. 3.5 Diagram of the relation between the magnitude of wave energy and sea level reefs, spur and groove features, and shelf-edge ridges along the Belize barrier reef. Wave force impact on the reefs is controlled by wind speed and direction and by the impedance and modifying effect of the offshore carbonate platforms (Figure redrawn from Burke 1982)

A combination of wind and waves is also responsible for the complex currents found in the back-reef lagoon of Ambergris Cay. Water accumulates in the lagoon behind the reef crest as swells force water over the reef. Constant northeast winds move some of the water southward into the shelf lagoon via longshore surface currents. However, most of the water is returned seaward as bottom currents through numerous breaks in the reef, building "ill-defined cusps of sediment along the western margin of the lagoon, opposite the reef breaks, and keeping the substrates in these channels highly mobile" (Ebanks 1967).

3.3   Chemical Parameters

Coral reef health and distribution are controlled primarily by light (Falkowski et al. 1990), nutrient level (Hallock and Schlager 1986), level of sedimentation, and temperatures. Reef communities are at their optimum in oceanic normal salinity

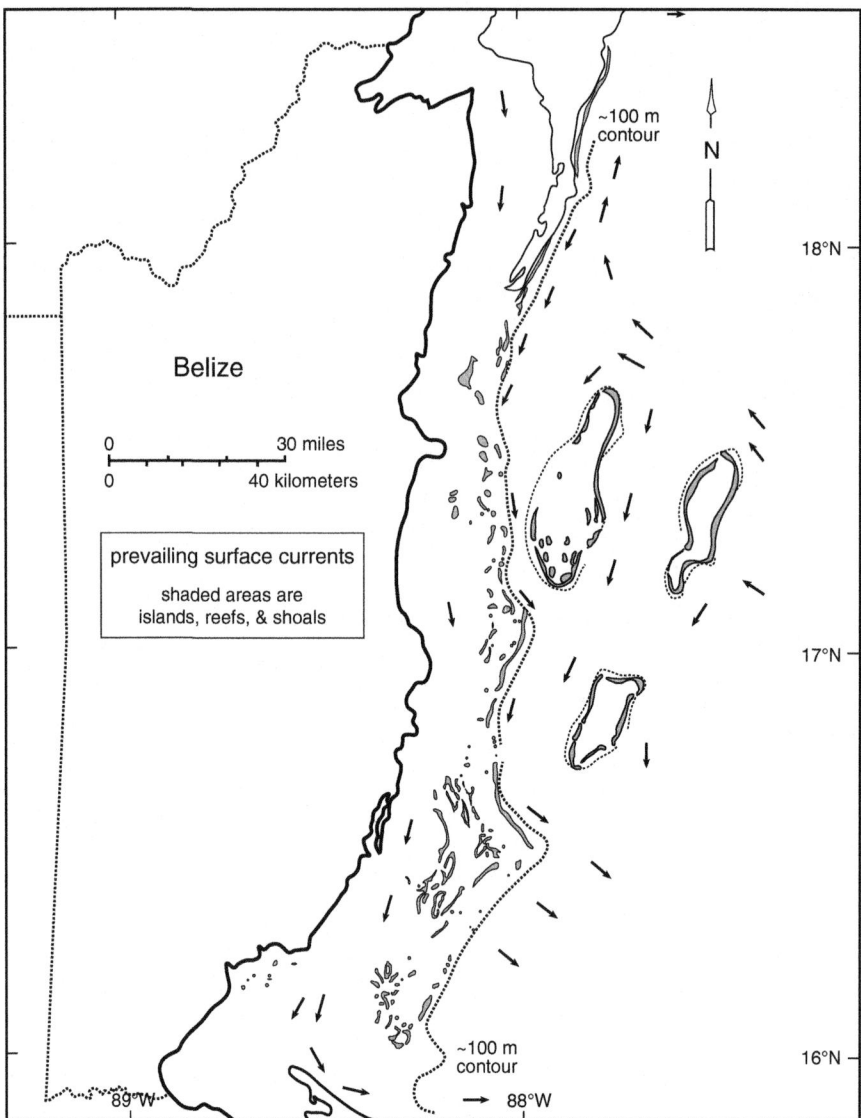

Fig. 3.6 A map illustrating the main surface circulation pattern and current velocities of the western Caribbean Sea (From Wust 1964 and reproduced from James and Ginsburg 1979). The two inserts at the left are water temperature and salinity profiles for a station (star) located between the barrier reef and Glovers Reef (Data from R. Molinari, reproduced from James and Ginsburg 1979)

(about 35 mg/l) but can survive over a salinity range of about 24–45 mg/l (Buddemeier and Kinzie 1976). However some corals have been reported in extremely high-salinity water of up to 47 mg/l in the Red Sea. On the other side of the spectrum, salinity maintained below 20 mg/l for longer than 24 h is lethal for most corals and reef organisms (Buddemeier and Kinzie 1976).

3.3.1 Salinity

Purdy et al. (1975) published a comprehensive analysis of salinity distribution in the Belize Lagoon. They reported surface-water and bottom-water salinity from June to August of 1961 and 1962 (Fig. 3.7). Two trends were apparent in their data: (1) a decrease in salinity northward into Chetumal Bay and southward into The Gulf of Honduras. Purdy et al. (1975) attributed the decreases to mainland runoff in the restricted bay, and to a substantial increase in freshwater runoff in The Gulf of Honduras, respectively; and (2) the tendency for nearshore surface salinities to be greater in the northern shelf lagoon than the southern shelf lagoon, related to rainfall differences and riverine inputs.

A comparison of the Purdy et al. (1975) surface and bottom salinities shows the mixing dynamics of the shallow northern lagoon and the deeper central and southern lagoon. In the north, the water column is more easily mixed through wind-wave agitation. In the south, the deeper water depth, coupled with

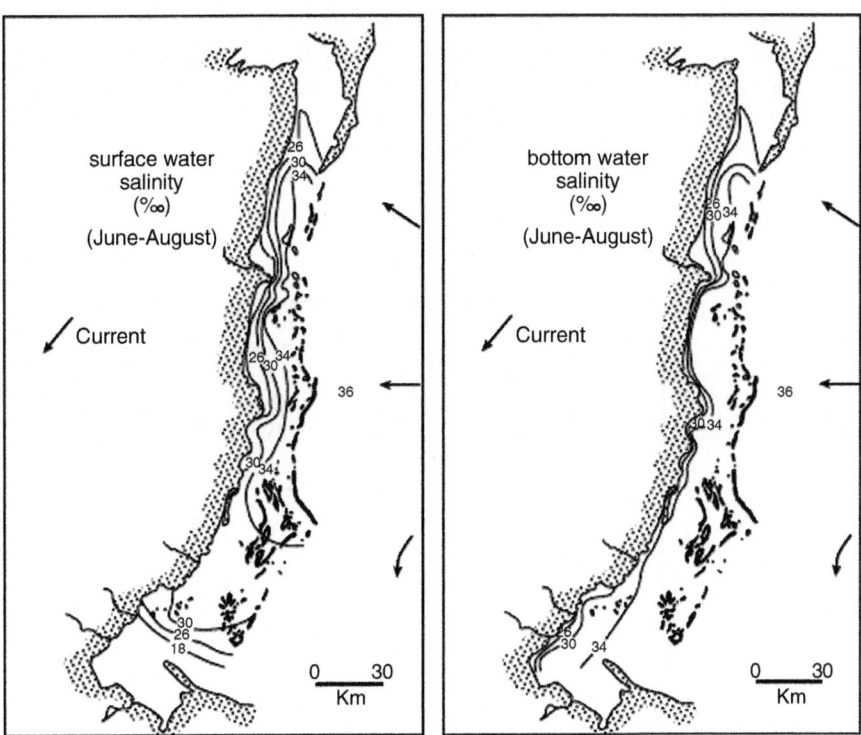

Fig. 3.7 A map illustrating the main surface circulation pattern and current velocities of the western Caribbean Sea (From Wüst 1964 and reproduced from James and Ginsburg 1979). The two inserts at the left are water temperature and salinity profiles for a station (*star*) located between the barrier reef and Glovers Reef (Data from R. Molinari, reproduced from James and Ginsburg 1979)

the greater freshwater runoff, favor the formation of freshwater wedges. The freshwater wedges originate at river outflows and float on top of the more dense seawater, and can travel great distances in the lagoon. Purdy et al. (1975) report freshwater reaching the barrier reef, and mud-laden plumes are often visible during the wet season reaching out to the barrier platform in the central lagoon. One event where a mud-laden plume reached the western side of the barrier platform (Fig. 3.8) was mapped by air in July 2000 (D.F. McNeill, unpublished observation). This salinity pattern controls to some extent the biotic and facies distribution of the lagoon.

In the northernmost lagoon, there is a salinity decrease north of Ambergris Cay (in Chetumal Bay), which acts as a physical barrier between the saline open ocean water and the bay that receives freshwater from mainland run-off. This effect is seasonal and is maximal during the rainy season, whereas, in the dry season, Chetumal Bay and the northern lagoon have increased salinity (Purdy et al. 1975). These high seasonal salinity variations (from 2 to 30 mg/l create a restricted marine fauna in Chetumal Bay and the northern shelf lagoon, except for algae, seagrass, and some benthic foraminifera.

The southern shelf lagoon displays a well-stratified water column. In the southern lagoon and The Gulf of Honduras, salinity is close to the normal open marine value. The salinity of the surface water is slightly decreased, whereas the bottom waters in the southern lagoon show normal salinity. The wedge of fresh water is "floating" on the denser, more saline deep water (Purdy et al. 1975). It is hypothesized that these surface salinity changes may partly explain why most of the reefs in the southern lagoon are not growing up to sea level. This freshwater wedge is measur-

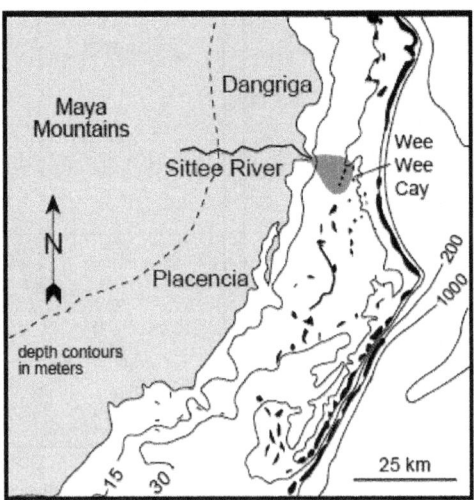

Fig. 3.8 A map of the Belize shelf around Dangriga showing the extent of a freshwater runoff plume on July 1, 1999. Surface water salinity in the vicinity of Wee Wee Cay was measured on June 29, 1999 at between 20–22 ppt (Unpublished observation and data by D.F. McNeill)

able as far eastward as the barrier reef (Purdy et al. 1975). The run-off of fresh water is the result of the Maya Mountains that provide a wide drainage area in association with significantly heavier rain in this southern part of Belize.

3.3.2 Aragonite-Calcite Saturation

Belize waters are supersaturated with respect to both calcite and aragonite. In contrast to Great Bahama Bank where most of the carbonate mud is aragonite, the mud in the northern shelf lagoon and Chetumal Bay is composed of calcite or high-magnesium calcite.

Where benthic foraminifera (like *Peneroplidae* and *Miliolidae*) dominate the skeletal grains, high magnesium calcite sediment occurs. Where *Halimeda*, and corals are the main sand size components, nearby mud tends to be more aragonitic.

Pusey (1975) and Reid et al. (1992) propose that the mud is produced by the breakdown of skeletal grains rather than by precipitation. Consequently, mud production and composition are not related to the difference in chemistry of water but to the mineralogy of the biota (magnesian calcite) it is derived from (Pusey 1975). Nevertheless, occurrences of whitings have been reported in the northern lagoon. Therefore, direct or bacterially induced precipitation cannot be excluded as a mud producer (Pusey 1975; Reid et al. 1992). Pusey (1975) suggests that bacterial reduction of organic matter might be an indirect cause of $CaCO_3$ precipitation by inducing pH differences in the water. This difference in northern shelf lagoon mud mineralogy might have implications for the understanding of fossil carbonate mudstones.

3.3.3 Holocene Dolomitization

In the northern shelf lagoon, syndepositional dolomitization has been documented in a 15-km²-shoal area, known as Cangrejo shoal (Mazzullo et al. 1995; Teal et al. 2000). This shoal has an average thickness of 3-m (with a maximum of 7-m) and sits on top of a karstic depression in the underlying Pleistocene limestone. Teal et al. (2000) show that early dolomite comprised as much as 30% of the Holocene shallow-marine transgressive sediment. The Cangrejo shoal dolomites are calcium rich (39.5–44.5 mol% $MgCO_3$) and are poorly ordered (Teal et al. 2000). The same authors show that the dolomite is predominantly cement, with an average crystal size of 7 μm. The stable-isotope composition (mean $ô^{18}O$ of 2.1‰ PDB and $\partial^{13}C$ of −5.2 to +11.6‰ PDB) of the dolomite suggests precipitation from normal salinity seawater and a dolomitization promoted by both bacteria sulfate reduction and methanogenesis in the pore water (Teal et al. 2000). The most widespread dolomitization is believed to have occurred during early transgression. This modern dolomite occurrence is one of the few documented examples of extensive syndepositional dolomitization in a near normal salinity seawater environment.

3.4 Fauna and Flora

Purdy et al. (1975) conducted a landmark study of the distribution of flora and fauna in the Belize shelf lagoon. They demonstrated that the biological system is organized in a regional north-south trend that follows the physiography of the shelf, as well as the depth and salinity changes associated with an east-west transition from the barrier to the lagoon to the mainland shoreline system.

The Belize shelf seems to have a biological organization that is dominated by two major trends. The first obvious organization is a barrier to lagoon to shore system. Superimposed on this trend, the physiography of the shelf induces a north-south transition in the biota due to differences in water-depth, energy setting and geography (rainfall, current, salinity). Additionally, some local environments exhibit hypersalinity or strong salinity variations due to river run-off and experience conditions ripe for development of atypical biota such as stromatolites or worm reefs (Burke et al. 1992; Rasmussen et al. 1993). The main biota of the Belize shelf lagoon is discussed below.

3.4.1 Corals

3.4.1.1 Coastal Reefs

Reefs adjacent to the Belize mainland occur in two general areas: the southern shelf lagoon between the village of Placencia and the Monkey River, and the embayment at the very southern part of the lagoon on the northwestern side of The Gulf of Honduras. The coastal reefs south of Placencia have been better explored relative to the less accessible southern coastal reefs. Cowan and McNeill (unpublished) have mapped out the reefal units in the Monkey River-Placencia corridor. They describe three main reef morphologies: the headland type, the open shoal reef, and individual reef mound shoals. The headland reefs are the most common, they occur on the seaward (open lagoon) facing side of mangrove islands along the coast. *Siderastrea siderea* head corals, and lesser amounts of *Diploria* sp., *Montastrea* sp., and some recently dead *Acropora palmata* dominate headland reefs. The open-shoal reef type is not attached to the shoreline or island, and consists of a shallow reef in the open embayment between headlands or islands. The third type of coastal reef is the small mound composed of branching finger-coral (*Porites divaricata*) mixed with red algae (*Amphiroa* sp.) and green algae (mainly *Halimeda* sp.).

The coastal reefs generate morphology distinctly different than that of the other lagoon and barrier reefs. They tend to form discontinuous elongate, shoreline parallel deposits, separated by either non-reefal sediment or open embayment. It is also worth noting that the coastal reefs in the southern shelf lagoon occur where there are abundant siliciclastics entering from rivers, weathering of the coastal plain, and redistributed by longshore drift. Whereas in the northern shelf lagoon, siliciclastics are relatively minor, but no coastal reefs occur.

3.4.1.2 Lagoon Patch Reefs

Patch reefs within the Belize lagoon are largely limited to the central and southern part of the shelf lagoon. No lagoon patch reefs are found north of about 16°50'N. Reefs do occur, however, on the barrier platform along the whole length of the shelf. The main lagoon patch reefs, detached from the barrier platform, start south of 16°50'N latitude and become more numerous to the south. Between the latitudes of 16°45'N and 16°30'N abundant patch reefs occur that have grown to near sea level. From 16°30'N southward, many of the lagoon patch reefs are submerged. The submerged reefs have built up to a level about 8 m below sea level (Fig. 3.9). The cause of the change from lagoon reefs that build upward to near sea level, and those submerged is not known, but it is speculated that either environmental influ-

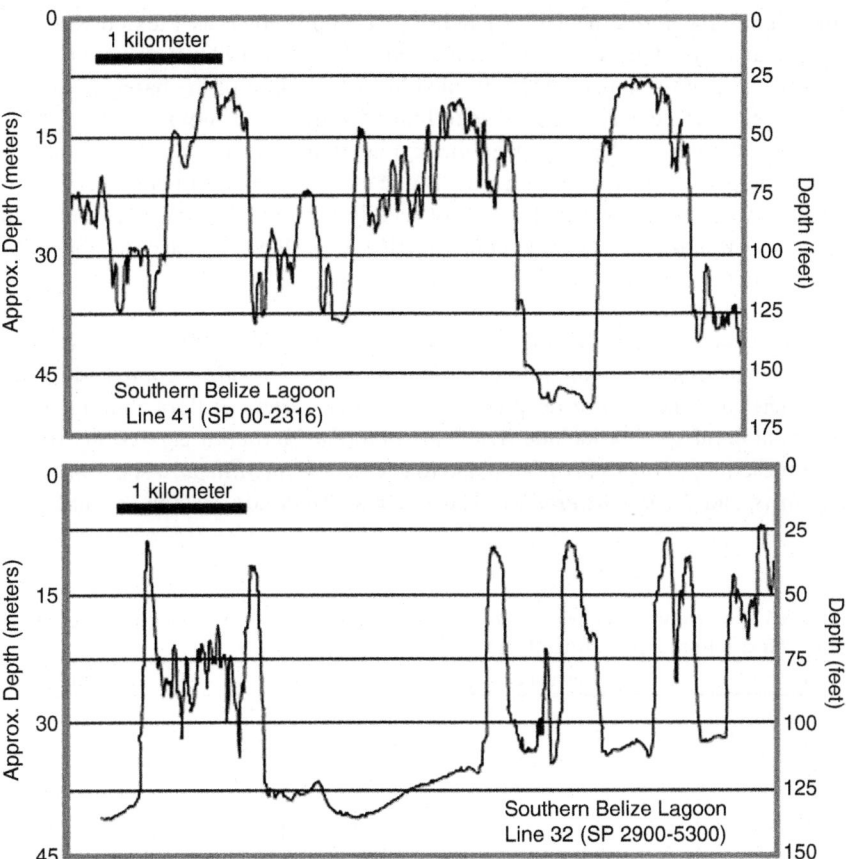

Fig. 3.9 Bathymetric profiles across the central southern Belize lagoon. The profiles show the two-dimensional morphology of pinnacle reefs. The reefs recorded in these profiles usually show either a composite, multi-pinnacle morphology (*top*) or a single-pinnacle morphology (*bottom, right*)

ences (salinity, temperature), neotectonism (rapid subsidence), or a combination of the two control the vertical aggradation. In addition, the role of upwelling of colder deep water from The Gulf of Honduras, perhaps rich in nutrients, has not been adequately studied for the southern shelf lagoon.

There are some striking differences between lagoon patch reefs from the north-central Belize shelf when compared to the southern ones. As the sea level rose during the Holocene, and because the Belize shelf was inclined southward, marine conditions progressively invaded the shelf from the south to the north. Southern patch reefs are older and have either kept up or caught up with sea-level or have been drowned, whereas, in the central and northern lagoon, patch reefs are younger and have experienced only slowly rising sea-level conditions.

As a result of the topographic slope of the lagoon and the likelihood of significant precursor topography, the style (and dimensional aspects) of the lagoon reefs varies from the southern lagoon (deep) to the central and northern lagoon (shallower). For example, a northern lagoon patch reef group, known as Mexico Rock has been studied in detail by several authors (Mazzullo et al. 1992; Burke et al. 1998). They found that these patch reefs formed by colonization of the basal Pleistocene limestone by single head corals. Due to limited vertical accommodation, growth of individual coral heads has expanded the reef laterally and the corals have coalesced to form larger composite reef masses. Burke et al. (1998) also noted that a biotic zonation within the patch reefs was not readily apparent. The domination by the head coral *Montastrea annularis* at Mexico Rocks is shown by the coral abundance measurements summarized in Table 3.5.

In the Southern lagoon, an extensive study has inventoried the biota of one of the rhomboid shoals, the Pelican Cays (Macintyre and Rützler 2000). These cays are peculiar because, here mangroves sit directly on a coral ridge (compared to mud elsewhere) and the interior deep ponds host a rich variety of flora and fauna. The dominant coral species include *Agaricia tenuifolia*, *Acropora cervicornis* (before 1998 bleaching), *Porites divaricata* and to a lesser extent *Millepora* sp., *Montastrea annularis* and *Porites asteroides*. The seagrass *Thalassia testudinum* and green

Table 3.5 Average weighted percent abundance of coral in a northern barrier lagoon patch reef group (Mexico Rocks) (Data from Burke et al. 1998)

Coral	Abundance (%)	Coral	Abundance (%)
Montastrea annularis	83	*Millepora alcicornis*	<1
Agaricia tenuifolia	4	*Montastrea cavernosa*	<1
Agaricia agaricites	4	*Agaricia agaritesvar. purpurea*	Rare
Porites porites	3	*Colpophyllia natans*	Rare
Porites asteroides	2.5	*Eusmillia fastiga*	Rare
Acropora palmata	1	*Isophyllia* sp.	Rare
Acropora cervicornis	<1	*Millepora complanate*	Rare
Dichocoenia stokesii	<1	*Mycetopyllia* sp.	Rare
Diploria sp.	<1	*Siderastrea siderea*	Rare

algae *Halimeda* sp. are also common. Similarly, the sponge community is rather extensive. Since the 1998 bleaching event, most of the corals are now dead, and have been replaced by algae and sponges.

In the southern lagoon, many living lagoon reefs are currently located about 8-m below sea level. Purdy et al. (1975) points out that salt wedges play an important role in the distribution of marine organisms, especially in the southern lagoon. Bathymetric data illustrate the 8-m growth limit of the patch reefs (Fig. 3.10). There is a simple, intuitive correlation between the occurrence of fresh water input and this pattern of reef growth. However, there are no direct data to sustain this hypothesis.

3.4.1.3 Barrier Reef

The Belize barrier reef system has been studied in detail at several locations, mainly in the central part of the barrier system. The most comprehensive study of the barrier ecosystem was carried out in and around Carrie Bow Cay and published by the Smithsonian Institution (edited by K. Rützler and I. Macintyre 1982b). The reader is referred to that publication for detailed studies of the physical environment,

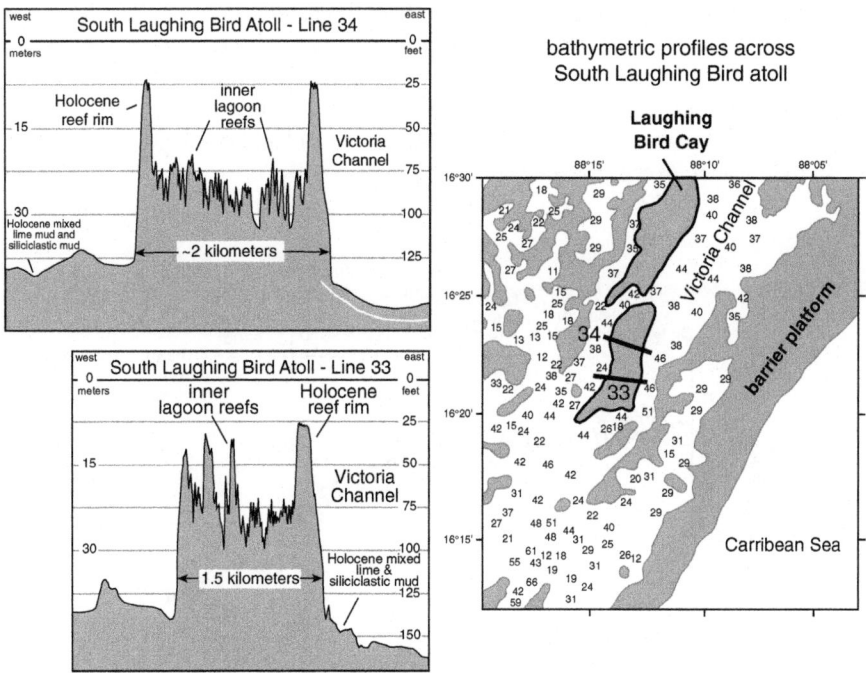

Fig. 3.10 Bathymetric profiles (*left*) across a submerged rhomboid reef shoal (South Laughing Bird Cay). The constructional reef rim sits about 8 m below sea level and the interior part (reefs?) of the shoal at about the 20–25 m depth

biogeological structure, flora and fauna, and morphology of the barrier reef complex. The seaward margin of the Belize barrier reef and atoll reefs were similarly characterized in a series of studies edited by N. James and R. Ginsburg (1979).

Studies in the compilation of Rützler and Macintyre (1982b) show that the main coral frame builders and carbonate sediment producers on the Belize barrier reef are generally similar to other reefal open-ocean facing Caribbean reef systems. More recently Macintyre et al. (1987) studied the barrier-reef crest and backreef apron of Tobacco Reef. Their detailed study showed that five different sediment-biozones occur (Fig. 3.11).

They proposed that the main control on sediment facies zonation on the outer barrier-platform is distance from the reef crest. The zonation described at Tobacco Reef from seaward to lagoon includes:

1. *Coralline-coral-Dictyota pavement.* This zone extends from the inner edge of the reef crest lagoonward for a distance of about 70 m. Scattered coral colonies of, for the most part, *Porites astreoides* and *Acropora palmata*, are attached to the crustose coralline algal encrusted pavement, which also supports an extensive cover of *Dictyota* sp. with lesser amounts of *Jania* sp., *Halimeda* and *Neomeris* sp.

2. *Turbinaria-Sargassum rubble.* This zone is about 40-m wide and the rubble and sand substrate supports a rich cover of fleshy algae, particularly the large brown algae *Turbinaria turbinata* and *Sargassum polyceratium*. Among the other common algal genera here are *Dictyota*, crustose corallines, *Gelidium*, *Laurencia*, *Halimeda*, *Penicillus*, *Neomeris*, *Gracilaria*, *Galaxaura*, *Rhipocephalus*, *Udotea*, *Dictyosphaeria*, *Liagora* and *Caulerpa*. Small colonies of the coral *Porites porites furcata* are also characteristic of this zone and all areas lagoonward of the study transect, except for the bare sand zone.

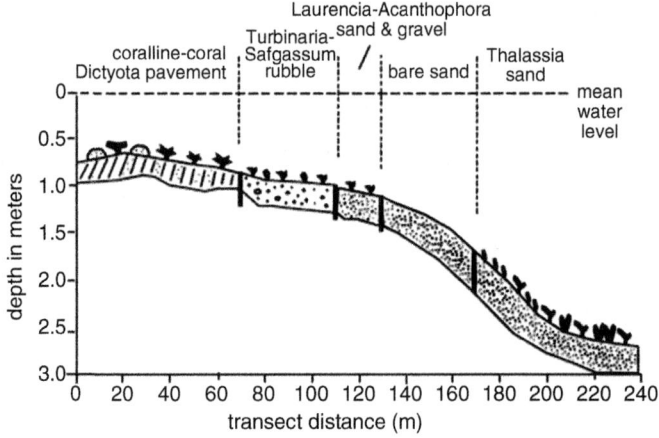

Fig. 3.11 Transect across the barrier reef sediment apron at Tobacco Reef. The authors mapped five distinct biogeological zones (Redrawn from Figure 10 of Macintyre et al. 1987)

3. *Laurencia-Acanthophora sand and gravel.* The loose substrate of this narrow (20 m wide) zone supports a variety of fleshy algae – most notably the red algae *Laurencia* and *Acanthophora.* Other common algae genera include crustose corallines, *Dictyota, Spyridia, Jania, Hypnea, Dictyosphaeria, Sargassum, Amphiroa, Neomeris, Halimeda, Turbinaria, Dictyota, Udotea, Liagora, Caulerpa* and *Penicillus.*

4. *Bare sand.* About 130 m leeward of the reef crest, the bottom consists of relatively open sand with scattered gravel. From the air, this zone appears as a distinct white band about 40 m wide. Plant cover is sparse, but here and there fleshy algae are present, primarily *Padina, Laurencia, Dictyota, Turbinaria, Hypnea, Halimeda, Udotea, Caulerpa, Rhipocephalus, Penicillus,* and *Ceramium.*

5. *Thalassia sand.* At the shoreward limit of the study transect, the sand bottom with occasional scattered gravel-sized fragments supports a rich cover of the seagrass *Thalassia testudinum.* Some fleshy algae are also present, primarily *Champia, Penicillus, Acanthophora, Dictyota, Udotea, Rhipocephalus, Avrainuillea, Caulerpa, Dictyosphaeria, Laurencia,* and *Halimeda.*

3.4.1.4   Northern Lagoon Foraminiferal Facies

Foraminifera are an important skeletal grain producer in the Belize lagoon. Foraminifera are especially abundant in the northern-most part of the lagoon, west of Ambergris Cay in Chetumal Bay (Pusey 1975; Wantland 1975). The northern Belize shelf, especially Chetumal Bay, experiences variations in salinity of between 10 and 35 mg/l (Pusey 1975). The salinity during the dry season (February–April) is considerably higher than normal seawater due to tidal restriction and evaporation.

The sediment facies mapped on the northern shelf show benthic foraminifers as a major grain component. Pusey (1975) mapped the marine sediment types north of Belize City (Fig. 3.12) and identified two widespread foraminifera-dominated facies (peneroplid sand and miliolid mud). The peneroplid sand facies occurs immediately west of, and in the lee of, Ambergris Cay. The miliolid mud dominates the inner part of the shelf from Belize City north to the Bulkhead shoal and island complex (Fig. 3.12).

Several workers have studied the distribution and ecology of benthic foraminifers on the central and southern parts of the Belize shelf. Cebulski (1969) was the first to identify many of the foraminifera faunas, their distribution, and their relative population densities. Cebulski (1969) divided the foraminiferal faunas into two main groups: reef tract and lagoon. The reef-tract fauna were further divided into three assemblages based on the absence or presence of marker species (main-reef assemblage, reef-margin assemblage, reef-channel assemblage). Wantland (1975) used transect sampling across the Belize lagoon, with wider geographic coverage (in the central and southern shelf lagoon and Gulf of Honduras) to characterize the foraminiferal facies. Ten benthic-foraminifera assemblages were distinct in the Belize lagoon surface sediments (Fig. 3.13).

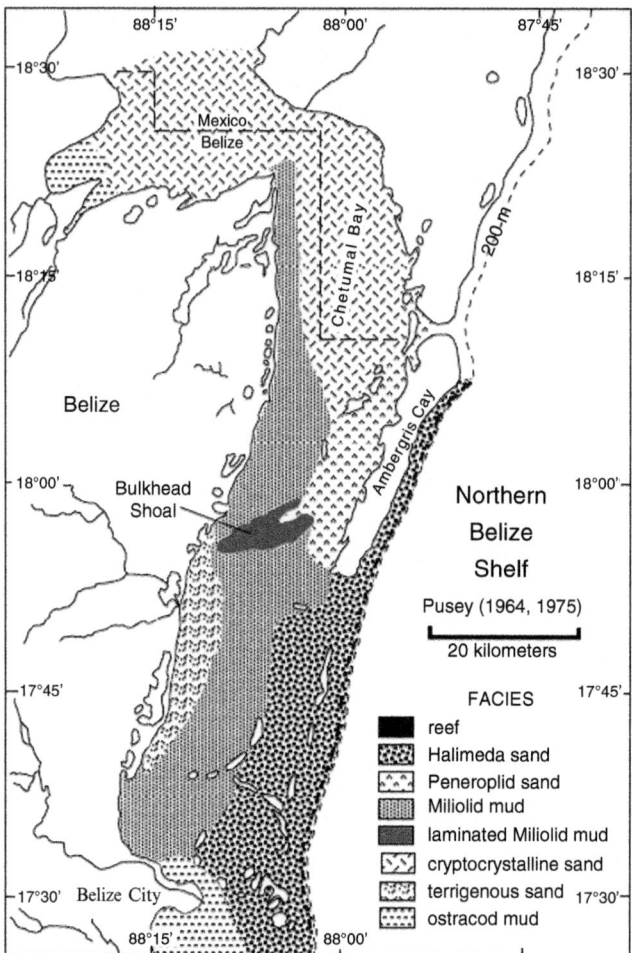

Fig. 3.12 Map of the main surface sediments on the northern Belize shelf and lagoon (Figure is redrawn from Figure 19 of Pusey 1975)

Wantland (1975) also showed that these assemblages are well defined based on faunal dominance within the total population and on faunal diversity. One of the key points promoted by Wantland (1975) was the idea that the assemblage distribution is largely controlled by bathymetry, hydrography, and sedimentation. Several significant environmental-faunal relations were proposed by Wantland: (1) miliolids are dominant in the shallow shelf (<10-m), the exceptions being the restricted coastal lagoons and fluvial settings that receive constant freshwater influx or quartzose sediments; (2) in deeper areas (>10-m) of the shelf lagoon calcareous perforate forms commonly dominate; and (3) with respect to the morphologic configuration of the Belize lagoon, the benthic foraminiferal assemblages are generally linear, elongate parallel to the trend of the shoreline and barrier reef.

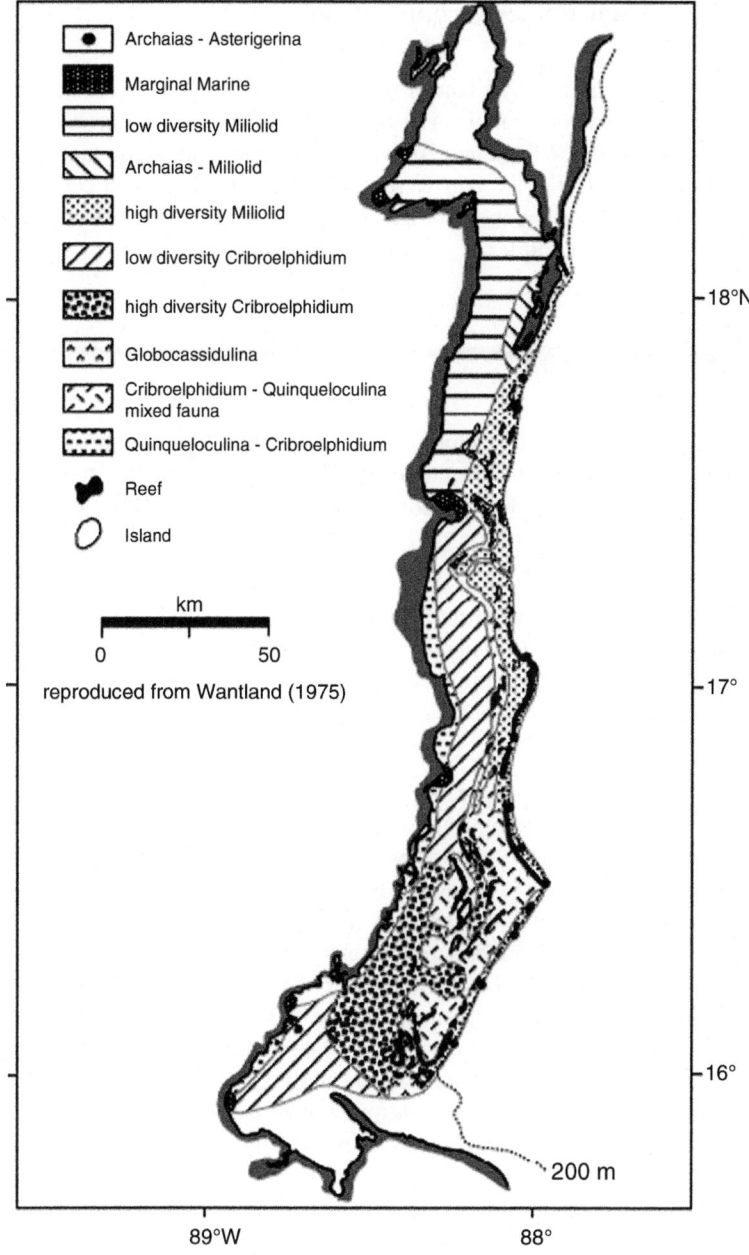

Fig. 3.13 Benthic foraminiferal zones of the Belize shelf lagoon (Figure reproduced from Wantland 1975)

3.5 Belize Shelf Facies Organization

The Belize shelf is rimmed by a barrier reef that has built up to sea level, composed of coarse skeletal sand and rubble. The Belize shelf lagoon and barrier reef ranges from 5 to 40 km wide (Wilson and Jordan 1983). Sediments grade landward from pure marine carbonate (atolls, barrier reef and platform) to mixed carbonate and terrigenous sediments (lagoon reefs and channels) to pure terrigenous sediment (parts of the shoreline complex).

3.5.1 Shelf Sediments

Three of the most important factors that influence the distribution of shelf sediments and the facies relationships are terrigenous input, carbonate production, and bathymetry (Purdy 1974a). Belize sediments can be split into three general shore parallel facies belts (Fig. 3.14):

1. A nearshore terrigenous belt consisting of quartz sand and mud
2. An axial marl belt (both carbonate and terrigenous fine-grained sediment) consisting of molluscs, *Halimeda*, *benthic foraminifers*, and ostracod grains
3. A marginal carbonate belt comprised of reefs, *Halimeda* sand, and other skeletal sands. A large barrier forms the border to the adjacent deep water to the east (Purdy 1974a; Fig. 3.14)

Superimposed on these shore-parallel facies belts there are significant changes in sediment type that occur from north to south. Bathymetry plays an important role in sediment distribution from north to south. North of Belize City (Northern Shelf Lagoon and Chetumal Bay) the lagoon is very shallow, less than 3–4 m, and the bathymetric change is small. As a result, sediment type is largely salinity dependent. Salinity decreases landward resulting in a distinct change from *Halimeda* sand to miliolid mud to non-skeletal peloidal grains (Purdy 1974a). Mudbanks and shoals occur at the transition from the northern shelf lagoon to Chetumal Bay and have recently been examined in detail by Mazzullo et al. (2003). The central and southern lagoon has a greater depth, which inhibits mixing of less saline surface waters and more saline deeper waters despite higher freshwater input from the land relative to the northern shelf lagoon. As a result, the southern shelf lagoon sediments are not generally salinity dependent (Purdy 1974a). The facies dimensions of two representative transects across the Belize shelf are shown in Table 3.6 (for the locations of the transects see Fig. 3.14). More recently, Purdy and Gischler (2003) reexamined the sediment facies (19 transects) of the Belize shelf and provided quantitative analysis of the components. The three general facies belts are now subdivided into eleven facies, and the north to south and east to west structural, bathymetric, siliciclastic source proximity, and salinity variations are reflected in the spatial distribution of these facies.

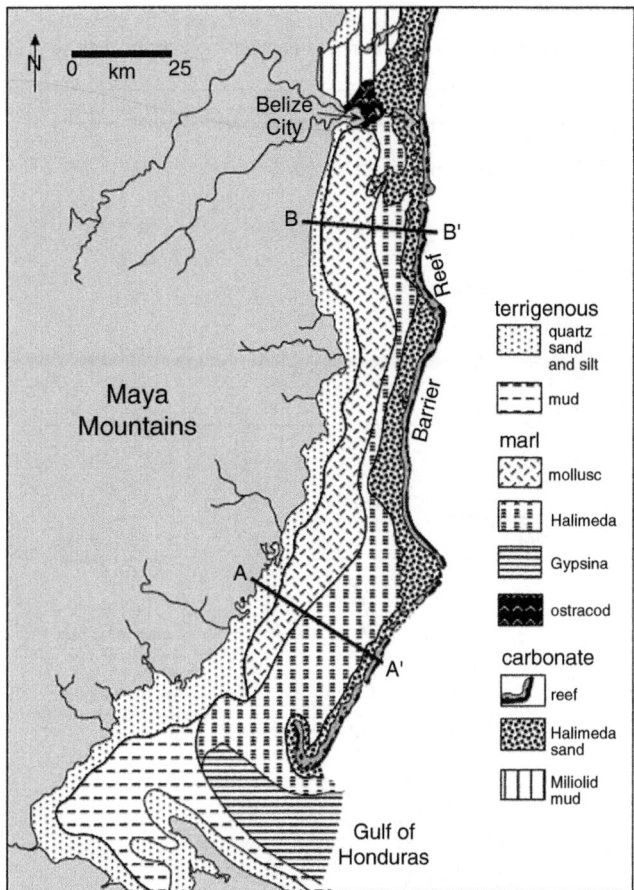

Fig. 3.14 The spatial distribution of surface sediment types on the southern part of the Belize shelf (After Wantland and Pusey 1971; Purdy 1974a). Principal components consist of quartz grains, molluscs, *Halimeda, Gypsina,* ostracods, miliolids and corals. The two transect lines refer to Table 5.1

3.5.2   *Barrier Reef*

The dominant feature of the Belize shelf is a barrier reef that runs 217 km along the shelf edge (James and Ginsburg 1979). Extending 250 km, the barrier-reef complex of Belize is the largest continuous reef system in the western Atlantic (Macintyre and Aronson 1997). The Belize barrier reef and fringing reefs form part of the Great Atlantic Reef Belt (Fig. 3.15). The barrier configuration changes from a fringing reef in the northern shelf (near Ambergris Cay) to a continuous barrier reef in the south (Fig. 3.16). The windward edge of the barrier also contains channels that open to the lagoon; carbonate shoals, algal covered pavement, and deeper reef communities (where the rim is topographically depressed).

Table 3.6 Lateral extension (in km) of facies belts across the Southern Belize shelf (For locations and sources of data see Fig. 3.14)

Facies	A–A' (km)	B–B' (km)
Terrigenous		
Quartz sand and mud	3.3	6.17
Marls		
Mollusc	11.7	8.2
Halimeda	8.3	16.7
Gypsina/ostracod	–	–
Carbonate		
Reef	1.7	1.5
Halimeda sand	3.3	2.2
Miliolid mud	–	–

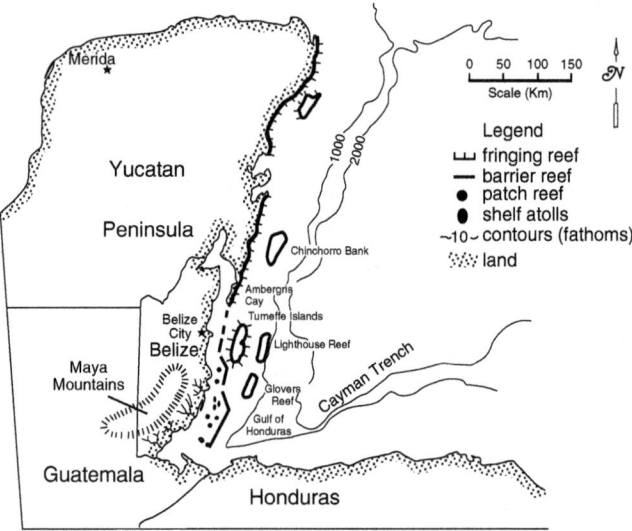

Fig. 3.15 General configuration of the Belize barrier-atoll complex. The system of reefs that range from The Gulf of Honduras to the northeastern tip of the Yucatan Peninsula comprises the Great Atlantic Reef Belt

Based on reef community structure and geomorphology, the barrier-reef platform and can be divided into three distinct provinces: the *northern*, *central*, and *southern* provinces (Macintyre and Aronson 1997). The northern province is characterized by 46 km of reefs that grow to sea level on the ocean-facing side of the shelf and is backed by the muddy shallows of Chetumal Bay. The reef crest in this northern part is dominated by crustose coralline algal mounds of *Millepora* sp. that also locally incrust the corals. Patch reefs in the shallow back reef east of Ambergris Cay are dominated by *Montastrea annularis*. Additionally, gently sloping pavement with only scattered coral colonies are present on the reef crest and inner fore reef areas of this province.

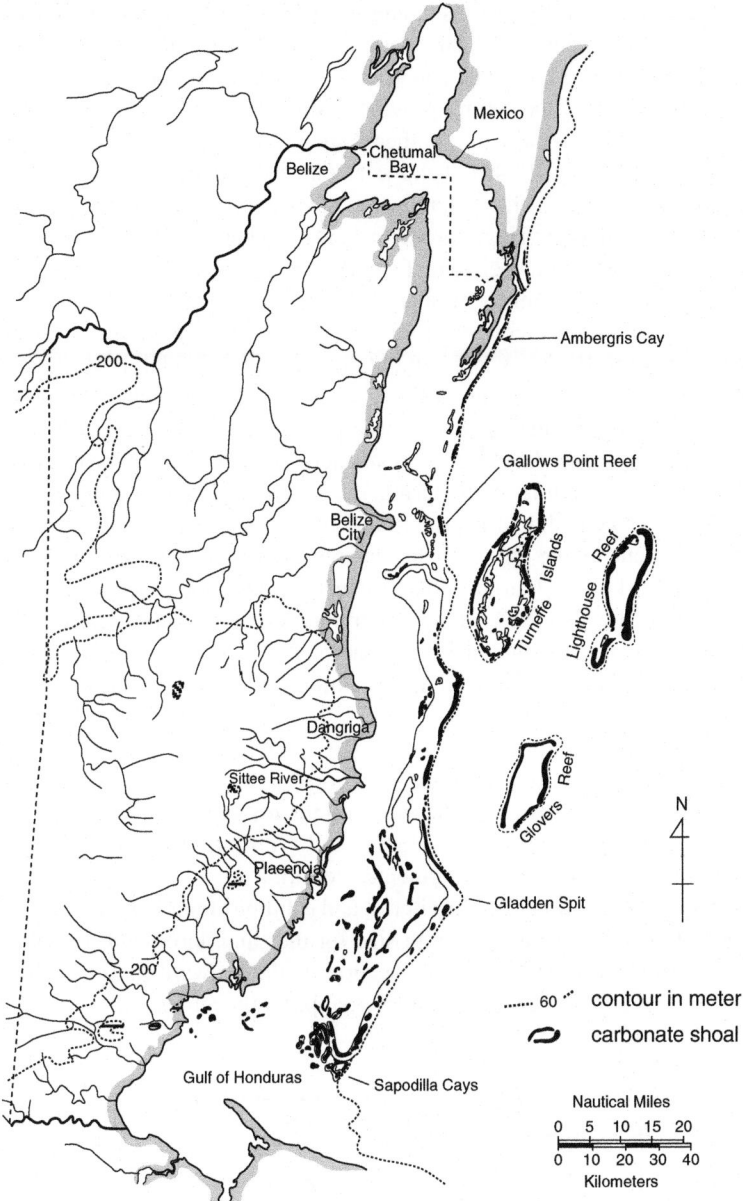

Fig. 3.16 Map showing the reefs that build up to near sea level and the nearly continuous nature of the reefs along the central and southern barrier margin (Figure reproduced from Burke (1982), his Figure 222 and additional data from our field observations and satellite data)

The central province covers the area between Gallows Point Reef and Gladden Spit (Macintyre and Aronson 1997). About 91 km of this interval are characterized by reefs close to sea level with long sections of uninterrupted barrier reef. Sand cays on the inner edge of the reef rims, mangrove islands, patch reefs, and grass flats are found toward the center and landward edge of the barrier platform. Numerous rhombohedral-shaped reefs and shoals in the shelf lagoon are characteristic of the back barrier region of the central province (Macintyre and Aronson 1997).

The southern province or southern lagoon extends from Gladden Spit to Sapodilla Cays (Purdy 1974a). It shows a zone only about 10 km long of reefs that grow close to sea level. These shallow-water reefs on the seaward edge of the barrier platform are mainly fringes around islands and passes. The lagoon behind these reefs is the deepest of the Belize reef system, for example reaching more than 27 m deep in the Victoria Channel. Similar to the central province, this back barrier part of the southern province contains rhomboidal reefs and shoals that continue northward to Laughing Bird Cay as incipient drowned rhomboidal, linear, and pinnacle reefs (Macintyre and Aronson 1997). The crests of these submerged reefs consist of crustose coralline algal mounds and coral rubble. The shelf depth is probably related to faulting and differential subsidence coupled with the progressive south-to-north flooding of this shelf during the Holocene. Additionally, variations in water chemistry related to terrestrial runoff, combined with wave energy, influenced the development of the Belizean barrier reef provinces (Burke 1993; Macintyre et al. 1995). Burke (1993) argued that the barrier morphology is controlled mainly by wave intensity; he also stressed the importance that Pleistocene and Holocene sea-level changes played in the barrier morphology.

The barrier reef can be divided into two morphologic zones: the reef flat and reef front (Fig. 3.17).

The reef flat or shallow reef is made up of the sand apron pavement or island. In the southern Belize lagoon sediments are mostly coarse-grained. The gravel fraction is made of mollusc shells, *Millepora* fragments and *Homotrema* tests and variable amounts of coral fragments (James and Ginsburg 1979). The sand fraction consists of *Halimeda*, coral fragments, coralline algae pieces, broken tests of *Homotrema* and *Gypsina* (Foraminifera), and echinoid spines and plates (James and Ginsburg 1979). The sediment grain size for the sand apron and reef pavement is illustrated in a morphological cross-section of the seaward margin of the barrier reef in Fig. 3.18. The reef pavement is often partially cemented.

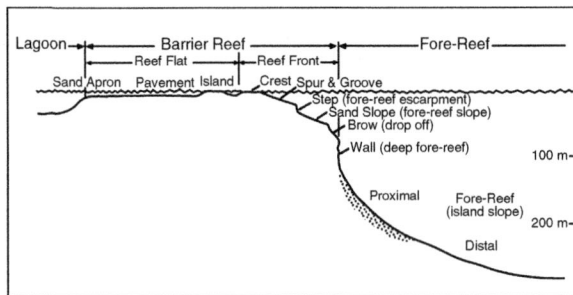

Fig. 3.17 Generalized morphologic zones of the Belize barrier reef (Figure redrawn from James and Ginsburg 1979) and includes the terminology of the reef front and fore-reef used by Land and Moore 1977

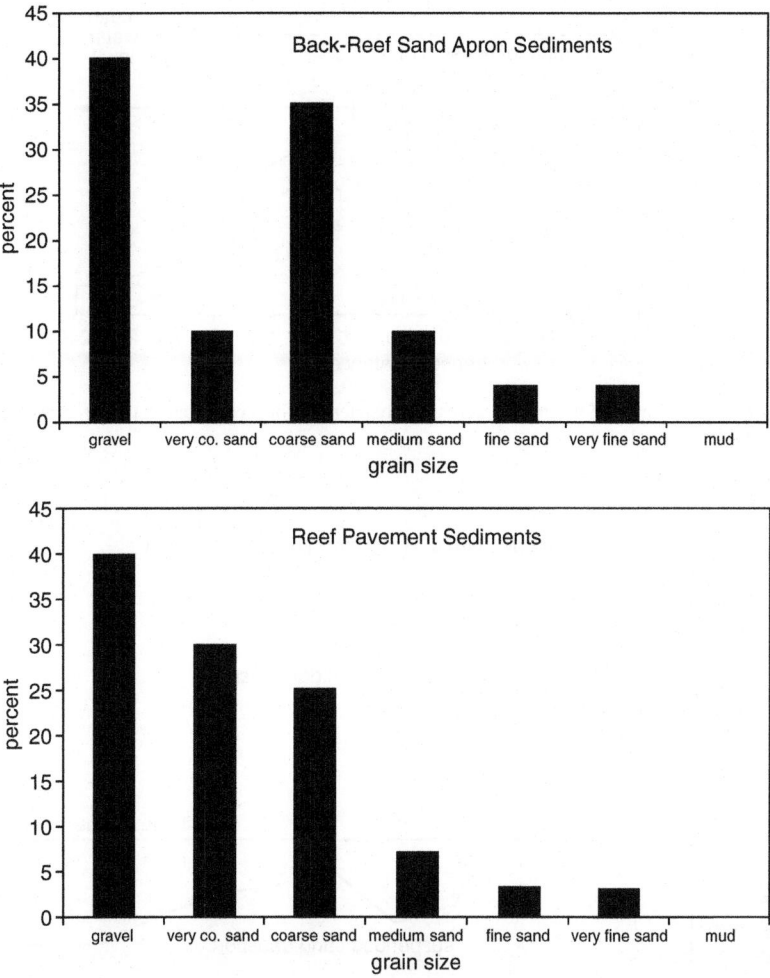

Fig. 3.18 Histograms of surface sediment grain size from the barrier reef sediment apron (Data are from James and Ginsburg 1979)

Within the reef flat waves serve to arrange the sedimentary facies in belts that parallel the barrier and reef crests. The facies zonation of the back-reef can be subdivided on the basis of biota, grain size, substrate and/or typical bottom water velocity, each of which are inherently related to wave energy and water depth. Here we use the scheme of Macintyre IG et al. (1987) because of their emphasis on wave related processes.

Wave energy controls: (1) which biological components exist on the reef; (2) the amount of mechanical erosion in the fore-reef and reef crest (the primary source region of back-reef sediments; Macintyre et al. 1987), and; (3) sediment types that are found in the back-reef.

Details of the sedimentary character of the back-reef belts of Macintyre et al. (1987) are summarized in Figs. 3.19 and 3.20.

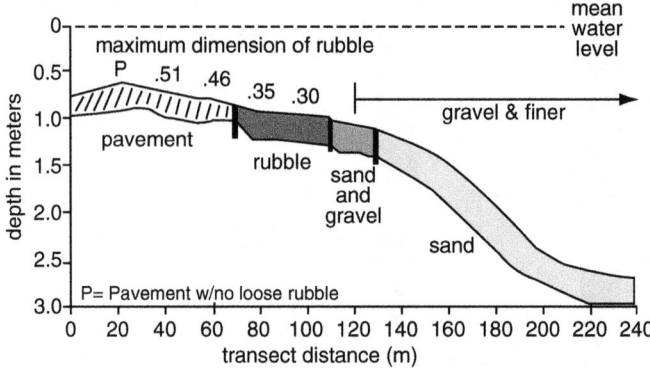

Fig. 3.19 Distribution of grain sizes across the barrier platform crest and shallow forereef (Figure from Macintyre et al. 1987)

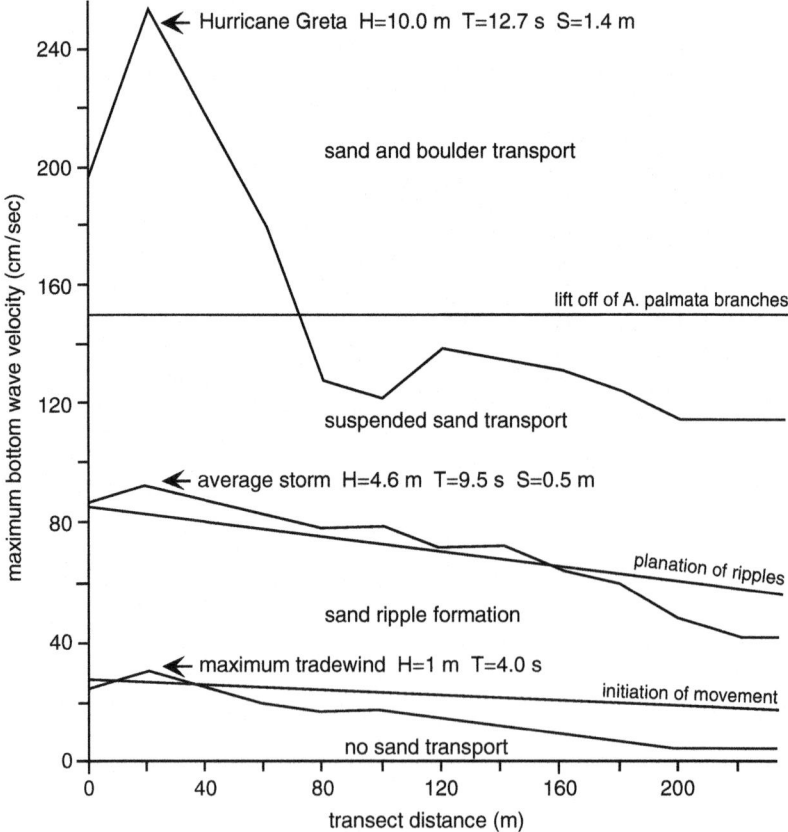

Fig. 3.20 Bottom current velocities across the Tobacco Reef (From Figure 8 of Macintyre et al. (1987)). H = wave height, T = wave period, and S = storm surge

The fore-reef, reef crest and back-reef regions are subdivided by various schemes depending on the author. Macintyre et al. (1987) made the following conclusions as a result of their study of the back-reef sediment apron of Tobacco reef:

1. Trade-wind conditions account for about 95% of the waves at the barrier reef crest.
2. Bottom-water velocities are too weak to initiate sand transport except for within 20 m of the reef crest and during high spring tides when the water depth is sufficiently increased to allow the propagation of larger waves into the back-reef.
3. Most sand transport occurs during episodic periods of storm activity (<1% frequency). Sand and finer than sand material are continually supplied to the back-reef from the fore-reef, but even during an average storm, transport to, and mobilization within the back-reef increases dramatically.
4. Sediments finer than sand-size are persistently winnowed from the back-reef in association with bioturbation.
5. The transport of boulders (>25.6 cm) requires hurricane force conditions. Fore-reef bottom water velocities must reach 4.0 m/s in order to fragment and transport *Acropora palmata* boulders to the back-reef, which generally is not reached under tropical storm conditions. Hurricanes, on the other hand, can generate bottom water velocities of up to 6.0 m/s (Macintyre et al. 1987).
6. The sediment size continuously decreases shoreward of the reef crest.
7. The size distribution is largely controlled by hydrodynamic conditions associated with storm and hurricane activity.
8. The size distribution and composition of the back-reef sediments indicates that for the most part, these sediments are derived from the fore-reef and reef crest.
9. The reef front reaches to depths of 70 m, and its morphological features include the crest, spur and groove, step and sand slope (James and Ginsburg 1979). The spur and groove area reaches to depths of 20 m. Spurs are 4.5–6 m wide and grooves are 2.5 m wide. The area is floored with coral rubble and cross-rippled sand (James and Ginsburg 1979). The crests contain between 15–40% living coral (including *Acropora cervicornis* and *Montastrea annularis),* sponges, *Pseudopterogorgia* sp., *Gorgonia ventialina,* and *Halimeda* sand (James and Ginsburg 1979). The barrier reef's spur and grooves also contain *A. palmata, P. porites, Agaricia agaricites, M. cavernosa, Porites asteroids, Millepora,* zooanthids and coralline algae (James and Ginsburg 1979).

The reef front "step" is at depths of 20–35 m, and the "sand slope" is from 35 to 45 m deep (James and Ginsburg 1979). The sand slope is made of gentle undulating swells and troughs 30–45 m wide (James and Ginsburg 1979). The troughs contain an accumulation of broken coral branches, conch shells, *Thalassia* grass, and rhodolites. The ridges contain fine-grained sand, burrow extrusion cones (~10 cm high), living *Penicillus,* and living *Halimeda* plants (3–5 plants/m^2) (James and Ginsburg 1979).

3.5.3 Offshore Atolls

Offshore atolls include Glovers Reef, Lighthouse Reef, the Turneffe Islands, and Banco Chinchorro (Mexico). All the atolls are located in a very similar geological setting in the open ocean but they consist of very different facies, indicating that small changes in ocean circulation and/or tectonic subsidence can result in widely varying facies patterns (James and Ginsburg 1979; Gischler 1994; Gischler and Lomando 1999; Purdy and Gischler 2003) (Fig. 3.21). Glovers Reef is an oval-shaped atoll 15 km east of the barrier reef. It is 28 km long and 10 km wide, and has a 6–18 m deep lagoon (James and Ginsburg, 1979). Its lagoon contains over 800 patch reefs. Lighthouse Reef has a reef rim with a shallow lagoon. In contrast, Turneffe consists of numerous mangrove islands with swamps (James and Ginsburg 1979). Glovers Reef has circular facies belts, whereas Lighthouse Reef and Banco Chinchorro have more linear facies belts. Table 3.7 gives a list of the areas and percentages of the different facies belts on the offshore atolls.

The reason for these variations in the different atolls is most likely the combined effect of ocean circulation and differential subsidence. Glovers Reef, Lighthouse Reef, and Banco Chinchorro are open to the Caribbean Sea and receive maximum wave force (Gischler and Lomando 1999). In contrast the Turneffe Islands is protected by the Lighthouse Reef to the east and exposed only to reduced wave force.

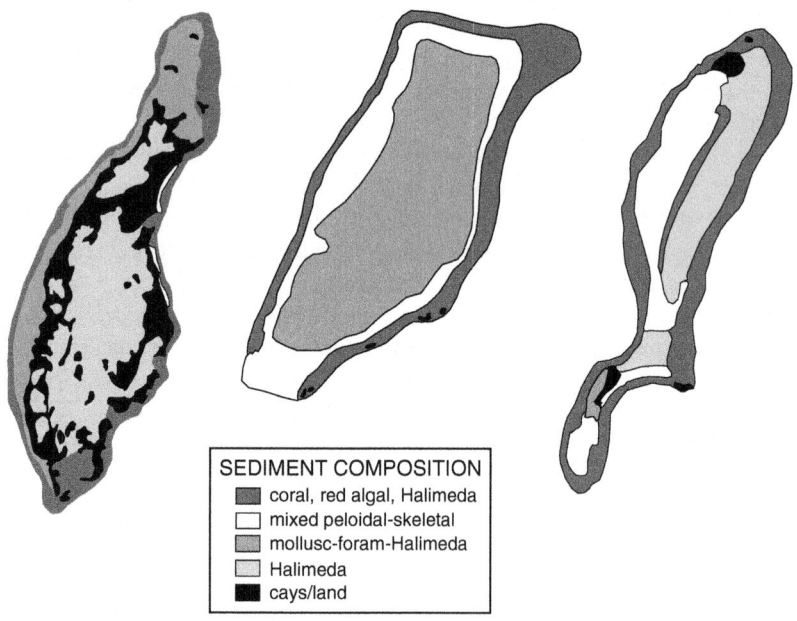

SEDIMENT COMPOSITION
- coral, red algal, Halimeda
- mixed peloidal-skeletal
- mollusc-foram-Halimeda
- Halimeda
- cays/land

Fig. 3.21 Sediment facies of the three offshore atoll-platforms (Figures are reproduced from Gischler and Lomando 1999)

Table 3.7 Geographical parameters of the three offshore atolls (Gischler 1994)

Atoll	Area km²	Land (%)	Lagoon (%)	Patch reef (%)	Reef rim (%)	Sand areas (%)	Tidalcuts	Maximum lagoon depth (m)	Lagoon setting
Glovers	260	0.2	74.5	9.8	5.3	10.2	3	17	Open
Light-house	200	2.9	56.1	5.7	8	27.3	4	8	Open
Turneffe	525	21.7	61	–	4.9	12.4	22	8	Restricted

Table 3.8 Occurrence of the four sediment types in the environments of the atolls (Gischler and Lomando 1999)

Environment	Coral-red-algae-Halimeda (%)	Mollusc foraminifer-Halimeda (%)	Mixed peloidal/skeletal (%)	Halimeda (%)
Island	8.6	0	0	0
Reef	22.4	0	4.1	0
Patch reef	5.7	0	0	0
Fore reef	17.8	0.9	4.1	0
Sand apron	26.4	2.7	22.8	0
Shallow lagoon	16.7	35.1	43.9	9.1
Deep lagoon	2.3	58.6	25.2	2.3
Restricted-turneffe lagoon	0	2.7	0	88.6

The wackestone in the interior of the Turneffe Islands is probably a direct result of this relatively protected position.

Antecedent topography might also explain the differences in the three ocean-exposed atolls. Lighthouse Reef and Banco Chinchorro have extensive grainstone belts, while Glovers is dominated by reefs in a muddy lagoon. Glovers Reef has the thickest Holocene sediment package (9.4–11 m) while both Lighthouse and Turneffe have a thinner Holocene sediment cover (6.4–7.9 m and 3.1–3.8 m, respectively) (Gischler and Lomando 2000). The difference in thickness is likely an indication of differential subsidence in the late Pleistocene and Holocene. The lower Pleistocene surface and a deep antecedent lagoon on Glovers resulted in an earlier initiation of reefs and thus the formation of a protective reefal rim that inhibited the formation of grainstone belts as found on the other two atolls.

The surface sediment of the offshore atolls can be divided into four facies types: coral-red algae-*Halimeda*, mollusk-foraminifera-*Halimeda*, mixed peloidal/skeletal, and *Halimeda*. Table 3.8 provides the composition and texture of the sediment types on the four atolls.

The coral-red algae-*Halimeda* type is most common in the reef, fore-reef, and sand apron environments (Gischler and Lomando 1999). The mollusc-foraminifera-*Halimeda* type occurs mostly in platform interior environments such as deep lagoonal areas (Gischler and Lomando 1999; Gischler 2003). The *Halimeda* type is primarily found in the restricted lagoons of the Turneffe Islands, and the mixed peloidal/skeletal type is found in leeward shallow lagoons of the platform interiors and windward sand aprons (Gischler and Lomando 1999).

3.5.4 Lagoon Reefs

The morphology and anatomy of Belize lagoon reefs has not been systematically studied, although several individual lagoon reefs have been examined in detail. Some examples of work on the lagoon reefs include studies by: (1) Westphall (1986) that examined the reef types and Holocene stratigraphy of deposits on Channel Cay; (2) Halley et al. (1977) that examined a back-barrier patch reef (Boo Bee patch reef); (3) Macintyre et al. (2000) that recently evaluated the Holocene stratigraphy of the Pelican Cays; and (4) Esker et al. (1998) that documented the seismic stratigraphy of both reef carbonates and surrounding siliciclastics.

A zonation of lagoon reefs was proposed by James and Ginsburg (1979) and is summarized below. The lagoon reefs exhibit a zonation of corals that is related to water depth and position. On the seaward-facing side, columnar *Montastrea annularis* dominates in the deep sloping areas and *Millepora* sp. and branching *Acropora palmata* dominates in shallow sloping areas. On the flat tops of the pinnacle reefs, coral rubble grades landward into the grass *Thalassia testudinium*. The leeward slope contains delicate branching coral *Acropora cervicornis* and *Porites porites*. The toe-of-slope contains *Halimeda* sand and coral rubble.

As part of the data collection for the Esker et al. (1998) study, analog bathymetric profiles were collected over the principal lagoon reefal areas. McNeill (in 1998) compiled these data as part of an unpublished report. Lagoon reefs and atolls of the Belize lagoon have been classified by morphology (two-dimensional bathymetric data) into three types: (1) single-pinnacle reefs that consists of an individual, steep-sided pinnacle with no laterally attached topographic features and are found in both deep and shallow areas of the lagoon; (2) multi-pinnacle reefs made up of two or more pinnacles that form a plateau-type feature relative to adjacent deeper water areas; and (3) shelf atolls with a central lagoon and steep sides. Reef types number 1 and 2 are equivalent to the "patch" reefs of James and Ginsburg (1979), while the shelf atolls encompass the rhomboid reefs in the southern shelf lagoon. James and Ginsburg (1979) have described a general zonation of the lagoon reefs. Westphall (1986) provides additional description of the surface fauna and sediments of Channel Cay a shelf atoll.

The dimensions of the three shelf lagoon reef types show distinct differences in both width and thickness (Fig. 3.22). The width of the multi-pinnacle reefs, as expected, is substantially wider (175–310 m range) than the single-pinnacle reefs (50–110 m range). Conversely, the mean thickness of the single-pinnacle reefs (~33 m) is almost double that of the multi-pinnacle reef structures (~17 m). If the depth to the base of the reef is taken as the point of initiation during the Holocene transgression, then the single-pinnacle reefs appear to have initiated earlier, lower on the shelf slope. The single-pinnacle reefs likely initiated earlier in the transgression during a period of relatively rapid sea-level rise (12–9 ka) (Fairbanks 1989). This earlier initiation likely explains the greater thickness of the single-pinnacle reefs. The precursor control of Pleistocene topography, is however, partially responsible for some of the differences in the reef attributes. The larger multi-pinnacle and shelf atoll reefs occur shallower on the shelf, and appears to have been initiated later (9–6 ka) during the transgression. The large multi-pinnacle and shelf atoll reefs are also more likely to require an existing template (Pleistocene) on which to establish themselves during the Holocene.

Lagoon coral reefs on the northern shelf are largely constrained to scattered, isolated patch reefs on the barrier platform (between the latitudes of Dangriga and Belize City), and become part of the fringing reef complex to the north between the

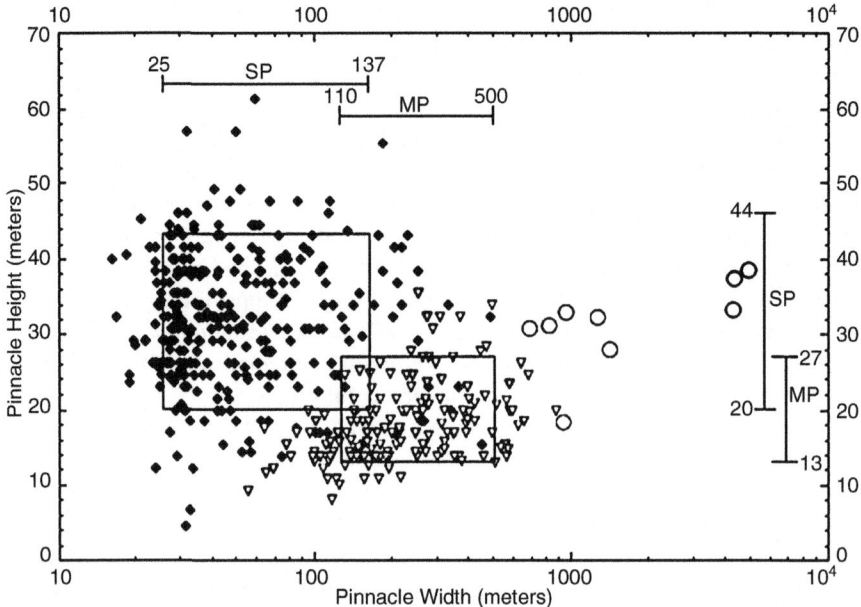

Fig. 3.22 Dimensional aspects of reefs from the Belize shelf lagoon. The data were compiled by D.F. McNeill from sounding charts associated with the collection of single-channel seismic data. The solid diamonds are single pinnacles, the open triangles are multi-pinnacles, and the open circles are shelf atolls (rhomboids). The boxes show the 10th and 90th percent quantiles, and their values are at the top and on the right side of the plot

latitudes of Belize city and the Mexican border). Besides coral reefs, small (~20–60 m), isolated reef shoals formed by agglutinated sabellariid worm tubes have been described by Burke et al. (1992) along the mainland shoreline.

The patch reefs behind the barrier/fringing reef complex north of Belize City have been partly explored and a general Holocene depositional history has developed. The best described northern lagoon patch reef in one found within the Mexico Rocks complex, located between the barrier reef and the Pleistocene limestone of Ambergris Cay (Mazzullo et al. 1992). Mazzullo et al. (1992) reported a pronounced zonation in coral fauna across Elmer Reef within the Mexico Rocks complex. *Montastrea annularis* dominated the reef crest, and *Acropora cervicornis* occurred on its windward and leeward flanks. A core through the Holocene section located within the *Montastrea* zone, shows a biotic composition consistent with a patch reef origin. Burke et al. (1992) dated the coral framework near the base of the core and obtained a carbon-14 age of 420 ± 130 bp. The relatively young age of this reef, and perhaps other nearby reefs, suggest that they likely developed during a static sea level regime. The limited vertical accommodation on this part of the shelf seems to have promoted lateral rather than vertical accretion, and the authors have termed these deposits "expansion reefs" (Mazzullo et al. 1992).

3.5.5   Forereef Slope

The forereef slope of the barrier reef and the offshore atolls have been thoroughly described in the publication compiled by James and Ginsburg (1979). Using mainly a manned submersible (James and Ginsburg 1979), and supplemented with seismic profiles in the offshore areas (Enos et al. 1979), the morphology, sediment composition, and organisms of the Belize barrier reef and atoll seaward margin were described. The reader is referred to that publication for details on the forereef setting, but a brief synopsis of the zonation is extracted below.

The forereef of the Belize barrier rim (and the rims of the offshore atolls) includes several morphologic zones: the forereef brow and wall, the proximal forereef, and distal forereef (Fig. 3.23).

The reefal margins of the barrier reef and atolls are either adjacent to a shallow basin (300–400 m deep), such as near Tobacco Cay and South Water Cay, or adjacent to the Cayman Trough, such as off of Gladden Spit and Glovers Reef (James and Ginsburg 1979) where the escarpment drops to several thousand meter depth. James and Ginsburg (1979) have summarized the variations in the reefal margins for each of the abovementioned (Table 3.9) settings.

James and Ginsburg (1979) describe the forereef: the forereef wall is made up of ledges, caves, and fissures. The ledges are 0.5–1.5 m wide and spaced 2–3 m apart. Fissures are near vertical, 1–3 m wide, <30 m long, and 5 m deep. Wall sediments consist of white *Halimeda* sand, coral plates, and mud on horizontal surfaces. Sediment bypasses the wall and accumulates at the toe-of-slope. Organisms along the wall show a transition between shallow water reefal communities to a

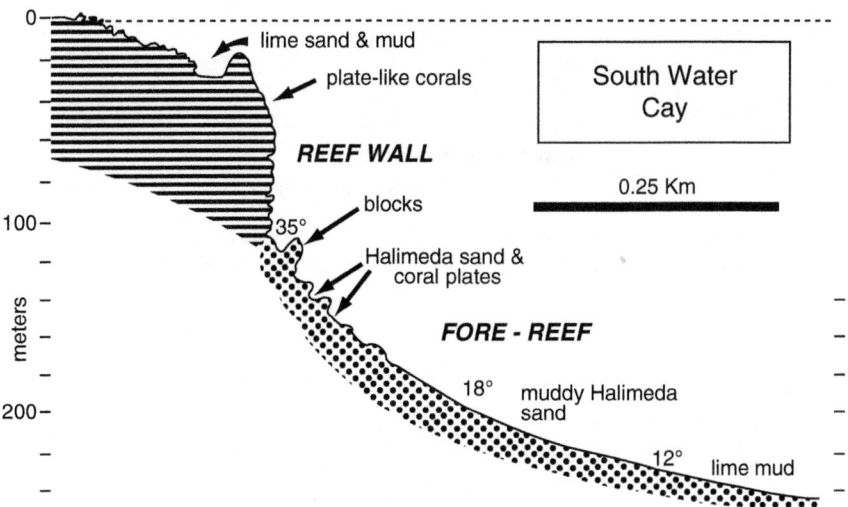

Fig. 3.23 A bathymetric profile of the barrier reef seaward margin near South Water Cay based on observations from submersible dives (Figure reproduced from Ginsburg and James 1976)

Table 3.9 Comparison of the reefal margins adjacent to shallow and deep basins (James and Ginsburg 1979)

| | Reefal margin | |
| | Adjacent to shallow | Adjacent to Cayman |
Morphological feature	basin	Trough
Wall	65 to 105–116 m	120–160 m
Vertical cliffs and gullies	–	150–200 m
Proximal forereef	To 200m	To 200m
Proximal forereef slope	30–45°	30–45°
Distal forereef	To 300 m	To 300 m
Distal forereef slope	10°	10°

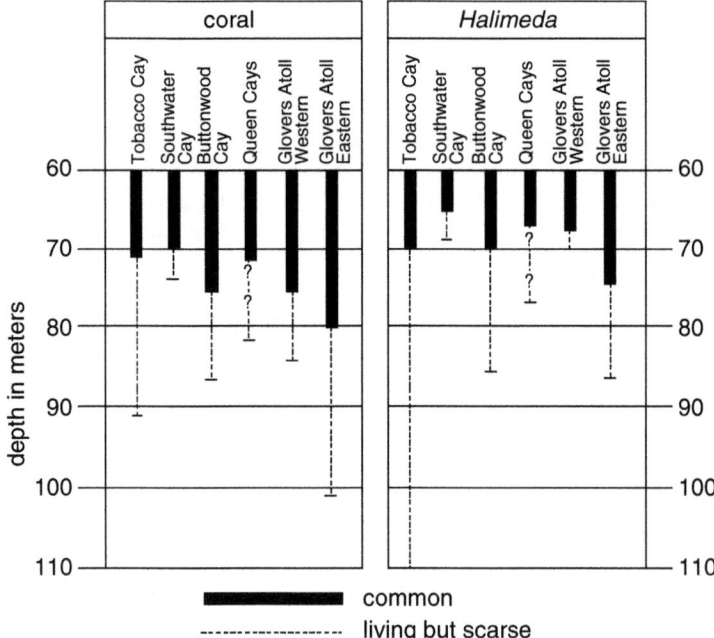

Fig. 3.24 A diagram summarizing the depth limits of the green alga *Halimeda* and reef building corals at several locations on the Belize barrier reef and offshore platforms (Figure from James and Ginsburg 1979)

deep-water community. The maximum depth at which various reefal organisms are found is summarized in Fig. 3.24.

The general forereef organism distribution (James and Ginsburg 1979) includes: from 67–80 m, *Agaricia fragilis, M. cavernosa, Madracis* sp., *Solenastrea* sp., *Stephanocenia* sp., and *Mycetophyllia reesii* make up 10% of the surface area. From 80–102 m, large hermatypic corals, *Madracis* sp., and *A. fragilis* dominate. *Halimeda* is only present down to ~70 m. In areas of vertical rock, 15% is covered

Table 3.10 Grain composition of sediment samples from the reef wall and sloping fore-reef (James and Ginsburg 1979)

Depth (m)	Location	Percentage of various particles														
		Halimeda	Mollusc	Coral	Coralline algae	Echinoids	Benthic foraminifera	Sponge chips	Worm tubes	Intraclasts	Peloids	Sponge spicules	Ascidian spicules	Alcyonarian spicules	Clay-size (<4 um)	Unknowns
48	Soft sediment, 4-m inside reef wall, glovers atoll	23	4	12	7	1	4	11	–	20	6	<1	–	2	2	7
100	Reef wall, inside cave, tobacco cay	39	6	6	6	3	11	7	2	6	5	<1	Tr	1	1	4
120	Top of sloping fore-reef, South water cay	50	6	11	6	3	9	1	2	4	2	<1	<1	<1	<1	1
177	Sloping fore-reef, tobacco cay	53	4	6	6	2	13	3	–	3	2	tr	tr	tr	1	5
190	Sloping fore-reef, tobacco cay	29	5	8	6	2	11	12	<1	2	9	Tr	Tr	Tr	3	9
200	Sloping fore-reef, South water cay	35	3	7	3	Tr	10	12	2	1	6	<1	<1	<1	2	11

Tr = trace amount

with yellow sponges and coralline algae, green algae, and alcyonians (James and Ginsburg 1979). Ledges contain sponges and 30–50% coralline algae. The percentage of the various grain types of the reef margin wall and sloping forereef is summarized in Table 3.10.

The proximal forereef exhibits fan-like features and contains blocks, coral debris, skeletal sand, fragments of plate-like corals, and *Halimeda* sand (James and Ginsburg 1979). The distal forereef contains muddy sand, corals, and blocks. Unlike the proximal forereef there is little downslope sediment movement in the distal forereef (James and Ginsburg 1979).

The forereef exhibits a grain size decrease from the sand-dominated proximal forereef to the mud-dominated basin (as shown in James and Ginsburg 1979). The proximal forereef sediment is bimodal with <5% mud. There is a 20% decrease in *Halimeda* sand and a fourfold increase in sponge chips (James and Ginsburg 1979). The transitional sediments between the proximal and distal forereef has 20–30% mud and <20% gravel fraction. The distal forereef sediment is 60% mud, and the fine sand fraction contains 20% Foraminifera, 5% pteropods, 25% shallow organism skeletal debris, 25% benthic foraminifera, 15% sponge chips, and 10% ovoid fecal pellets. The basin sediments consist of >70% mud (James and Ginsburg 1979).

The Belize barrier forereef morphology and sediment character are similar to other steep forereef and upper slopes described from the Caribbean (Ginsburg et al. 1991; Grammer et al. 1991; Grammer and Ginsburg 1992).

3.5.6   Backreef Sand

Landward of the Belize barrier reef a skeletal sand body, approximately 100–200 m wide runs the 160-km length of the barrier. There are unbroken stretches of this sand apron with lengths of 30 km (Halley et al. 1983). The sand body is nearly flat and ranges in depth of 1.5 m nearest to the reef and deepens lagoonward. The backreef sand flat is generally wider, better developed along the southern barrier reef. This southern barrier platform, outside of the lee of Glovers and Turneffe offshore atolls, is less protected than the central parts of the barrier reef. Backreef sand aprons are also well developed on the offshore atolls, especially on the eastward (seaward facing side.

3.5.7   Lagoon

The lagoon proper includes the area landward of the barrier reef and seaward of the mainland shoreline. There is both an west-to-east and north-to-south trend in the types shelf lagoon sediment. In the northern shelf lagoon there is a west-to-east trend from quartz sand nearshore to miliolid mud to *Halimeda* sand to reef front

Table 3.11 Carbonate content, grain size, and constituent composition of sediments in Belize (Purdy et al. 1975)

Purdy Facies	Mean grains size			Mean vol% of fraction >1/16 mm^3									No. of thin sections	Total no. of samples
	Mean wt% CaCO$_3$	wt% 1/8–1/16 mm	wt% <1/16 mm	Coralline Algae	Halimeda	Foraminifera	Corals	Mollusks	Pteropods	Ostracods	Crypto, grains	Terrigenousclastic		
Terrigenous														
Northern quartz sand	79(4)	8(4)	10(4)	–	–	13	–	28	–	–	22	19	4	4
Southern quartz sand	8(14)	6(16)	8(23)	–	–	–	–	–	–	–	–	99	14	23
Sandy mud	30(28)	7(27)	61(24)	–	3	–	–	14	–	–	–	76	17	28
Mud	23(8)	3(8)	88(8)	–	–	23	–	12	10	–	–	50	3	8
Marl														
Ostracod	73(5)	7(6)	89(6)	–	–	16	–	21	–	20	26	5	6	6
Mollusk	54(36)	5(32)	83(37)	–	16	11	–	37	–	–	–	32	22	37
Halimeda	70(46)	7(41)	72(47)	–	43	5	21	14	–	–	–	4	28	47
Pteropod	81(8)	3(8)	94(8)	–	22	14	4	8	39	–	–	–	8	8
Gypsina	76(5)	11(5)	36(5)	12	20	52	9	3	4	–	–	–	4	5
Carbonate														
Cryptocrystalline sandy mud	90(15)	19(15)	34(15)	–	–	10	–	16	–	–	67	–	15	15
Miliolid sandy mud	89(17)	20(17)	50(17)	–	–	34	–	22	–	4	22	–	17	17
Halimeda sandy mud	94(30)	14(30)	62(31)	–	40	6	22	17	–	–	–	2	33	33
Halimeda sand	97(30)	8(34)	14(36)	2	50	10	18	16	–	–	–	–	35	36
Coralgal sand	98(28)	3(24)	4(28)	4	26	6	37	9	–	–	–	–	28	28
Peneroplid sand	95(13)	6(13)	10(13)	–	–	56	–	18	–	–	20	–	13	13

(Purdy et al. 1975). In the central and southern shelf lagoon the sediment type from west-to-east transition from quartz sand to mollusc marl to *Halimeda* marl to *Halimeda* sand to reef. Likewise, the carbonate content increases from ~30% to ~90% west to east (Purdy et al. 1975). The Holocene marine facies mapped by Purdy et al. (1975) (Fig. 3.14) have also had their carbonate content, grain size, and textural composition tabulated (summarized in Table 3.11, the Table 4 of Purdy et al. 1975).

For the northern shelf, Pusey (1975) mapped and characterized the sediment facies. The facies assignment of Pusey WC (1975) are the same as the more regional ones used by Purdy et al. (1975). The exception being the addition of "carbonate shoal" as a facies to describe the mud-rich deposits of Cangrejo and Bulkhead shoals in the northern shelf lagoon. Pusey (1975) described in detail the composition of the northern shelf marine facies. One of the most extensive facies is the broad (~12 km wide) and elongate (45 km) *Halimeda* sand deposit. The composition of the "*Halimeda*" facies is shown in Table 3.12 (from Pusey 1975 his Table 2).

The amount of terrigenous sediment is greater in the central and southern shelf lagoon due to the input of siliciclastics from the Maya Mountains. The clay mineral assemblages for the Belize shelf have been described and mapped by Scott (1975). Five principal assemblages were assigned based on the relative abundance of montmorillonite (smectite) versus kaolinite and illite. Chetumal Bay and the northern shelf lagoon (area north of Belize City), is characterized by montmorillonite-rich clay minerals. These montmorillonite clays are derived from the lime-enriched soils (less weathered) of northern Belize. The southern shelf clay assemblage is one of mixed kaolinite, illite, and montmorillonite. The more kaolinite/illite-rich clay mineral suite likely reflects the proximity to the Maya Mountains and more thoroughly leached soils of southern Belize (Scott 1975). Clay mineral distribution along the central and southern shelf lagoon is governed by differences in settling characteristics of these minerals, the influence of currents, and the general bathymetry (Scott 1975). Illite and kaolinite are found in the nearshore areas and on the barrier platform. Montmorillonite, with a slow setting velocity, is preferentially transported to the deeper parts of the shelf lagoon and to the offshore basin.

Carbonate mud originates predominantly from the breakdown of carbonate skeletons and nannoplankton. From the north to the south in the lagoon skeletal particles grade from mollusc and foraminifera to *Halimeda* to encrusting *Gypsina* fragments (Purdy et al. 1975). Coccoliths make up 5–20% of the bulk mud in the northern lagoon (Purdy et al. 1975). The southern shelf lagoon also has up to 20% coccoliths and coccolith fragments (Scholle and Kling 1972).

3.5.8   Islands

The Belize shelf is characterized by islands ranging from small (<100 m) mangrove cays to the large (kilometers wide) peninsula island of Ambergris Cay. The islands

Table 3.12 Composition, grain size class, and carbonate mineralogy of the *Halimeda* facies (Pusey 1975, his Table 2)

Facies type	Mean	Observed range	Standard deviation
Coralline algae	0.5	0.0–3.0	0.8
Melobesia	2.5	0.4–10.7	3.0
Halimeda	41.4	7.3–79.4	22.4
Miliolidae	3.2	0.6–7.7	2.1
Peneroplidae	3.3	0.5–10.1	2.6
Other foraminifera	4.5	1.4–9.9	2.5
Corals	0.7	0.0–2.3	2.1
Mollusks	26.5	11.7–49.0	11.4
Ostracods	1.6	0.0–11.8	2.9
Misc. skeletal	2.7	0.6–10.1	2.8
Unknown skeletal	5.7	1.3–19.6	4.2
Total skeletal	92.7	80.9–99.8	6.2
Fecal grains	4.7	0.0–16.6	4.6
Crypto. grains	1.6	0.0–6.1	2.5
Crypto. Skeletal	6.5	0.0–47.7	11.6
Total >50% crypto.	8.0	0.0–50.4	12.3
Country rock	0.8	0.0–9.2	2.1
Terrigenous minerals	0.3	0.0–3.5	0.8
wt% <0.125 mm	40.7	1.8–80.6	27.7
wt% <0.062 mm	26.8	1.2–68.9	23.6
Aragonite as wt% carb.	71	54–89	10
Mg carb in calcite wt%	12	7–15	2

of the shelf fall into four general categories with respect to their formation: (1) Pleistocene foundation with a Holocene veneer; (2) coral rubble islands along the barrier, offshore atolls, lagoon reefs, and shoreline reefs; (3) mangrove cays founded on a carbonate shoal; or (4) islands founded on mangrove-peat deposits. For example, Ambergris Cay and some of the islands in Chetumal Bay have a shallow Pleistocene foundation and only a thin veneer of Holocene sediment (Ebanks 1975; Ambergris Cay, Cay Corker, Blackadore Cay). Many of the smaller cays (or sand bores) on the barrier platform are formed by the accumulation of coral rubble and redeposition by storm-driven wave energy (Stoddart 1965, 1969; Stoddart et al. 1982). Examples of coral rubble islands include; Tobacco Cay, the numerous sand bores at tidal passes on the barrier reef, islands on the barrier platform (Shinn et al. 1982), Long Cay on Glovers atoll, and most of the headland reefs along the Placencia-Monkey River shoreline. The third type of island consists of mangrove colonization of a carbonate shoal, either reef rubble, sand shoal, or mudbank deposit. For example, islands along the Placencia-Monkey River shoreline, lagoon reefs such as the Channel Cay rhomboid shoal where mangrove islands have established (Westphall 1986), and the Pelican Cays rhomboid shoal islands where underlying reef shoals may have been initially controlled by karst topography (Macintyre et al. 2000). The other island type that is known from the Belize shelf lagoon is that formed from the long-lived deposition of mangrove peat. The best example of these peat islands is along the north side of the Tobacco Range cored and described by Macintyre et al. (1995).

3.6 Structural and Sedimentological Controls on Facies Distribution

The modern facies distribution in the Belize lagoon is largely determined by antecedent topography derived from the underlying structure (Fig. 3.25). The following sections summarize the key points related to the regional tectonics and their associated control on the shelf topography.

3.6.1 Tectonics

The continental margin of Belize is believed to have formed through left-lateral strike-slip motion between the North American Plate and Caribbean Plate that resulted in extension and normal faulting (Dillon and Vedder 1973; Lara 1993). These tectonic movements have subsequently determined the architecture, the extent of the shallow-water area, and the regional facies distribution. The large-scale structures, i.e., the regional transform faults and en-echelon faults, mark the boundaries of the lagoon, and in the offshore area define the margin for the four isolated platforms (Fig. 3.26).

The Tertiary tectonic history is marked by two major transtensional episodes (Lara 1993) (Table 6.1). The first episode during the Paleocene formed Southwest-northeast trending en-echelon normal faults as Cuba moved in a left-lateral strike-slip motion parallel to the eastern coast of the Yucatan peninsula (Lara 1993). The en-echelon faults created a pattern of ridges and troughs, which would later influence sedimentation in the basin. A comparison of the deep structures and the present bathymetry of the lagoon showed that portions of the barrier platform and the elongate to rhombohedral platform atolls, such as Laughing Bird Cay, are located over ancient structural highs, whereas bathymetric lows, such as the Victoria Channel, are long-lived troughs (Fig. 3.26; Lara 1993). These structures suggest that Quaternary reef foundations are deeper than previously thought. These deep structures, and overlying Miocene-Pliocene structures exercised a major control on sedimentation throughout the Tertiary evolution of the central and southern shelf lagoon (Lara 1993; Esker et al. 1998).The second transtensional episode during the Pliocene (and/or later) formed southwest-northeast trending high-angle normal faults due to movement along the Caribbean/North American left-lateral transform boundary (Lara 1993). The normal faults formed half-grabens that run along three southwest northeast oriented depressions (Fig. 3.27, Esker 1998). The transtensional structures influence the deposition of the modern sediments in the lagoon and on the three offshore "atolls". These three isolated, atoll platforms, Turneffe,

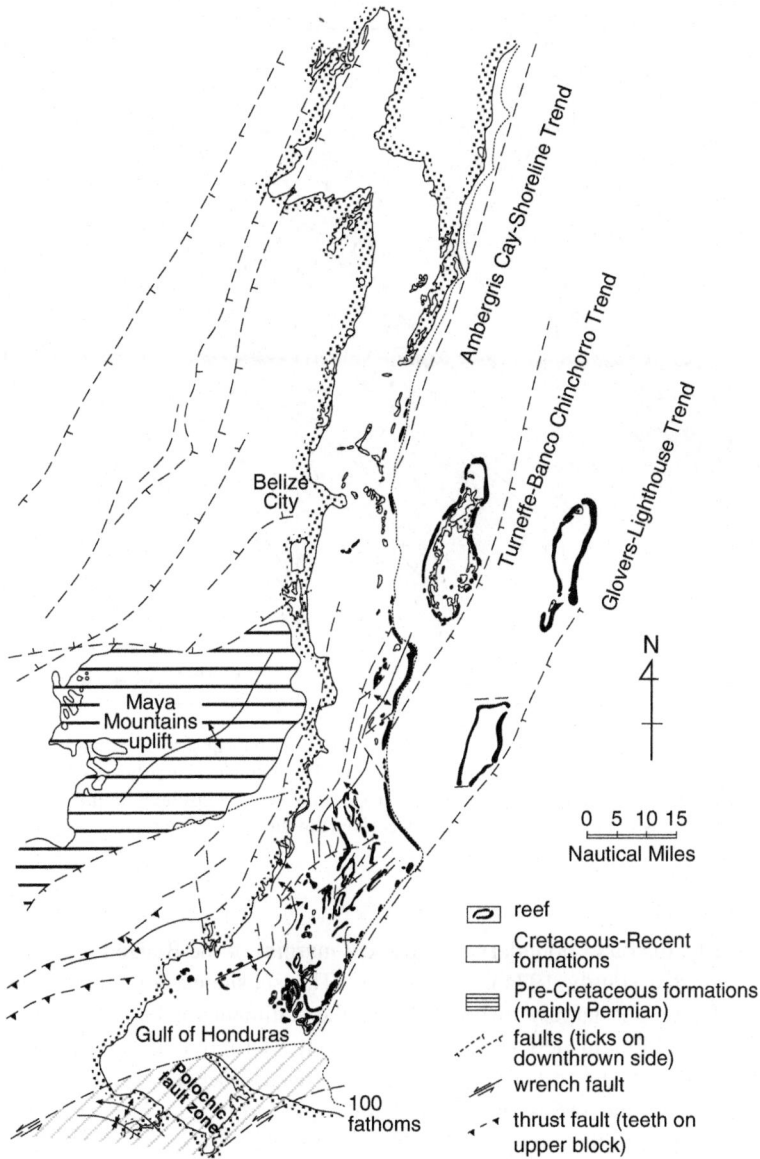

Fig. 3.25 Structural composition of Belize by Purdy (1974b), compiled from several published and unpublished sources. Note the northeast-diverging structural lineaments that contribute to the Belize reef margin and offshore platforms

Lighthouse, and Glovers, are all surrounded by water ranging from 800–2,000 m deep. The basement of the platforms is made of continental-derived fault blocks that are bounded on either side by deep-rooted faults of a left-lateral transform boundary (Dillon and Vedder 1973).

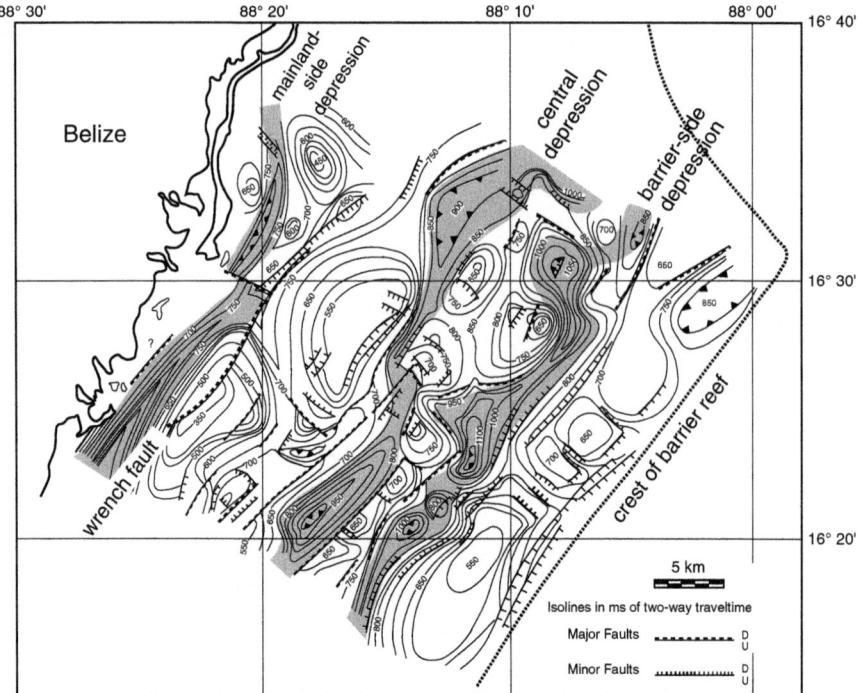

Fig. 3.26 Isochron map of a lower Eocene unconformity in the subsurface of the Belize lagoon (Figure from Lara (1993)). Contour interval is 50 ms. The map shows northeast-southwest elongate structural trends. The highs and lows are proposed to have controlled the initiation of reefs (*highs*) and the positioning of the deep channels (*lows*) in the modern lagoon

Similarly, the outer boundary (barrier reef margin) of the Belize lagoon is likely fault controlled (Purdy 1974a; Lara 1993). The subsidence of these large-scale tectonic features is largely responsible for the sediment thickness and the water depth in the different depositional environments. Differential subsidence of the offshore atolls results in variable thickness of the Holocene sediments, which in turn determines the facies on the individual platforms (Gischler and Lomando 1999). The fastest subsiding platform, Glovers Reef, is also the atoll with the greatest number of patch reefs and the most well developed marginal reefs. The age of Pleistocene bedrock under the Belize shelf and offshore atolls varies from between 280 and 120 ky (Gischler et al. 2000). The differences in Pleistocene bedrock elevation (and associated sedimentation patterns) likely reflect the combination of differential subsidence along the fault block. This subsidence, combined with differential karstification, may be the two important processes controlling the sedimentation (Gischler et al. 2000). A comprehensive overview of the antecedent topography and tectonic controls is presented in Purdy (1998) and Purdy et al. (2003) (Table 3.13).

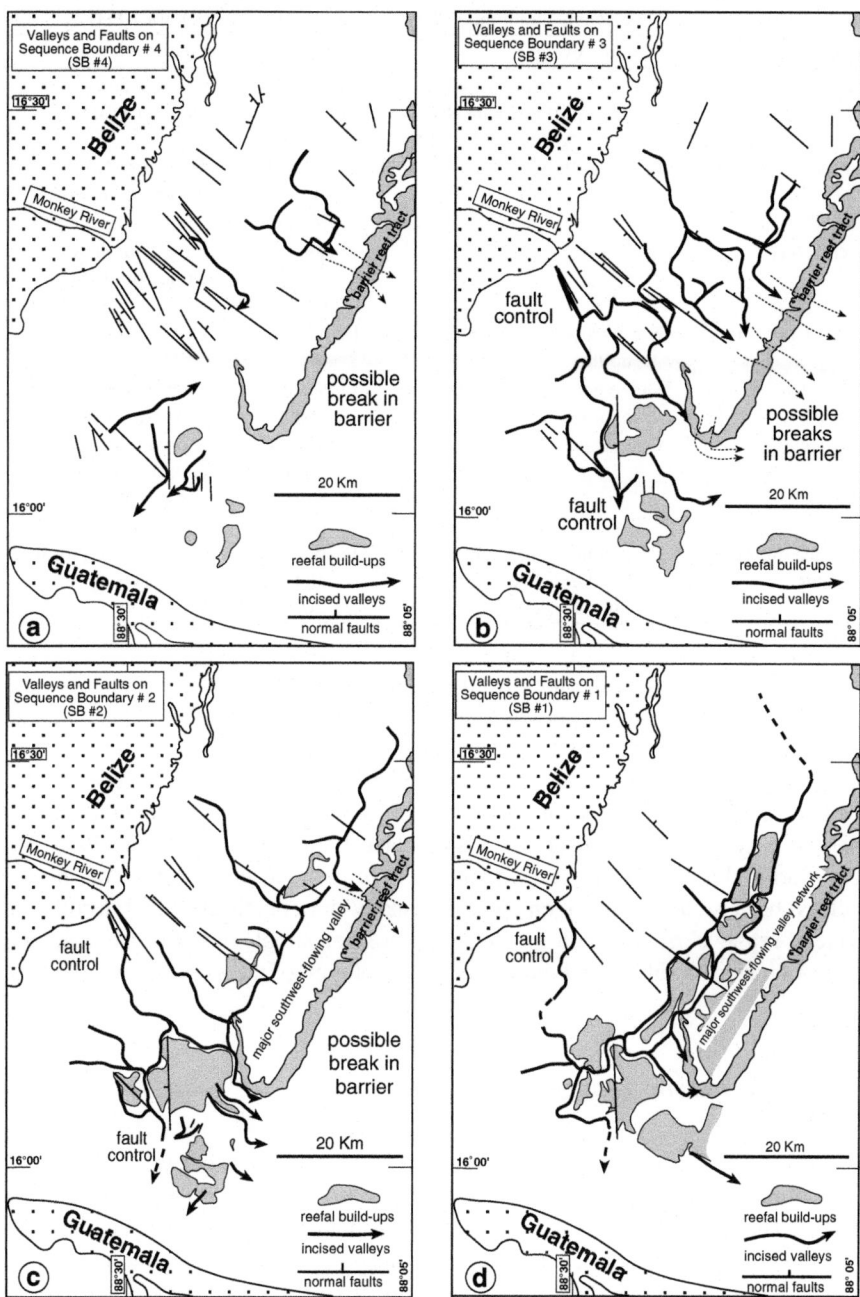

Fig. 3.27 Routes of incised-valleys and reef development during four depositional (sea level) events. The interpreted record shows drainage in the lagoon deflected to the south through time (SB 4 – SB 1). The development of reef bodies shows that they are relatively young features and that they reoccupy the existing topographic highs to form stacked deposits (Reproduced from Figure 18 of Esker et al. 1998)

Table 3.13 Age of tectonic events along the Belize margin (Modified from Lara 1993)

Age	Tectonic event	Regional tectonic setting
Late Permian-Triassic	Maya Mts: deformation and uplift of Santa Rosa Group granitic intrusion	North-south compression along the Sierras of Northern Central America
Jurassic	Block faulting	Yucatan Block rifting from North America; development of Gulf of Mexico
Cretaceous	Subsidence	Yucatan Block rifting from South America; development of Proto-Caribbean Sea
Latest Cretaceous	Northward compression and uplift	Development of Motagua suture zone
Paleocene	Divergent wrench faulting parallel to the coast	Left-lateral strike-slip motion of Cuba along Yucatan eastern coast
Oligocene	Compression and uplift	Left-lateral strike-slip motion along inshore portion of the North American-Caribbean plate boundary
Pliocene and/or later	Divergent wrench faulting parallel to the coast. Subsidence	Eastward migration of the Caribbean plate. Spreading at the Cayman trough

In the Belize lagoon, the long-lived structures and the distribution of reefs and lagoonal sediments can be thought to act as a depositional template. This template effectively provides a feedback mechanism for subsequent sedimentation (i.e., reef reoccupy the highs, siliciclastics reoccupy the lows). To form this template, structural features likely influenced the initial position of reefs and incised valleys in the Belize lagoon, as suggested by subsurface seismic data (Lara 1993; Esker et al. 1998). At least during the Pleistocene, incised valleys were prone to inhabit ancestral incised valleys, because accommodation space is usually not completely filled during one cycle of sea level rise and fall. Figure 3.28 presents a simple model of the structural control on incised valleys and the subsequent evolution of the reefs and basin facies.

3.6.2 Precursor Topography

The structural control on sedimentation patterns occurs on two levels. One is the relatively old underlying tectonic framework that was created during the evolution of the margin. The other is the more recent tectonic activity that has shaped the margin during the Neogene. A brief tectonic summary is presented below for the more recent events.

The basement of the Belize shelf contains NNE-tilted fault blocks created by Tertiary tectonic activity (Dillon and Vedder 1973; Purdy 1974a; Lara 1993). The resultant series of ridges and troughs provided the structural template that influenced the development and distribution of both the Quaternary carbonate reefs and siliciclastic systems (Lara 1993).

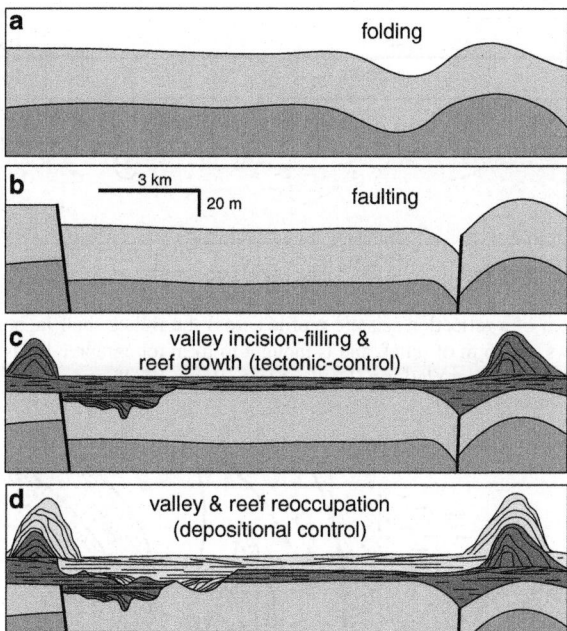

Fig. 3.28 Schematic model of the evolution of incised valleys and carbonate buildups during the late Pleistocene in the Belize lagoon. Structural control is likely the main factor for positioning of the valleys and reefs early in the depositional history. Later, the topographic highs influence each subsequent highstand and promote valley and reef reoccupation (Figure redrawn and modified from Esker et al. 1998)

Similarly, the Pleistocene reefs likely grew on bathymetric high areas that originated from structural features, highs formed from karst dissolution (Fig. 3.29), and perhaps topographic highs formed from channel banks and bars of a fluvial and deltaic system (Fig. 3.30). As a result of this precursor topography, subsequent siliciclastic deposition on the Belize shelf was constrained to the topographic lows (Figs. 3.26 and 3.27).Holocene reefs formed on the topographic highs of the Pleistocene carbonates as well as on topographic highs of the latest siliciclastic deposits along the mainland coast (Choi and Ginsburg 1982; Lara 1993; Esker et al. 1998). The topographic lows became incised valleys that were reactivated by each of the Quaternary sea level falls (Esker et al. 1998). These valleys are u-shaped to irregularly shaped and measure 1.5–5 km across (Choi and Ginsburg 1982). The depth to the underlying Pleistocene appears to increase from the north to south in the Belize lagoon and perhaps beneath the Belize barrier reef (Fig. 3.31) (Burke 1993).

Purdy (1974a, b) proposed that structurally influenced karst processes were largely responsible for much of the relief found today on the Belize shelf. Purdy (1974b) argues that relief is initially developed by dissolution and then accentuated by reef growth. Purdy et al. (2003) and Purdy and Gischler (2003) provide subsurface and surface facies maps that further support the importance of antecedent

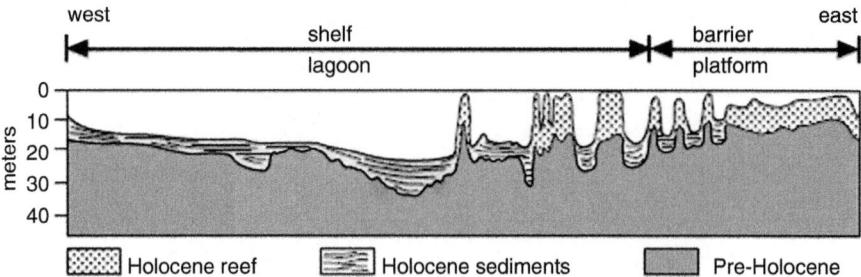

Fig. 3.29 A sketch of a reflection seismic profile across the Belize shelf lagoon. The pre-Holocene surface shows 10–15 m of relief, and is thought to form topographic highs for reoccupation of the lagoon reefs and barrier platform (Data are from Purdy 1974b and redrawn from James and Ginsburg 1979)

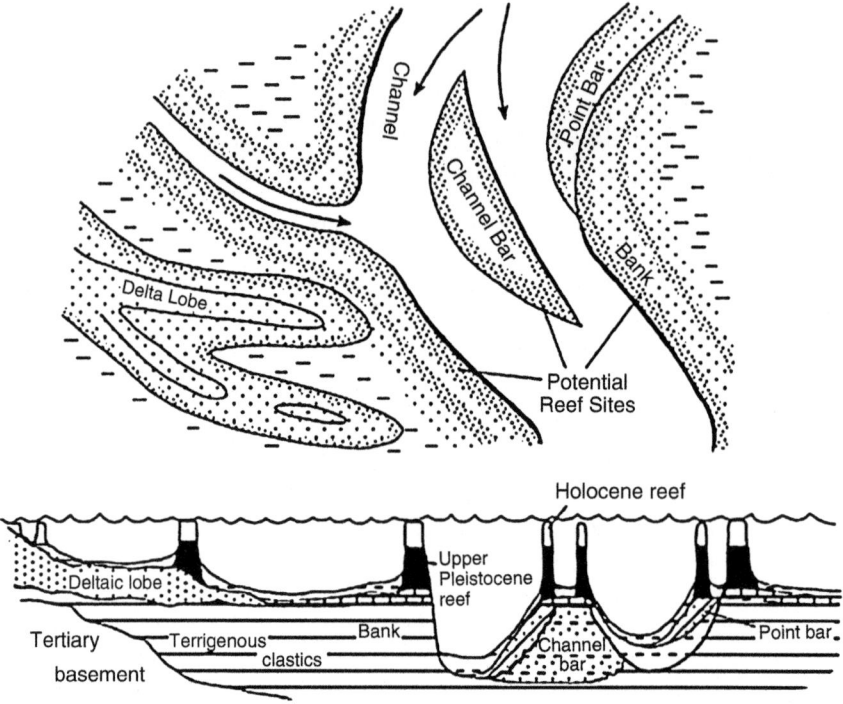

Fig. 3.30 Schematic map and cross-section portraying the suspected development of reefs on localized elevation of fluvial and deltaic deposits. Choi and Holmes (1982) propose that modern reef geometry is directly related to underlying (Pleistocene) fluvial morphology (Data to test this hypothesis has not been collected)

topography and a long-lived control on sedimentation. In fact, the new facies map presented in Purdy and Gischler (2003) argues that the distribution and type of Holocene facies is strongly conditioned by the antecedent topography derived from the structural features and influences.

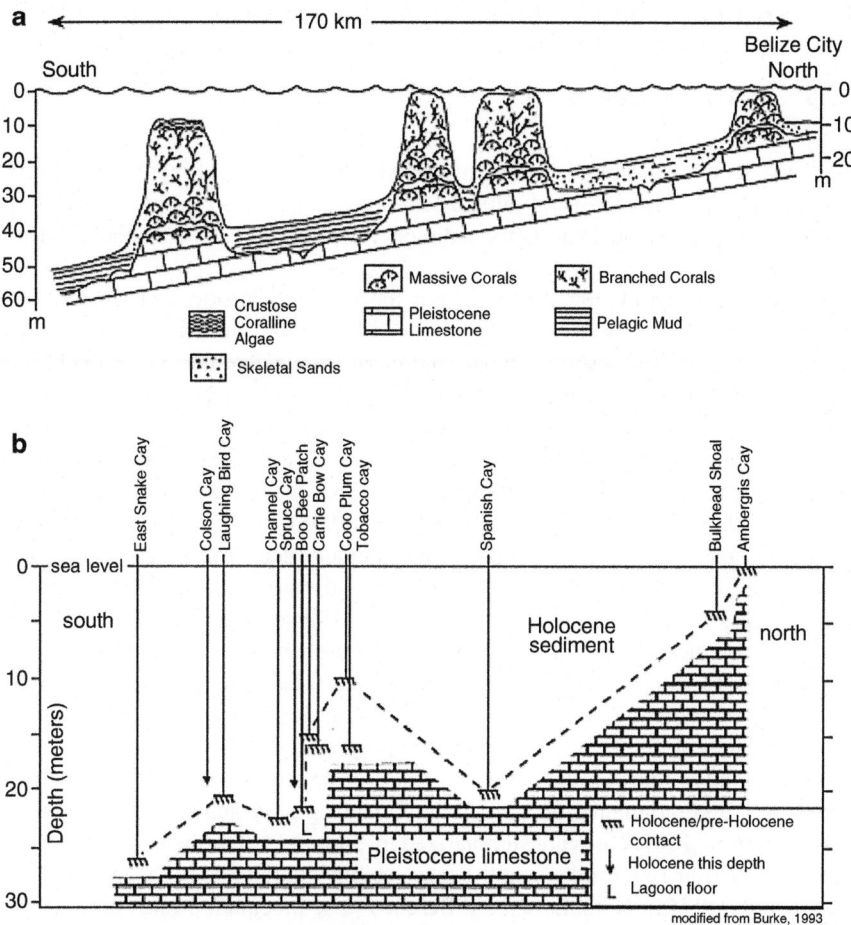

Fig. 3.31 (**a**) Schematic diagram of the inferred succession of reef builders along a north-south transect in the Belize lagoon. (**b**) A figure redrawn from Burke (1993) showing the known Holocene-Pleistocene contacts along the Belize barrier reef. This general southward decrease in elevation of the Pleistocene was confirmed with 11 new cores collected and analyzed by Gischler and Hudson (2004)

Present-day Belize lagoon geomorphic features include reefal plateaus and various submarine channels that run sub-parallel to the barrier reef (Fig. 3.27; Esker 1998). Esker (1998) identified three seismic facies units within the incised-valley fills of the Belize lagoon. The units indicate that the valley fills increase in thickness with time due to increase in accommodation space, reefs become more numerous and widespread through time, with reef development forcing incised valleys to flow around them.

3.7 Stratigraphic Evolution: Holocene Deposits

3.7.1 Accumulation Rates

Holocene accumulation rate data from various depositional settings on the Belize shelf are available from several published studies. Purdy (1974a) provided one of the most comprehensive datasets for sediment accumulation on the barrier platform and shelf lagoon (Table 3.14).

Additional accumulation data for both the barrier platform and shelf lagoon were provided by the study of Shinn et al. (1982) based on age dates from peat recovered in vibrocores (shelf lagoon) and coral from rotary coring on the barrier platform (Table 3.14). Gischler and Hudson (2004) provide a comprehensive data-set from 11 core borings on the accumulation of Holocene deposits along the barrier reef. They concluded that initiation of the reef was nonsynchronous due to variation in elevation of antecedent topography. They also calculated a mean barrier reef accumulation rate of 3.25 m/ky. Esker et al. (1998) provided sediment accumulation rate information for the southern shelf lagoon (deeper channels and off-reef areas) from their carbon-14 age dates on peat and marine molluscs (Table 3.14). More recently, a very comprehensive data set on Holocene accumulation has been published for the offshore atolls (Table 3.14)(Gischler and Hudson 1998, 2004; Gischler and Lomando 2000). Several sediment accumulation rate values also exist for the lagoon rhomboid reef (1.5–3.1 m/ky) at Channel Cay (Westphall 1986) and the peat islands at Tobacco Range (1.4–3.5 m/ky)(Macintyre et al. 1995).

Sedimentation rates on the Belize barrier reef derived from carbon-14 dates of coral heads, and sedimentation rates in the Belize lagoon taken from peat/sediment recovered from piston cores are summarized in Table 3.14. The mean sedimentation rate on the barrier is 2.7 m/ky, and the mean sedimentation rate in the lagoon is 0.2 m/ky (Purdy 1974a). Notable is the difference in sedimentation rates between the barrier reef and shelf lagoon, on the order of 10–1, respectively.

3.7.2 Sea-Level History

In the southern shelf lagoon and in The Gulf of Honduras, reefs may have been drowned probably due to a high rate of sea-level rise (give-up reefs) (Burke et al. 1998; Westphall 1986). Burke et al. (1998) suggest that those reefs may not have reached a sufficient diversity of fast-growing species that would have allowed them to follow the rapid increase in accommodation space. According to the Fairbanks (1989) sea-level curve, later combined with local data by Esker (1998) (Fig. 3.32) for the southern Belize lagoon, the fastest sea-level rise in the Holocene would have been too fast for corals to keep up during two periods around 12 and 9.5 ka (Meltwater Pulses I and II). A recent summary of Belize sea level data from corals and mangrove peat for the past 10 ka (calendar years) details this period of rapid rise and the

Table 3.14 Summary of carbon-14 age dates and accumulation rates for the Belize shelf

Locality/sample	Sample type	Sediment thickness	C-14 age (C-14 years)	Accumulation Rate	Reference
Barrier platform and lagoon reefs					
RC1a	Coral	5.79	5625 ± 85	1.03	Shinn et al. (1982)
RC1b	Coral	8.23	6165 ± 90	1.33	"
RC1c	Coral	10.97	6140 ± 90	1.79	"
RC1d	Coral	17.67	7175 ± 100	2.46	"
RC4	Coral	15.54	6960 ± 110	2.23	"
Spanish cay	Coral	6.3	4025 ± 125	1.5	Purdy (1974a)
Tobacco cay	Coral	5.6 (mean)	5100 ± 135	1.1	"
Spruce cay	Coral	17.4 (mean)	3925 ± 125	4.4	"
Laughing bird cay-a	Coral	5.1 (mean)	3950 ± 125	1.3	"
Laughing bird cay-b	Coral	23.9 (mean)	4775 ± 130	5.0	"
Laughing bird cay-c	Coral	29.6 (mean)	6725 ± 155	4.4	"
Colson cay	Coral	6.7 (mean)	5550 ± 140	1.2	"
CC B-11953	Coral	5.00	2130 ± 70	2.4	Westphall (1986)
CC B-11954	Coral	8.48	3160 ± 60	2.7	"
CC B-11955	Coral	10.35	3370 ± 80	3.1	"
CC W-5596	Coral	4.55	2890 ± 250	1.6	"
CC W-5598	Coral	1.84	800 ± 200	2.3	"
CC W-5600	Coral	4.55	3010 ± 250	1.5	"
CC W-5602	Coral	2.38	1290 ± 450	1.8	"
BBPR	Coral	~14	SL curve	1.6	Halley et al. (1977)
Elmer reef	Coral	1.5	420 ± 130	3.5	Mazzullo et al. (1992)
BBR 1–2.8	Coral	2.5	4640 ± 70	0.5	Gischler and Hudson (2004)
BBR 1–8.1	Coral	7.8	6020 ± 60	1.3	"
BBR 1–13.9	Coral	13.6	6800 ± 60	2.0	"
BBR 2–2.0	Coral	1.7	4770 ± 70	0.4	"

(continued)

Table 3.14 (continued)

Locality/sample	Sample type	Sediment thickness	C-14 age (C-14 years)	Accumulation Rate	Reference
BBR 2–4.4	Coral	4.1	6230 ± 80	0.7	"
BBR 3–3.5	Coral	2.6	6200 ± 80	0.4	"
BBR 3–8.4	Coral	7.5	7220 ± 90	1.0	"
BBR 3–20.0	Coral	19.1	7290 ± 60	2.6	"
BBR 4–1.7	Coral	1.1	3380 ± 70	0.3	"
BBR 4–6.0	Coral	5.4	6170 ± 80	0.9	"
BBR 4–9.3	Coral	8.7	7220 ± 60	1.2	"
BBR 5–2.8	Coral	1.9	1420 ± 60	1.3	"
BBR 5–7.3	Coral	6.4	5740 ± 70	1.1	"
BBR 5–14.0	Coral	13.1	7180 ± 90	1.8	"
BBR 6–3.2	Coral	2.6	4310 ± 70	0.6	"
BBR 6–7.6	Coral	7.0	5450 ± 90	1.3	"
BBR 7–3.0	Coral	2.4	4530 ± 70	0.5	"
BBR 7–6.4	Coral	5.8	5480 ± 80	1.1	"
BBR 8–2.9	Coral	2.6	3170 ± 70	0.8	"
BBR 8–8.3	Coral	8.0	5580 ± 80	1.4	"
BBR 8–14.5	Coral	14.2	6500 ± 70	2.2	"
BBR 8–20.0	Coral	19.7	7790 ± 60	2.5	"
BBR 9–2.9	Coral	1.7	1100 ± 60	1.5	"
BBR 9–10.7	Coral	9.5	4990 ± 80	1.9	"
BBR 9–15.5	Coral	14.3	6570 ± 70	2.2	"
BBR 9–19.8	Coral	18.6	7570 ± 60	2.5	"
BBR 10–2.2	Coral	1.9	5010 ± 80	0.4	"
BBR 10–4.2	Coral	3.9	5810 ± 80	0.7	"
BBR 10–7.0	Coral	6.7	6940 ± 60	1.0	"
BBR 11–2.0	Coral	1.7	4990 ± 80	0.3	"

BBR 11–44	Coral	4.1	6090 ± 60	0.7	"
BBR 11.68	Coral	6.5	6480 ± 60	1.0	"
Shelf lagoon					
Deep lagoon sediments					
CC W-5601	Coral	1.90	9190 ± 450	0.2	Westphall (1986)
23	Peat	1.8	5775 ± 140	0.3	Purdy (1974)
31	Peat	1.5	6125 ± 145	0.2	"
255	Peat	0.9	5600 ± 140	0.2	"
92	Peat	2.1	8100 ± 170	0.2	"
139	Peat	2.0	10,075 ± 210	0.2	"
165	Peat	1.5	9375 ± 195	0.2	"
300	Peat	2.1	8400 ± 180	0.2	"
308	Peat	1.0	6000 ± 145	0.2	"
330	Peat	1.7	8250 ± 175	0.2	"
331	Peat	1.0	8250 ± 175	0.1	"
UM92 03.80	Gastropod	0.80	4740 ± 60	0.17	Esker et al. (1998)
UM92 03.105	Peat	1.05	7335 ± 70	0.14	"
UM92 04.228	Peat	2.25	7705 ± 75	0.29	"
UM92 07.154	Peat	1.54	8855 ± 75	0.17	"
UM92 07.204	Gastropod	2.04	8490 ± 75	0.24	"
UM92 11.193	Gastropod	1.93	7920 ± 70	0.24	"
UM92 25.168	Seagrass roots	1.68	5670 ± 75	0.30	"
UM92 26.175	Peat	1.75	8865 ± 85	0.20	"
UM92 26.233	Mollusc	2.33	6260 ± 75	0.37	"
UM92 27B.104	Worm tube	1.04	8685 ± 90	0.12	"
UM92 27B.142.5	Barnacle	1.43	8980 ± 85	0.16	"

(continued)

Table 3.14 (continued)

Locality/sample	Sample type	Sediment thickness	C-14 age (C-14 years)	Accumulation Rate	Reference
UM92 28.154.5	Oyster	1.55	11,185 ± 90	0.14	"
UM92 28.273.5	Oyster	2.74	11,230 ± 90	0.24	"
UM92 30.169.5	Mollusc	1.70	9230 ± 120	0.18	"
Mangrove peat island					
TR 1	Peat	6.94	4830 ± 80	1.43	Macintyre et al. (1995)
TR 2	Peat	7.18	4840 ± 95	1.48	"
TR 3	Peat	5.20	3660 ± 100	1.42	"
TR 4	Peat	1.20	780 ± 110	1.54	"
TR 5	Peat	5.12	1450 ± 100	3.53	"
TR 6	Peat	2.41	1060 ± 110	2.27	"
TR 7	Peat	5.69	2220 ± 110	2.56	"
BBPR	Peat	~3.2	8780 ± 100	0.4–0.5	Halley et al. (1977)

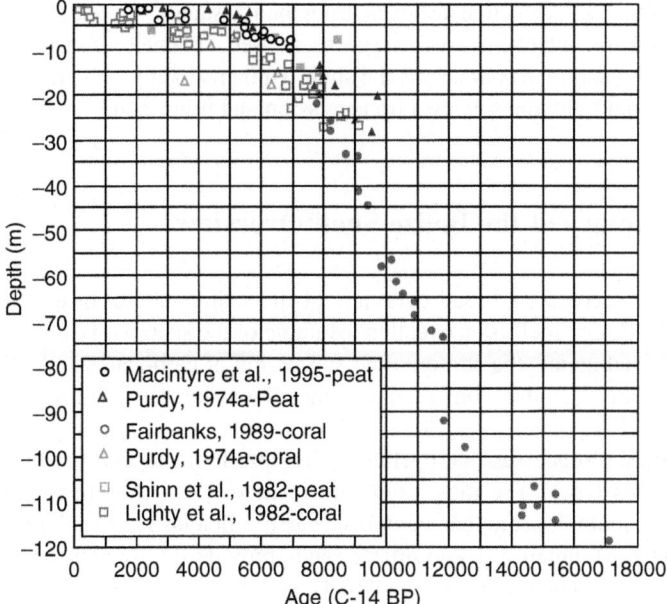

Fig. 3.32 Compilation of carbon-14 dated samples from the Caribbean, Florida, and Belize that are indicative of the position of sea level (plus any water depth) over the past ~18 ky (Data source as indicated on the figure. Some of these data and other new data have recently been compiled by Toscano and Macintyre (2003) and Gischler and Hudson (2004) to refine the western Atlantic sea-level curve for the past 11 ky)

deceleration starting at about 6,000 years before present (Gischler and Hudson 2004). In addition, the southern lagoon appears to be subsiding considerably faster than the central and northern parts of the shelf. Purdy (1998) has attributed structural controls to the morphology of the southernmost "hook" part of the barrier reef.

The modern topography of the Belize lagoon is most likely a combination of structure and differential sediment accumulation related to the high-amplitude Quaternary sea level fluctuations. In the north, the combination of freshwater influx, shallow water depth, and relatively recent sea level transgression has produced shelf lagoon deposits that are generally within a few meters of modern sea level. Limited ocean circulation and copious freshwater runoff contribute to salinity-stressed eco-logic conditions and sedimentary facies. These deposits are often either mud-rich or skeletal sands dominated by benthic foraminifers. A fringing reef occurs seaward of the large Pleistocene island (Ambergris Cay). In the central and southern lagoon, the deeper lagoon and shallow reef contrast becomes distinct with pinnacle reefs and rhomboid shelf atolls surrounded by water depths of about 20–30 m. The reefs reach sea level in the northern part of the southern shelf lagoon but dip to a depth of 7–8 m below sea level in the southernmost part of the southern shelf lagoon.

This shallow reef and deep lagoon morphology has effectively created a deposi-tional template. During the highstands, the inter-reef areas accumulate a mixture of lime mud and mud-size siliciclastics. During the lowstands, the reefs are subaerially exposed and the inter-reef areas serve as incised channels, transporting fluvial material (sand and mud) out across the shelf. Upon reflooding, the reef "highs" are reoccupied and track the sea level transgression and highstand.

3.8   Summary of the Belize Shelf Overview

This chapter serves as a broad overview of the sedimentation in the Belize shelf lagoon. The Belize shelf is an excellent example of the spatial mixing of carbonate-siliciclastic sediments, the stratigraphic succession of carbonates and siliciclastics, and the relative control on such sedimentation by an antecedent structural template.

The physical processes that operate within the Belize lagoon produce a diverse suite of facies types. The most dominant include: (1) a shoreline system in the central and south lagoon that contains coarse siliciclastics weathered from the nearby Maya Mountains. These quartz sands provide the foundation for fringing reefs, but also smother modern coastal reefs, and they are currently both accumulating and being reworked along the mainland shoreline; (2) the lagoon configuration allows the suspended transport of mud-sized siliciclastics to the middle and outer shelf lagoon where they mix with lime mud and settle adjacent to the shallow reef structures. This middle and outer lagoon is largely biologically dominated with the accumulation of reef sediment. The prevalent calm sea conditions in the shelf lagoon limit the physical transport of coarse-grained carbonate sediments and thus the reefs have established and vertically tracked sea level during the Holocene transgression; (3) the barrier platform and barrier reef are dominated by coralgal organisms but are subjected to considerably greater wave energy due to normal wave conditions as well as by waves of an occasional tropical storm system. The majority of sediment transport occurs during these periods of tropical-force winds and associated wave energy; and (4) the offshore atolls provide a key example of the influence of tectonics and structure on morphology and sedimentation. The atoll platforms all show a distinct windward-leeward rim development as well as a diverse set of lagoon facies. These atoll lagoon facies have been hypothesized to result from differences in subsidence and precursor depth of their Pleistocene foundations.

Physical and biological sedimentation in the Belize lagoon has been relatively well characterized; and most of the physical and chemical processes that influence or control that sedimentation are also well understood. The remaining challenges in this mixed carbonate-siliciclastic system include: (1) quantifying the influence of tectonics on the sedimentary architecture, especially in the southern shelf lagoon; (2) the development of a lagoon-wide understanding of the freshwater impacts on reefs and the influence of these freshwater events on the suspended transport of mud-sized material from the mainland; and (3) the relative contribution of precursor topography to the modern lagoon shelf morphology.

Bibliography

Aronson RB, Precht WF (1997) Stasis, biological disturbance, and community structure of a Holocene coral reef. Paleobiology 23:326–346

Bateson JH, Hall IHS (1977) The geology of the Maya Mountains, Belize. Institute of Gelogical Sciences Overseas Mem No. 3, p 44

Buddemeier RW, Kinzie RA (1976) Coral growth. Oceanogr Mar Biol Ann Rev 14:183–225

Burke RB (1982) Reconnaissance study of the geomorphology and benthic communities of the outer barrier reef platform, Belize. In: Rützler K, Macintyre IG (eds) The Atlantic barrier reef ecosystem at Carrie Bow Cay, Belize I. Smithsonian contribution to the marine sciences 12:509–526

Burke RB (1993) How have Holocene sea level rise and antecedent topography influenced Belize barrier reef development? In: Ginsburg RN (ed) Proceeding of the colloquium on global aspects of coral reefs: Health, Hazards and History. University of Miami, pp 14–20

Burke CD, Mazzullo SJ, Bischoff WD, Dunn RK (1992) Environmental setting of Holocene sabellariid worm reefs, Northern Belize. Palaios 7:118–124

Burke CD, McHenry, TM, Bischoff WD, Mazzullo SJ (1998) Coral diversity and mode of growth of lateral expansion patch reefs at Mexico Rocks, northern Belize shelf, Central America. Carbonates and Evaporites 13:32–42

Cebulski DE (1969) Foraminiferal populations and faunas in barrier-reef tract and lagoon, British Honduras. In: McBirney AR (ed) Tectonic Relations of northern Central America and the Western Caribbean-The Bonacca Expedition. AAPG Mem 11:311–328

Choi DR (1981) Quaternary reef foundations in the southernmost Belize Shelf, British Honduras. Proceedings of the 4th international coral reef symposium, Manilla, vol 1, pp 635–642

Choi DR, Ginsburg RN (1982) Foundations of Quaternary reefs in the Belize Lagoon, British Honduras. GSA Bull 93:116–126

Choi DR, Holmes CW (1982) Foundations of Quaternary reefs in south-central Belize Lagoon, Central America. AAPG Bull 66:2663–2671

Cowan CA, McNeill DF Mixed carbonate and siliciclastic facies along the southern Belize coast, Central America (unpublished manuscript)

Dengo G, Bohnenberger O (1969) Structural development of northern Central America. In: McBirney AR (ed) Tectonic Relations of northern Central America and the Western Caribbean-The Bonacca Expedition. AAPG Mem 11:203–220

Dill RF (1977) The blue-holes – geologically significant submerged sink holes and caves off British Honduras and Andros, Bahama Islands. In: Proceedings of the 3rd international coral reef symposium, Rosenstiel School of Marine and Atmospheric Science, University of Miami, vol 2, pp 237–242

Dill RF, Land LS, Mack LE, Schwarcz HP (1998) A submerged stalactite from Belize: petrography, geochemistry, and geochronology of massive marine cementation. Carbon Evapor 13: 189–197

Dillon WP, Vedder JG (1973) Structure and development of the continental margin of British Honduras. GSA Bull 84:2713–2732

Dixon CG (1956) Geology of southern British Honduras with notes on adjacent areas. British Honduras Gov Printing Office, Belize, p 85

Ebanks WJ (1967) Recent carbonate sedimentation and diagenesis, Ambergris Cay, British Honduras. Ph. D. thesis, Rice University, p 189

Ebanks WJ (1975) Holocene carbonate sedimentation and diagenesis, Ambergris Cay, Belize. In: Wantland KF, Pusey WC (eds) Belize Shelf-carbonate sediments, clastic sediments, and ecology. AAPG Stud Geo 2:234–296

Enos P, Koch WJ, James NP (1979) The geophysical anatomy of the southern Belize continental margin and adjacent basins. In: James NP, Ginsburg RN (eds) The Seaward margin of Belize barrier and Atoll reefs. Int Assoc Sedimentol Spec Publ 3:15–24

Esker D (1998) The interplay between tectonics and eustasy in a modern mixed carbonate-siliciclastic system, southern Belize lagoon. Ph.D. thesis, University of Miami, Coral Gables, FL, p 281

Esker D, Eberli GP, McNeill DF (1998) The structural and sedimentological controls on the reoccupation of Quaternary incised valleys, southern lagoon of Belize. AAPG Bull 82:2075–2109

Fairbanks RG (1989) A 17,000-year glacio-eustatic sea level record: Influence of glacial melting rates on the Younger Dryas event and deep-ocean circulation. Nature 342:637–642

Falkowski PG, Jokiel PL, Kinzie RA (1990) Irradiance and corals. In: Dubinsky Z (ed) Coral reefs – ecosystems of the world, vol 25, pp 89–107. Elsevier Science, Amsterdam

Ferro CE, Droxler AW, Anderson JB, Mucciarone D (1999) Late Quaternary shift of mixed siliciclastic-carbonate environments induced by glacial eustatic sea-level fluctuations in Belize. In: Harris PM, Saller AH, Simo JA (eds) Advances in carbonate sequence stratigraphy; application to reservoirs, outcrops and models. SEPM Spec Publ 63:385–411

Fuglister FC (1947) Average monthly sea surface temperatures of the western North Atlantic Ocean. M.I.T. and Woods Hole Oceanographic Institute, Papers in physical oceanography and meteorology 10, p 25

Gentry RC (1971) Hurricanes, one of the major features of air-sea interaction in the Caribbean Sea. In: Symposium on Investigation and Research of the Caribbean Sea and Adjacent region, Curacao, 18–26 Nov 1968. UNESCO, Paris, pp 80–87

Ginsburg RN, James NP (1976) Submarine botryoidal aragonite in Holocene reef limestones, Belize. Geology 4:431–436

Ginsburg RN, Harris PM, Eberli GP, Swart PK (1991) The growth potential of a bypass margin, Great Bahama Bank. J Sed Petrol 61:976–987

Gischler E (1994) Sedimentation on three Caribbean atolls: Glovers reef, Lighthouse reef and Turneffe islands, Belize. Facies 31:243–254

Gischler E (2003) Holocene lagoonal development in the isolated carbonate platforms off Belize. Sed Geol 159:113–132

Gischler E, Lomando AJ (1997) Holocene cemented beach deposits in Belize. Sediment Geol 110:277–297

Gischler E, Hudson JH (1998) Holocene development of three isolated carbonate platforms, Belize, Central America. Mar Geol 144:333–347

Gischler E, Lomando AJ (1999) Recent sedimentary facies of isolated carbonate platforms, Belize-Yucatan system, Central America. J Sed Res 69:747–763

Gischler E, Pisera A (1999) Shallow water rhodoliths from Belize reefs. N Jb Geol Paläont Abh 214:71–93

Gischler E, Lomando AJ (2000) Isolated carbonate platforms of Belize, Central America: sedimentary facies, late Quaternary history and controlling factors. In: Insalaco E, Skelton PW, Palmer TJ (eds) Carbonate platform systems: components and interactions. Geol Soc (Lond) Spec Publ 178:135–146

Gischler E, Lomando AJ, Hudson JH, Holmes CW (2000) Last interglacial reef growth beneath Belize barrier and isolated platform reefs. Geology 28:387–390

Gischler E, Zingeler D (2002) The origin of carbonate mud in isolated carbonate platforms of Belize, Central America. Int J Earth Sci 91:1054–1070

Gischler E, Hudson JH (2004) Holocene development of the Belize barrier reef. Sed Geol 164:223–236

Grammer GM, Ginsburg RN, McNeill DF (1991) Morphology and development of modern carbonate foreslopes, Tongue of the Ocean, Bahamas. In: Larue DK, Draper G (eds) Transactions of the 12th Caribbean geological conference, Miami Geological Society, pp 27–32

Grammer GM, Ginsburg RN (1992) Highstand vs. lowstand deposition on carbonate platform margins: insight from Quaternary foreslopes in the Bahamas. Mar Geol 103:125–136

Grammer GM, Ginsburg RN, Harris PM (1993) Timing of deposition, diagenesis, and failure of steep carbonate slopes in response to a high-amplitude/ high-frequency fluctuation in sea level, Tongue of the Ocean, Bahamas. In: Loucks RG, Sarg JF (eds) Carbonate sequence stratigraphy. AAPG Mem 57:107–131

Greer JE, Kjerfve B (1982) Water currents adjacent to Carrie Bow Cay, Belize. Smithsonian Contri Mar Sci 12:53–58

Halley RB, Shinn EA, Hudson JH, Lidz B (1977) Recent and relict topography of Boo Bee Patch
   Reef. In: Proceedings of the 3rd international coral reef symposium, Rosenstiel School of
   Marine and Atmospheric Science, University of Miami, vol 2, pp 29–37
Halley RB, Harris PM, and Hine AC (1983) Bank margin environment. In: Scholle PA, Bebout
   DG, Moore CH (eds) Carbonate depositional environments. AAPG Mem 33:463–506
Hallock P, Schlager W (1986) Nutrient excess and the demise of coral reefs and carbonate plat-
   forms. Palaios 1:389–398
High LR (1969) Storms and sedimentary processes along the northern British Honduras coast.
   J Sed Petrol 39:235–245
High LR (1975) Geomorphology and sedimentology of Holocene coastal deposits, Belize. In:
   Wantland KF, Pusey WC (eds) Belize shelf-carbonate sediments, clastic sediments, and ecol-
   ogy. AAPG Studies in Geology 2:53–96
International Marine Tide Tables (1998). International Marine Publishing Company, Rockport,
   Maine
James NP, Ginsburg RN, Marzalek DS, Choquette PW (1976) Facies and fabric specificity of
   early subsea cements in shallow Belize (British Honduras) reefs. J Sed Petrol 46:523–544
James NP, Ginsburg RN (1979) The seaward margin of Belize barrier and Atoll reefs. Int Assoc
   Sedimentol Spec Publ 3, p 191
Kjerfve B (1978) Diurnal energy balance of a Caribbean barrier reef environment. Bull Mar Sci
   28:137–145
Kjerfve B, Rütlzer K,Kierspe GH (1982) Tides at Carrie Bow Cay, Belize. In: Rützler K,
   Macintyre IG (eds) The Atlantic barrier reef ecosystem at Carrie Bow Cay, Belize I.
   Smithsonian Contri Mar Sci 12:47–52
Kjerfve BJ, Dinnel SP (1983) Hindcast hurricane characteristics on the Belize Barrier Reef. Coral
   Reefs 1:203–207
Kling SA (1975) A lagoonal coccolithophore flora from Belize (British Honduras).
   Micropaleontology 21:1–13
Land LS, Moore CH (1977) Deep fore-reef and upper island slope, north Jamaica. In: Frost SH,
   Weiss MP, Saunders JB (eds) Reefs and related carbonates: ecology and sedimentology.
   AAPG Stud Geo 4:53–67
Lara ME (1993) Divergent wrench faulting in the Belize Southern Lagoon: implications for ter-
   tiary Caribbean plate movements and Quaternary reef distribution. AAPG Bull 77:
   1041–1063
Lighty RG, Macintyre IG, Stuckenrath R (1982) Acropora palmata reef framework: a reliable
   indicator of sea level in the western Atlantic for the past 10,000 years. Coral Reefs 1:125–130
Macintyre IG (1984) Extensive submarine lithification in a cave in the Belize barrier reef platform.
   J Sed Petrol 54:221–235
Macintyre IG, Burke RB, Stuckenrath R (1981) Core holes in the outer fore reef off Carrie Bow
   Cay, Belize: a key to the Holocene history of the Belizean barrier reef complex. Proceedings
   of the 4th international coral reef symposium, Manilla, vol 1, pp 567–574
Macintyre IG, Graus RR, Reinthal PN, Littler MM, Littler DS (1987) The barrier reef sediment
   apron: tobacco reef, Belize. Coral Reefs 6:1–12
Macintyre IG, Aronson RB (1997) Field guidebook to the reefs of Belize. Proceedings of the 8th
   international coral reef symposium, vol 8, pp 203–221
Macintyre IG, Littler NM, Littler DS (1995) Holocene history of Tobacco Range, Belize, Central
   America. Atoll Res Bull 430:1–18
Macintyre IG, Precht WF, Aronson RB (2000) Origin of the Pelican Cays ponds, Belize. In: Macintyre
   IG, Rützler K (eds) Natural history of the Pelican Cays, Belize. Atoll Res Bull 467:1–11
Macintyre IG, Goodbody I, Rützler K, Littler DS, Littler MM (2000) A general biological and
   geological survey of the rims of ponds in the major mangrove islands of the Pelican Cays,
   Belize. In: Macintyre IG, Rützler K (eds) Natural history of the Pelican Cays, Belize. Atoll
   Res Bull 467:15–34
Macintyre IG, Rützler K (eds) (2000) Natural history of the Pelican Cays, Belize. Atoll Res Bull
   467:333

Matthews RK (1965) Genesis of Recent lime mud in Southern British Honduras. J Sed Petrol 36:428–544

Mazzullo SJ, Anderson-Underwood KE, Burke CD, Bischoff WD (1992) Holocene coral patch reef ecology and sedimentary architecture, Northern Belize, Central America. Palaios 7:591–601

Mazzullo SJ, Bischoff WD, Teal CS (1995) Holocene shallow-subtidal dolomitization by near-normal seawater, northern Belize. Geology 23:341–344

Mazzullo SJ, Teal CS, Bischoff WD, Dimmick-Wells K, Wilhite BW (2003) Sedimentary architecture and genesis of Holocene shallow-water mud mounds, northern Belize. Sedimentology 50:743–770

Miller JA, Macintyre IG (1977) Field guidebook to the reefs of Belize. Atlantic Reef Committee, University of Miami, Miami, p 36

National Ocean Service (1997) Tide tables 1998. High and low water predictions, East Coast of North and South America. International Marine Publishing, Washington DC, p 279

Portig WH (1976) The climate of Central America. In: Schwerdtfeger W (ed) Climates of Central and South America. World survey of Climatology 12, pp 405–488. Elsevier, New York

Purdy EG (1974a) Karst-determined facies patterns in British Honduras: Holocene carbonate sedimentation model. AAPG Bull 58:825–855

Purdy EG (1974b) Reef configuration: cause and effect. In: LaPorte LF (ed) Reefs in Time and Space. SEPM Spec Publ 18:9–76

Purdy EG (1998) Structural termination of the southern end of the Belize Barrier Reef. Coral Reefs 17:231–234

Purdy EG, Pusey WC, Wantland KF (1975) Continental shelf of Belize-regional shelf attributes. In: Wantland KF, Pusey WC (eds) Belize shelf-carbonate sediments, clastic sediments, and ecology. AAPG Stud Geo 2:1–39

Purdy EG, Gischler E (2003) The Belize margin revisited 1: Holocene marine facies. Int J Earth Sci 92:532–551

Purdy EG, Gischler E, Lomando AJ (2003) The Belize margin revisited 2: Origin of Holocene antecedent topography. Int J Earth Sci 92:552–551

Pusey WC (1964) Recent calcium carbonate sedimentation in northern British Honduras. Ph.D. thesis, Rice University, Houston, Texas, p 247

Pusey WC (1975) Holocene carbonate sedimentation on Northern Belize Shelf. In: Wantland WF, Pusey WC (eds) Belize shelf-carbonate sediments, clastic sediments, and ecology. AAPG Stud Geol 2:1–52

Rao RP, Ramanathan R (1991) Tectonics and petroleum potential of Belize. In: Larue DK, Draper G (eds) Transactions of the 12th Caribbean geological conference. Geol Soc Miami, pp 520–527

Rasmussen KA, Macintyre IG, Prufert L (1993) Modern stromatolite reefs fringing a brackish coastline, Chetumal Bay, Belize. Geology 21:199–202

Reid PR, Macintyre IG (1998) Carbonate recrystallization in shallow marine environments: a widespread diagenetic process forming micritized grains. J Sed Res 68:928–946

Reid PR, Macintyre IG, Post JE (1992) Micritized skeletal grains in Northern Belize Lagoon: a major source of Mg-calcite mud. J Sed Petrol 62:145–156

Romney DH (1959) Land in British Honduras, report of the British Honduras land use survey team. London, Colonial Res Publ 24, p 327

Rützler K, Ferraris JD (1982) Terrestrial environment and climate, Carrie Bow Cay, Belize. In: Rützler K, Macintyre IG (eds) The Atlantic barrier reef ecosystem at Carrie Bow Cay, Belize I. Smithsonian Contri Mar Sci 12:77–91

Rützler K, Macintyre IG (1982a) The habitat distribution and community structure of the barrier reef complex at Carrie Bow Cay, Belize. In: Rützler K, Macintyre IG (eds) The Atlantic barrier reef ecosystem at Carrie Bow Cay, Belize I. Smithsonian Contri Mar Sci 12:9–45

Rützler K, Macintyre IG (eds) (1982b) The Atlantic barrier reef ecosystem at Carrie Bow Cay, Belize. Smithsonian Contri Mar Sci 12:539

Scholle PA, Kling SA (1972) Southern British Honduras: lagoonal coccolith ooze. J Sed Petrol 42:195–204

Scott MR (1975) Distribution of clay minerals on Belize Shelf. In: Wantland KF, Pusey WC (eds) Belize shelf-carbonates sediments, clastic sediments, and ecology. AAPG Stud Geol 2:97–130

Shinn EA, Hudson JH, Halley RB, Lidz B, Robbin DM (1979) Three-dimensional aspects of Belizean patch reefs. AAPG Bull 63:528

Shinn EA, Hudson JH, Halley RB, Lidz B, Robbin DM, Macintyre IG (1982) Geology and sediment accumulation rates at Carrie Bow Cay, Belize. In: Rützler K, Macintyre IG (eds) The Atlantic barrier reef ecosystem at Carrie Bow Cay, Belize. Smithsonian Contri Mar Sci 12:63–75

Stoddart DR (1962) Three Caribbean atolls: Turneffe islands, Lighthouse reef, and Glover's reef, British Honduras. Atoll Res Bull 87:151

Stoddart DR (1963) Effects of hurricane Hattie on the British Honduras reefs and cays, 30–31 Oct 1961. Atoll Res Bull 95:1–142

Stoddart DR (1965) Comprehensive overview of Belize islands. Inst British Geogr Trans Pap 36:131–147

Stoddart DR (1969) Post-hurricane changes on the British Honduras reefs and cays: Re-survey, 1965. Atoll Res Bull 131:1–31

Stoddart DR (1974) Post-hurricane changes on the British Honduras reefs: Re-survey, 1972. Proceedings of the 2nd international coral reef symposium, vol 2, pp 473–483

Stoddart DR, Fosberg FR, Spellman DL (1982) Cays of the Belize reef and lagoon. Atoll Res Bull 256:1–76

Teal CS, Mazzullo SJ, Bischoff WD (2000) Dolomitization of Holocene shallow-marine deposits mediated by sulfate reduction and methanogenesis in normal-salinity seawater, northern Belize. J Sed Res 70:649–663

Teeter JW (1975) Distribution of Holocene ostracoda from Belize. In: Wantland KF, Pusey WC (eds) Belize shelf-carbonates sediments, clastic sediments, and ecology. AAPG Stud Geo 2:400–499

Tide Tables (1998) High and low water predictions, East Coast of North and South America including Greenland. Int Mar (formally published by National Ocean Service)

Toscano MA, Macintyre IG (2003) Corrected western Atlantic sea-level curve for the last 11,000 years based on calibrated 14C dates from Acropora palmata framework and intertidal mangrove peat. Coral Reefs 22:257–270

Trewartha GT (1961) The earth's problem climates. University of Wisconsin Press, Madison, Wisconsin, p 334

U.S. Naval Oceanographic Office (1963) Yucatan channel and approaches: Chart H.O. 966, 67th edn (revised), scale 1, p 906/530

Vermeer DE (1963) Effects of hurricane Hattie, 1961 on the cays of British Honduras. Zeitschr. für Geomorphologie, N. Folge, 7th edn, H. 4, pp 332–354

Wallace RJ, Schafersman SD (1977) Patch-reef ecology and sedimentology of Glovers Reef Atoll, Belize. AAPG Stud Geo 4, Reefs and Related Carbonates – Ecology and Sedimentology, pp 37–52

Wantland KF, Pusey WC (1971) A guidebook for the field trip to the Southern Shelf of British Honduras. New Orleans Geological Society, p 87

Wantland KF, Pusey WC (eds) (1975) Belize shelf-carbonate sediments, clastic sediments, and ecology. AAPG Stud Geo 2, p 599

Wantland KF (1975) Distribution of Holocene benthonic foraminifera on the Belize Shelf. In: Wantland KF, Pusey WC (eds) Belize shelf-carbonates sediments, clastic sediments, and ecology. AAPG Stud Geol 2:332–399

Westphall MJ (1986) Anatomy and history of a ringed-reef complex, Belize, Central America. M.S. thesis, University of Miami, Coral Gables, FL, p 159

Wilson JL, Jordan C (1983) Middle shelf. In: Scholle PA, Bebout DG, Moore CH (eds) Carbonate depositional environments. AAPG Mem 33:297–344

Wright ACS, Romney DH, Arbuckle RH, Vial VE (1959) Land in British Honduras. Colonial Res Publ 24, Colonial Office, London

Wüst G (1964) Stratification and circulation in the Caribbean-Antillean Basins, Part I. Columbia University Press, New York, p 201

Chapter 4
The Gulf: Facies Belts, Physical, Chemical, and Biological Parameters of Sedimentation on a Carbonate Ramp

Bernhard Riegl, Anthony Poiriez, Xavier Janson, and Kelly L. Bergman

4.1 Introduction

The Holocene of The Gulf, also referred to as the Arabian or Persian Gulf, is frequently cited as a classic example of a mixed carbonate-siliciclastic ramp system for an arid climate. This notion of a ramp is supported by the recognition that The Gulf area has a dominant shallow water carbonate/evaporite basin fill from the Permian to today despite a complex tectonic history (Alsharhan and Kendall 2003). The current depositional setting is that of a proximal foreland ramp (Burchette and Wright 1992; Evans 1995; Kirkham 1998). Walkden and Williams (1998), however, argue that since The Gulf has been above sea level for over much of the past 2.5 Ma, and since it is in tectonic, eustatic and depositional disequilibrium it should not be considered a ramp. Despite this controversy, the Holocene sedimentary fill of the current Gulf has been and will continue to be used as a model for a carbonate ramp.

B. Riegl (✉)
National Coral Reef Institute, Nova Southeastern University Oceanographic Center, Dania, Florida, USA

A. Poiriez
Rosenstiel School for Marine and Atmospheric Sciences, University of Miami, Miami, Florida, USA

X. Janson
Bureau of Economic Geology, Austin, Texas, USA
and
Rosenstiel School for Marine and Atmospheric Sciences,
University of Miami, Miami, Florida, USA
e-mail: Xavier.Janson@beg.utexas.edu

K.L. Bergman
ETC Chevron Corporation, San Ramon, California, USA
and
Rosenstiel School for Marine and Atmospheric Sciences,
University of Miami, Miami, Florida, USA
e-mail: KBergman@chevron.com

H. Westphal et al. (eds.), *Carbonate Depositional Systems: Assessing Dimensions and Controlling Parameters*, DOI 10.1007/978-90-481-9364-6_4,
© Springer Science+Business Media B.V. 2010

This interest in the area is hightened by the fact that is one of the few places in which Holocene dolomite and evaporites form.

Due to its geographic location, the sedimentological research history of The Gulf has not been as intense as that of Florida, Bahamas or Belize, although the region's vibrant economy has recently attracted much new scientific investigation. Motivated by exploration for hydrocarbons and metallic ores, intensive research was intiated in the region. This started with mostly physical geographical work undertaken in the 1940s (Lees 1948; Lees and Falcon 1952). From then on, the research direction was rapidly dominated by sedimentological issues, begun with the classical early work of Emery (1956). Concerted, large scale efforts by the Imperial College of London mainly within the United Arab Emirates, were directed by G. Evans and D.J. Shearman and led to a series of theses and publications (Evans and Shearman 1964; Evans 1966, 1970; Evans et al. 1964, 1969; Kinsman 1964a; Murray 1965a; Shearman 1963). Kiel University, directed by E. Seibold concentrated on the northern Gulf and The Gulf of Oman and published in the "Meteor Forschungsergebnisse" (1969–1972). Shell Research investigated the area around Qatar in the 1950s and 1960, yielding numerous publications (Houbolt 1957; Wells 1962; Illing et al. 1965; Taylor and Illing 1969). These results combined with research by the Swiss Institute of Technology Zürich (Hsü and Schneider 1973), the Central Geological Survey Prague and the University of Baghdad (Kukal and Saadallah 1973), the Museum National d'Histoire Naturelle (Loreau and Purser 1973) were collected by Purser (1973a) into the well known volume of Gulf sedimentology. More recent overviews of coastal carbonate and evaporite facies of The Gulf were written by Alsharhan and Kendall (2003), and Gischler and Lomando (2005).

Besides the classical marine sedimentary systems, The Gulf intertidal, in particular the sabkha, or coastal salt flats, raised early interest that has been sustained to this day. Models concerning its geomorphology, hydrology and evaporite formations date back to early works by Butler (1969), Evans et al. (1969), Kendall and Skipwith (1968) and Kinsman (1969), classical models were developed by Hsü and Siegenthaler (1969), Hsü and Schneider (1973) and McKenzie et al. (1980). Later work is that of Patterson and Kinsman (1977, 1981), Alsharhan and Kendall (1994), Sanford and Wood (2001) and Wood et al. (2002) up to Alsharhan and Kendall (2003). Detailed geophysical work in the area dating to the 1970s is reviewed in Ross et al. (1986).

Since the late eighteen hundreds The Gulf has also been the focus of biological investigations, which are relevant to carbonate sedimentaology. Mollusks received early attention (Smith 1872; von Martens 1874; Fischer 1891). F.W. Townsend made collections when cleaning cables for the Indo-European Telegraph Department on the steamer "Patrick Stewart". The mollusks were scientifically treated by J.C. Melville (1897, 1898, 1899, 1904, 1917, 1928 and many others). H.E.J Biggs (1973) collected from 1911–1935. The 1950 Yale Peabody Museum Harvard Expedition to the Near East yielded Haas (1952). The history of malacological research was exhaustively reviewed in Bosch et al. (1995). Oil exploration greatly accelerated the pace of biological research. Foraminifera, a most important group for carbonate sedimentologists, were studied by J.W. Murray (1965a, b; 1966a, b, c; 1970a, b), the German Meteor expedition led by E. Seibold (Lutze et al. 1971) and others (Basson and Murray 1995; Cherif et al. 1997). Largely funded by oil companies, detailed overview work on the biological component of the sedimentary system was published (e.g. Basson et al. 1977). Early work on corals of The Gulf was by Burchard (1979), Downing (1985), Coles (1988), and Sheppard (1988). Sheppard produced several exhaustive reviews of the area's fauna and ecology (Sheppard and Sheppard 1991; Sheppard et al. 1992). Saudi Arabian corals received detailed treatment by Fadlallah (1996) and Fadlallah et al. (1992, 1995a, b). Many other studies followed (George and John 1999, 2000a, 2000b, 2002; Riegl 1999, 2002, 2003; Purkis and Riegl 2005; Purkis et al. 2005; Burt et al., 2008; Riegl and Purkis 2009). Early phycological work in The Gulf goes back to Endlicher and Diesing (1845), Borgensen (1939), and Newton (1955a, b). The most exhaustive recent overview was provided by De Clerck and Coppejans (1996).

The 1991 Gulf War and the ensuing ecological problems led to several international expeditions that included the American Mt. Mitchell expedition (ROPME 1993; Abuzinada and Krupp 1994; Downing and Roberts 1993; Krupp et al. 1996). Since then, the amount of sedimentological and biological research in the area has increased with the influx of interest and money centered on oil and recently real-estate and industrial devlopment, with the result of attracting an increasingly diverse and active research community. Thus growing scientific output and results from this interesting region can be expected.

4.2 Sedimentary and Tectonic Setting

The Gulf is a marginal, epicontinental sea of approximately 1,000 km length, 200–300 km width with an area of 226,000 km^2 (Purser and Seibold 1973). It is connected to the Indian Ocean by the 60-km-wide Straits of Hormuz. An unstable Tertiary fold belt system lines the Iranian side, opposed to the stable Arabian foreland on the Arabian side (Purser and Seibold 1973). The Gulf is a distal foreland basin that is periodically flooded through the Straits of Hormuz (Figs. 4.1 and 4.2).

The Gulf basin is asymmetric with a gently inclined floor that has a slope of 175 cm/km on the Iranian side and 35 cm/km on the Arabian side (Purser and Seibold 1973). Because The Gulf is surrounded by land and lies in an arid subtropical climate, no buffering system (e.g. riverine input or the rapid turn-over of oceanic waters) exists for its temperature and salinity. Thus temperature variability is high (Kinsman 1964b) and evaporation reaches up to 124 cm/year.

Important structural elements of The Gulf region include the Arabian platform, the Zagros Mountains, the Musandam Peninsula, and the Oman line, at which the Zagros Mountains terminate (Fig. 4.3). The Arabian platform underlies The Gulf and the Arabian Peninsula (The Gulf being the flooded portion thereof) and exhibits anticlines formed in response to uplift of the Oman Mountains and possibly salt diapirism (Ross et al. 1986). The Zagros Mountains border the northern Gulf coast and consist of folded Paleozoic to Cenozoic shelf carbonates (Ross et al. 1986). The Musandam Peninsula consists largely of Mesozoic carbonates and juts into

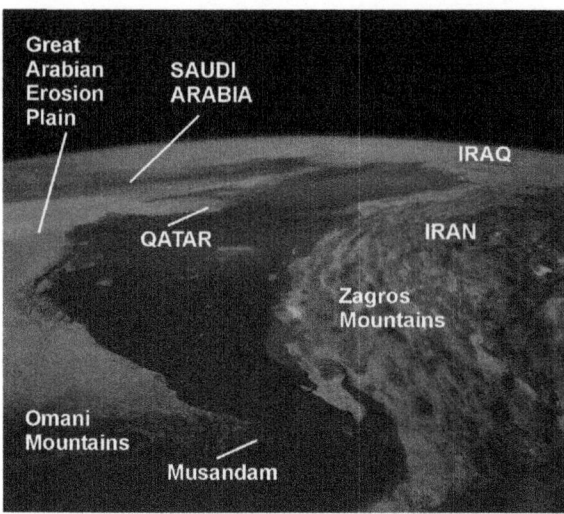

Fig. 4.1 An East to the West astronaut's view of The Gulf and the major features that control the distribution of the carbonate sediments. The epicontinental character of The Gulf is clearly visible and represents a flooded proximal foreland basin. The flat Arabian side is in stark contrast to the Zagros fold belt in Iran (Image courtesy NASA)

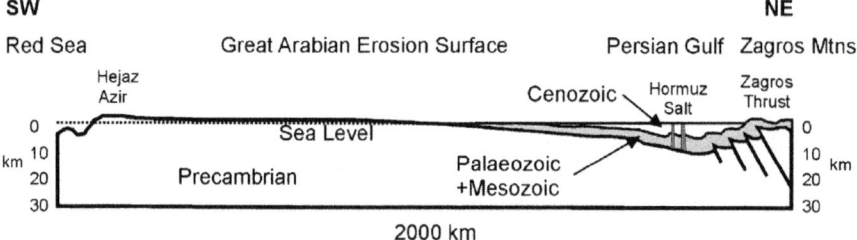

Fig. 4.2 Schematic NE-SW section through the Arabian plate with The Gulf representing a flooded foreland basin (From Walkden and Williams 1998. Permission by Geological Society of London)

The Gulf at its entrance. It is presently subsiding while the Oman Mountains were uplifted to a rate of 60 m in the last 10,000 years (Ross et al. 1986). The Oman line marks the SE end of the Zagros range and marks the end of the post-paleozoic sequence of The Gulf. It separates the carbonate platform to the west and a deep-water flysch and radiolarite facies to the east (Alsharhan and Nairn 1997). It also signifies a change in the tectonic margin between the Arabian and Eurasian plates, and changes from a continent-continent collisional margin in the west to a subduction margin in the east (Ross et al. 1986).

The tectonics of the coastline surrounding The Gulf and the antecedent topography of the basin exert control on the distribution of facies belts. The Iranian coastline consists of steeply dipping anticlines trending NW-SE that formed during the Plio-Pleistocene Zagros Orogeny (Kassler 1973). The anticlines' dips decrease from 50° on the mainland to 10–20° near the coast (Kassler 1973). As a result of

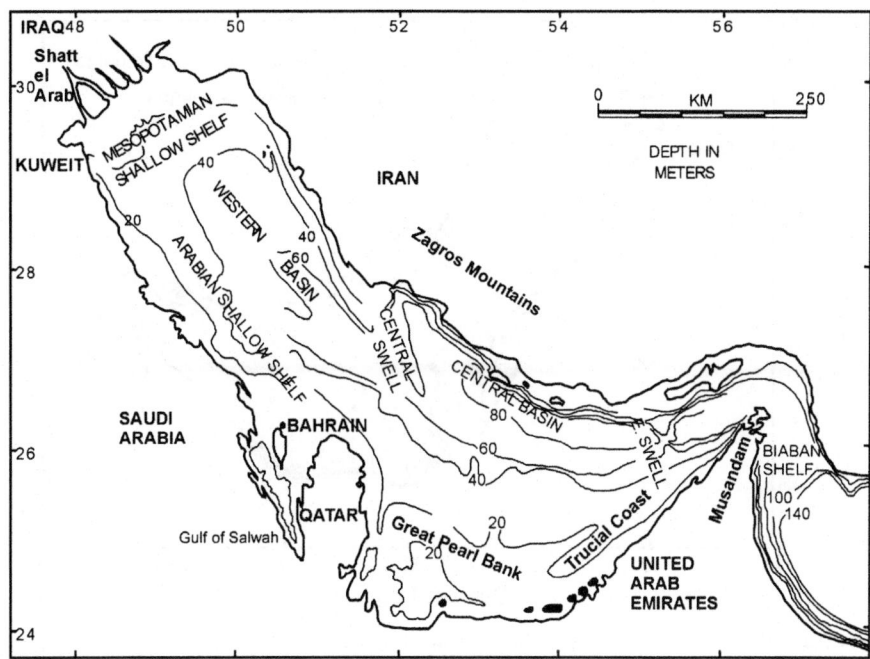

Fig. 4.3 Basin subdivisions of The Gulf (Purser and Seibold 1973). The Central Swell divides the Western Basin from the Central Basin (Permission by Springer)

mountain building, the Iranian shelf is narrow and is bound by a discontinuous island chain on its seaward side (Ross et al. 1986). Rivers import mud and fine sand, and the resulting sediment off the Iranian coast is dominated by marl. Grain size and carbonate content increases away from this coast (Seibold et al. 1973).

The Arabian coast consists of N-S to NE-SW trending, gently dipping anticlines (Kassler 1973) and the flooded portion of the Arabian Shield. The most recent uplift event of the Oman Mountains began in the late Tertiary and continues until the Recent (Kassler 1973). The Arabian marine shelf is wider and gentler than the Iranian shelf and contains flat-topped banks and shoals produced largely by salt diapirism and erosional relicts of the Quaternary (Kassler 1973). As a result, the Arabian coast is characterized by low sandy islands, beaches with shallow channels, and coastal salt flats in which evaporites are precipitated, locally known as sabkhas (Ross et al. 1986).

4.2.1 Overview of Ramp Sediments

The Iranian side of The Gulf receives terrigenous sediments from the Zagros Mountains of Iran and the Tigris-Euphrates delta of Iraq (Enos 1983) while its Arabian side is gently sloping and fits the model of a typical carbonate ramp setting. There, the coastal area is formed by wide sabkhas (tidal flats composed mainly of fine, wind-blown material and in situ evaporites that prograde into the shallow coastal sea),

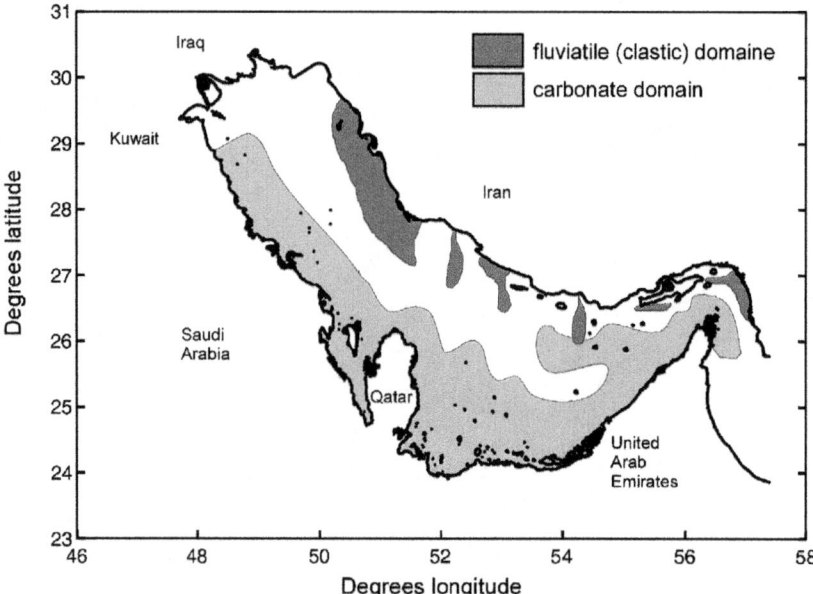

Fig. 4.4 Major sedimentary domains of The Gulf. The northern domain is characterized by high fluviatile input (grey = influence by fluviatile sedimentation), while the southern domain is characterized by carbonate sedimentation (Combined from Diester-Haas 1973; Uchupi et al. 1999. Permissions by Springer and Elsevier)

tidal lagoons and flats, channels, sand bars, and coral buildups. Thus, The Gulf is separated into two major sedimentary realms: a northern, Iranian, domain strongly influenced by fluviatile sedimentation, and a southern, Arabian, carbonate domain of mainly authochthonous, pure carbonates with aeolian-derived siliciclastic admixture (Fig. 4.4; Diester-Haas 1973; Uchupi et al. 1996, 1999).

4.2.2   Bathymetric Topography

The southern Gulf is marked by numerous bathymetric highs that range from tens of meters to kilometers in diameter and occur at variable depths from around 20 m to depths shallow enough to allow reef growth or exposure at low tide (Purser 1973b). Bathymetric highs can be categorized as basin central, intermediate, or coastal (Table 4.1; Purser 1973b).

Sedimentological trends on bathymetric highs include: (1) foraminiferal sands on more deeply submerged highs towards the basin axis and coral-algal reefs on the less deeply submerged highs towards the Arabian shoreline, and (2) increased sand size with increasing water energy near the Arabian shoreline (Wagner and van der Togt 1973; Purser 1973b). Around salt domes "exotic" material, pushed to the surface by the salt's action, can be reworked and incorporated into the sediment. Thus, Paleozoic dolomite, volcanic rocks and heavy minerals such as magnetite can be locally found (Purser 1973b).

Table 4.1 Facies of bathymetric highs in the Persian/Arabian Gulf (From Purser 1973b. Permission by Springer)

	Bathy-metric high	Morphology	Sedimentology	Examples
1	Outer homocline (Basin center)	Surrounded by deep water (>50 m); incl. salt diapirs, islands, submerged shoals	Variable with water depth (differential wave action and biota)	
1A	Deeply submerged		Coarser grains on crest incl. large foraminiferal sands, compound grains, lithified lumps; muddier sediments at foot of high	Shoals near Halul Island
1B	Shallow submerged		Crests: reefs and coral-algal sands; flanks: pelecypod muds; storms sweep off sediment on highs and deposit sediment in adjacent environment; reefs best developed on shallower leeward sides	Shah Allum shoal, Abu Thama, Naiwat Arragie
1C	Emerged highs		Coral-algal reefs on top; reef growth on all sides and highest on windward; sediment arranged concentrically around highs with more fine sand and silt on downcurrent side, tails as thick as 20 m, tail up to 20 km long, longest tails in shallower settings	Abu Musa, Farsi and Arabi islands, Halul, Das, Abu Nu'air, Dalma, Zirko, Yas
2	Inter-mediate homocline highs	100s m to > 50 km; incl. salt dome emergent islands, sand cays; depth < 36 m	Because of high energy, sediments on lows and highs are similar	
2A	Submerged below wave base		Highs: pelecypod sands and compound grains, pelecypod muddy sands around highs	Umm Shaif
2B	Crests culminating above wave base		Shallow sediments incl. coral-algal gravels, pelecypod sands, compound grains accumulate on windward sides, pelecypod muddy sands and muds on leeward sides	Rig az Zakum
3	(3) Inner homocline (coastal) highs	Near axis of Great Pearl Bank barrier, occur as low islands, shoals; cores of pre-Holocene limestone	Sediment tails extend in leeward direction from highs, some oolitic sand bars form, sediment tails may form coastal peninsulas or tombolas	Abu Dhabi

4.3   Physical Parameters

Physical constraints to carbonate sedimentology in The Gulf region are the semi-enclosed, shallow nature and the arid setting with resulting hypersalinity and, among other phenomena, sabkha sedimentation (Sheppard 1993). Also typical are the high annual temperature variations, and the complicated circulation patterns (John 1992a). Much of the seafloor is located within the photic zone, allowing for an expansive photozoan tropical carbonate factory (Schlager 2005). Restriction of oceanic circulation due to the narrow Straits of Hormuz results in a relatively low rate of water exchange with an estimated 90% flushing time of 5.5 years (Hughes and Hunter 1979).

4.3.1   Climate

The Gulf is situated in a strictly arid climatic zone (Grasshoff 1976) characterized by low rainfall (Emery 1956) and high evaporation rates (estimates vary from 144–5,000 cm/year Seibold 1973; Johns et al. 2003), the resulting net loss of water leading to a Mediterranean-style, reverse-flow, estuarine-type circulation (Reynolds 1993). The region is mainly subjected to extra-tropical weather systems that are strongly influenced by orography (Murty and El-Sabh 1984). The Straits of Hormuz form a boundary between the generally east to west travelling tropical weather systems south of The Gulf and the west to east travelling extra-tropical weather systems within The Gulf basin (Murty and El-Sabh 1984). The arid climate, in the context of sediment transport, translates to low stream discharge with consequently slow deposition of clastic sediments that do not mask the carbonate sediments in the southern Gulf at all (Emery 1956). Additionally, because rainfall is so rare, few perennial streams, other than the Shatt el Arab, ever reach The Gulf. Since most of the approximately 4–24 cm/year of rainfall occur in just a few days during the winter months, markedly pulsed delivery of clastic sediment occurs during this period (Table 4.2).

4.3.2   Wind

Among the most notable wind patterns in The Gulf region is the Shamal. The word "Shamal" means "north" in Arabic and refers to seasonal northwesterly winds that occur most dramatically during winter. Alsharhan and Kendall (2003) consider the Shamal as equally important sedimentological drivers as the Caribbean hurricanes. Winter Shamals occur from November through March and tend to follow cold fronts. At most locations, winds exceed 20 knots (~10 m/s) only for less than 5% during winter (Perrone 1981). Shamals tend to set in with great abruptness and force (Murty and El-Sabh 1984), often with wind speeds of 40–50 km/h and gusts

Table 4.2 Summary of physical parameters of The Gulf

Parameter	Datum	Source
Waves	4–5 m (storm)	Murty and El-Sabh (1984)
	6 m (storm)	Shinn (1976)
	Wave base 20 m	Purser and Seibold (1973)
Wind	Speed: 5–12 m/s	John et al. (1990)
	Direction: N-WNW	
	Maximum sustained speed:	
	40–50 km/h	Murty and El-Sabh (1984)
	65 km/h	Shinn (1976)
	Maximum gust speed: 100 km/h	John et al. (1990)
Tides	1–3 m	Lehr (1984)
	2–4 m	Jones (1986), Sheppard (1993)
	0.5 m (Gulf of Salwah)	Sheppard (1993)
Currents	Tidal: 50+ cm/s (even at 0–4 m from bottom)	Seibold (1973)
	100–200 cm/s (through restrictions and past islands)	Sheppard (1993)
	Residual: 10 cm/s (bottom current exiting strait)	Koske (1972), Grasshoff (1976)
	10–40 cm/s	U.S. Navy Hydrographic Office (1960)
Water temperature	20–32°C (offshore)	Hughes Clarke and Keij (1973)
	15.9–35.5°C (W Gulf)	John et al. (1990)
	15–40°C (lagoons)	Purser and Seibold (1973)
Light penetration	20 m (shallow-deep water transition depth based on biota)	Hughes Clarke and Keij (1973)
	30+ m (Gulf axis)	Purser and Seibold (1973)
Storms	Frequency: several Shamals per year	Murty and El-Sabh (1984)
Rainfall	3–8 cm	Reynolds (1993)
Evaporation	140–500 cm	Reynolds (1993)

up to 100 km/h (John et al. 1990). Both summer and winter Shamals are known, those in summer being less powerful.

The northwest to southeast airflow of the region is attributed to orography. Sharply rising mountains lie to the east and north while gently rising mountains lie to the west and southwest effectively funneling the low-level airflow. This low-level channeling of air also affects southerly to southeasterly winds known as "Kaus" in Arabic and "Shakki" in Farsi that are strongest on the eastern side of The Gulf where the Zagros mountains in western Iran intensify the flow. The Kaus can precede the Shamal and increase in intensity as the Shamal-bearing cold fronts approach. These Kaus-events may generate galeforce winds (62–74 km/h) (Murty and El-Sabh 1984).

The Shamal has two characteristic durations, short (24–36 h) or long (3–5 days). The interaction between upper and lower airflows determines duration. Longer Shamals are associated with a large pressure gradient between The Gulf of Oman Low and the Saudi Arabian High making the Shamal strongest in the south and southeast ranging from 30–40 knots (~15–20 m/s) and peaking at over 50 knots (~25 m/s). Typical speed is about 5–15 knots (~2.5–7.5 m/s), which is greater than the average wind speed in the northern Gulf (Fig. 2.3).

Table 4.3 Wind speed and directions of western Gulf stations (From John et al. 1990)

Month	Speed and direction		
	Dhahran	Ras Tanura	Safantyah
January	19 NW	17 NW	15 NW
February	21 NNW	17 N	15 N
March	21 N	18 N	17 N
April	21 N	17 N	15 N
May	22 N	17 NNW	15 NNW
June	24 NNW	18 NNW	15 NNW
July	19 N	15 N	12 N
August	19 NNW	15 NNW	13 NNW
September	19 N	13 NNW	10 NNW
October	16 N	13 NNW	13 WNW
November	19 NW	15 WNW	15 WNW
December	19 NW	17 WNW	17 WNW
No. of years	24	6	6

Upon the Shamal's onset, the wind direction at a given locality depends on coastal orography. In the northern part of The Gulf the wind blows from between N and WNW. Winds in the middle Gulf blow from between WNW to NW. On the southeast coast, the Shamal blows from the West. Around the Strait of Hormuz the air flows from the SW (Table 4.3). The Shamals are among the most important wave-generating winds in The Gulf.

The dry Shamal transports sediment from desert regions in both Iran and Arabia towards The Gulf and consequently aeolian transport accounts for an appreciable proportion of the delivery of siliciclastic material (Emery 1956). Airborne dust can become so thick that visibilities of less than 55 m have been reported (Table 4.4). Vessels passing during Shamals become covered by a fine red deposit of dust and sand. Three samples of these shipboard deposits were analyzed by Emery (1956) and consisted of 83% calcite. In the northern Gulf, dust fall-out from southern Iraq amounted to 6.9 g/m^2/year (ROPME 1987). Elsewhere in Arabia, dust precipitation values of up to 22 g/m^2/year have been recorded (Behairy et al. 1985).

The Gulf is also characterized by strong local seabreeze-landbreeze systems (Table 4.3). The effects of local winds appear to be more important to the marine biota, especially in terms of exerting stress, than the broad scale wind systems (Sheppard 1993). During summer afternoons along the coastlines, the sea breeze, caused by convection associated with heating of the nearby land, can reach tens of meters per second. These breezes can affect shallow sediments as well as biota like offshore reefs. They also cause mixing of the water layers, removing stratification and thermal stress, which is important for the persistence of corals. At night, the breezes reverse and can have a powerful desiccating effect on intertidal biota and coastal vegetation (Sheppard 1993). Dew caused by the local seabreeze-landbreeze systems of the United Arab Emirates encourages the development of the local vegetation on the nearshore dune systems, that protects sediment from the wind but sparsens out away from the shoreline (Kendall et al. 2003). Such vegetation is important for the formation of phytodunes (nebkhas).

Table 4.4 Frequency of dust storms in 1967 and average for 1943–1967 in central (Baghdad) and southern (Basrah) Iraq (From Kukal and Saadallah 1973)

	Baghdad 1967	Baghdad average	Basrah 1967	Basrah average
January	1	1.1	0	0.3
February	1	2.1	0	0.6
March	1	2.5	2	1.0
April	4	2.3	4	1.3
May	3	2.3	3	1.4
June	1	2.2	0	2.9
July	1	3.6	4	3.1
August	0	1.6	3	1.7
September	0	0.6	1	1.2
October	3	1.3	1	0.8
November	1	1.1	0	0.3
December	2	0.8	1	0.1
Total	17	21.5	19	14.7

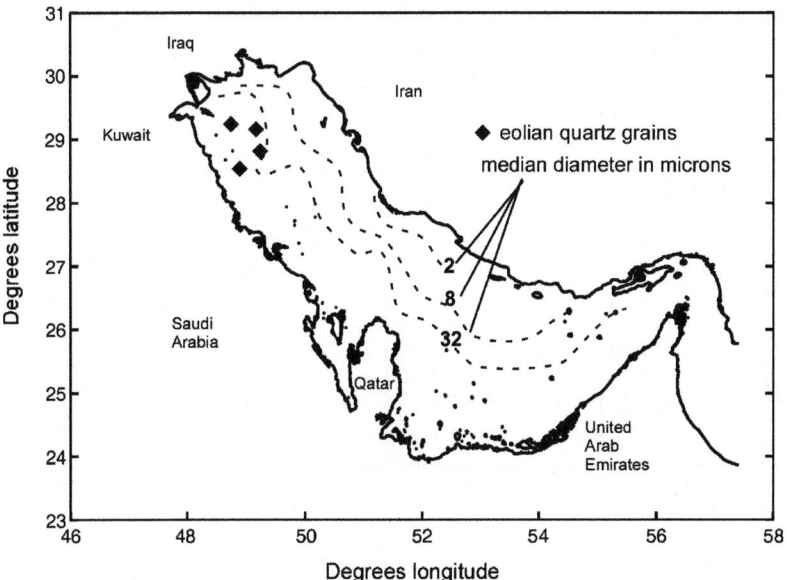

Fig. 4.5 Grain size distribution of insoluble residue. Eolian derived quart grains (*black diamonds*) are believed to be introduced into The Gulf by migrating dunes and subsequently transported offshore (From Emery 1956. Permission by AAPG)

Besides the contribution of airborne sediments to The Gulf, the wind-driven dune migration also brings sediments to the coastal areas. In a study between Ras Tanura (Saudi Arabia) and Kuwait (Fig. 4.5), the eolian lag deposit was found to consist of sand grains and fragments that migrate as dunes toward the sea at about 20 m/year . At this location, dune migration has led to shore progradation and contributed to the sediment load of the nearshore currents (Emery 1956).

A similar mechanism is found in SE Qatar in the Khor al Odaid region (Shinn 1973). Volumetrically, the contribution of these dunes to Gulf sediment budget is unknown, but there is significant potential for aeolian sand transport (potential sand movement is 2.1 t.m/year in the coastal lowlands of Saudi Arabia's Jubail region. Thus, ~50% of sands are transported by winds blowing only 4.5% of the year; Barth 2001). The readily identifiable sediments have been found in sea-floor samples as far as 75 km offshore (Fig. 4.5; Emery 1956).

Kirkham (1998) and Alsharhan and Kendall (2003) suggest a constant dominant wind direction throughout the Holocene and Quaternary. Even throughout the Pleistocene, wind-direction was generally comparable though during the glacials winds were stronger than during the interglacials (Glennie 1996).

4.3.3   Wave Energy

Waves and surface currents are the most important mechanisms of sediment transport in The Gulf (Purser and Seibold 1973). Since the strongest winds are the northwesterly Shamals, waves and surface currents are, for much of the year, also mainly oriented towards the SE. This makes the southeastern Gulf the most exposed environment and the northwestern Gulf, near the Shatt el Arab, the most sheltered (Fig. 4.6). The sediments on the Arabian (southern) side of The Gulf are clearly

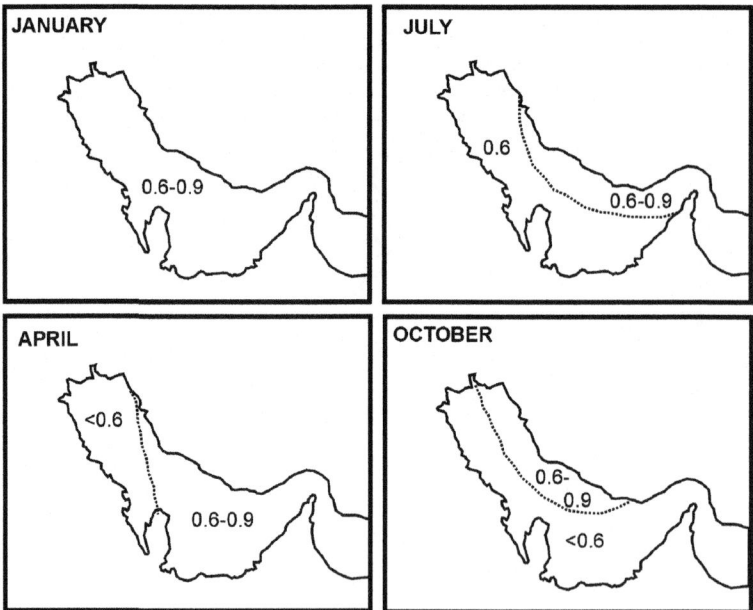

Fig. 4.6 Wave amplitudes in meters in The Gulf for January, April, July and October (Modified from Murty and El-Sabh 1984. Permission by UNESCO)

more of the "exposed" type (i.e. bioclastic and oolitic) and can occur to depths of about 20 m in its southeastern portion. Waves and surface currents produced by Shamals can cause strong water column mixing to a depth of about 30 m (Sarnthein 1970). This strong mixing and associated heat transfer has important implications for carbonate producing fauna (such as reefal organisms).

The strongest wave generating winds in The Gulf region are certainly the Shamals. Relationship between wind speed and wave heights developed for deep water are likely not to hold true for the shallow Gulf and Perrone (1981) suggested that the shallow depths of The Gulf, as well as its stratification, produces wave amplitudes much larger than those expected under similar conditions in deeper water. Observations made from oil rigs indicate that persistent gale force winds blowing for up to 12 h during Shamal conditions can generate waves with amplitudes of 4–5 m (Murty and El-Sabh 1984). Shinn (1976) described a 1964 winter Shamal that lasted several days as packing winds of 65 km/h and driving waves of up to 6 m. During Shamals, the entire Gulf experiences gale force winds, with the strongest winds in the southern portion, and it can take several days for the swell to decay (Murty and El-Sabh 1984).

The northern portion of The Gulf typically experiences lower wave heights than the southern portion due to the Shamal's limited fetch. Perrone (1981) suggested three factors as generating high waves (up to 5 m) in the southern part of The Gulf during Shamals of longer duration:

1. Increase in wind speed in the southern Gulf contributing to locally generated seas
2. Longer duration of gale force winds over the whole Gulf, the northern part of The Gulf generating swell that travels into the southern part
3. No fetch limitation in the southern Gulf

Seibold (1973) reports that the Central Swell (Fig. 4.3, on the Iranian side) should be at least partially attributed to wind generated wave transport of material parallel to the coast which built the swell to a height from 10–40 m with lateral dimensions of roughly 150 km length and 50 km width. The Central Swell is partially situated in that part of The Gulf identified by Murty and El-Sabh (1984) as a region of stronger than usual northwesterly winds and higher seas (Fig. 4.7). In contrast, Uchupi et al. (1996) suggest fluvial input from Iran to have formed at least the basis of the Central Swell, Seibold et al. (1973) cite as evidence in support of longshore and wave dominated sedimentation:

1. Regional net transport of the bipolar tidal currents is small and is reflected in (a) facies boundaries that are relatively sharp, (b) limited dispersion of provenance-specific sediments, and (c) plant and vertebrate remains away from the coast at river mouths.
2. Wave base is recognized by a maximum concentration of echinoderm fragments, a sediment component that dominates the central swell. Wave base is at a depth of 15–50 m, depending on fetch (Sarnthein 1970).
3. Net transport in the western basin is basinward (SE) along the central swell as a result of wind-induced longshore currents, wave action and tidal currents which is evidenced by (a) total carbonate content of several size fractions, (b) calcite:aragonite

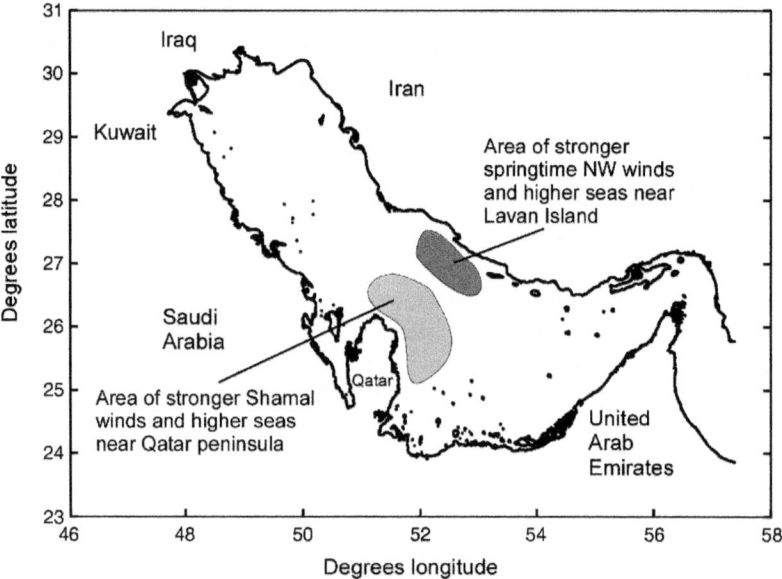

Fig. 4.7 Areas of stronger than average winds and higher waves as indicated by simulation (From Murty and El-Sabh 1984. Permission by UNESCO)

ratios in the 2–6 μm and 20–63 μm fractions, (c) quartz:dolomite ratios in the 2–6 μm fraction, and (d) the quantitative component analysis of the coarse fraction.

Also the northeast coast of the Qatar Peninsula is characterized by wave-induced longshore sediment transport and as a result by extensive carbonate cheniers, bars and spits (Shinn 1973). It also coincides with Murty and El-Sabh's (1984) southern region of higher than normal seas and winds (Fig. 4.7).

Along the northeast coast of the UAE (Fig. 4.8) longshore transport is important. The region is characterized by major spit system developments (Purser and Evans 1973) in the Abu Dhabi, Umm al Qawein and Al Hamra regions. Although not described by Murty and El-Sabh (1984) as having unusually high seas, Purser and Evans (1973) argue that the coast of the UAE is affected by maximum wave fetch since it faces directly toward the long axis of The Gulf.

Murty and El-Sabh (1984) simulated Shamal conditions to investigate their effect on sea surface elevations at various points within The Gulf. They found significant positive and negative storm surges that could reach nearly 4 m in the southeastern portion of The Gulf (Fig. 4.9).

4.3.4 Tides: Interactions with the Sedimentary System

The tides of The Gulf are complex and regionally variable. They range from diurnal to semi-diurnal to mixed. Tidal ranges are relatively large, exceeding 1 m everywhere

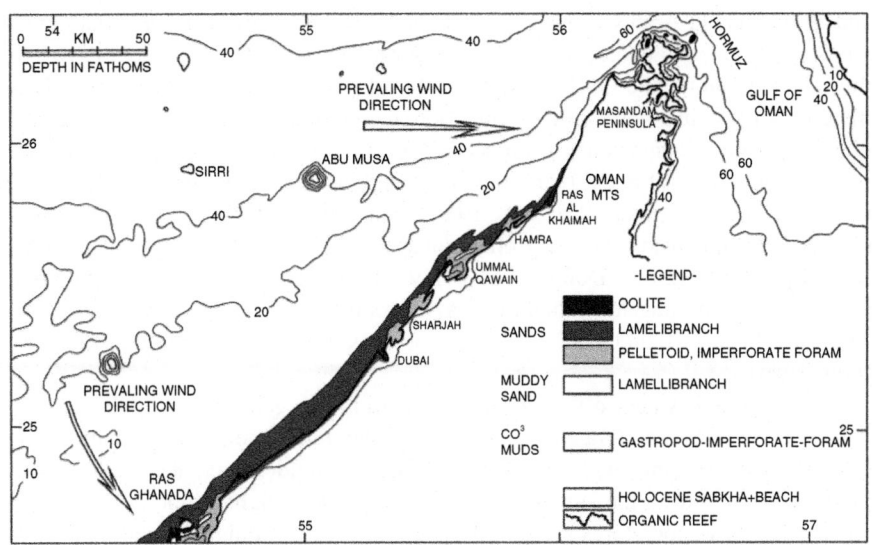

Fig. 4.8 Morphology and distribution of sedimentary units in UAE in response to wave-action and exposure to fetch. Note spit development in Umm al Qawein and Al Hamra regions (From Purser and Evans 1973. Permission by Springer)

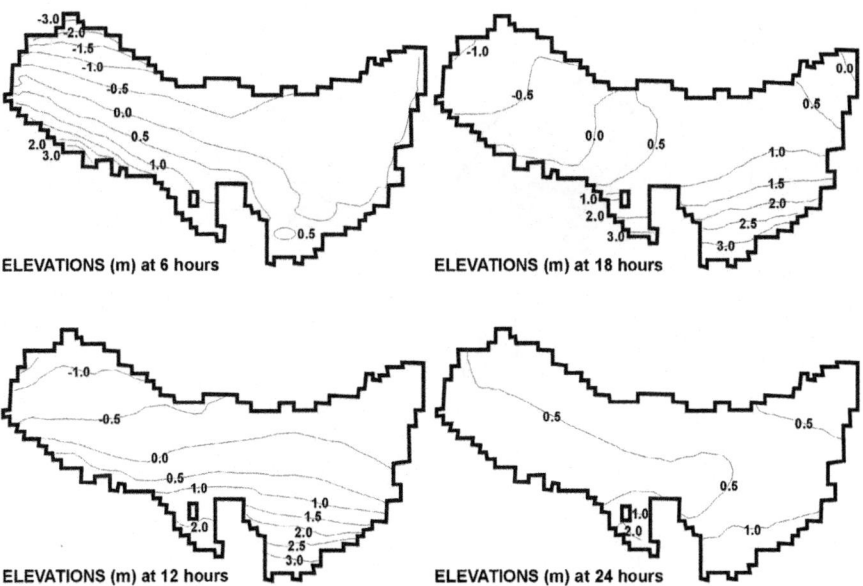

Fig. 4.9 Distribution of positive and negative storm surge heights (Murty and El-Sabh 1984) for the peak hours of a long duration winter Shamal (Permission by UNESCO)

and at the Shatt el Arab surpassing 3 m (Lehr 1984). Off Kuwait, spring tidal ranges are 2 m in the south and up to 4 m in the north (Jones 1986; Sheppard 1993). The diurnal tide enters The Gulf through the Strait of Hormuz and progresses along the coast of Iran in a northwesterly direction, then turning to the southeast and, hugging the Saudi Arabian coast. It proceeds back toward the Straits in a Kelvin wave fashion with a central nodal point, the tidal range increasing progressively from the node to the coast (Defant 1961; Lehr 1984; John 1992b). The diurnal component of the tides has dimensions not far from resonance (Hughes and Hunter 1979), with a natural period of 22.6 h (Defant 1961). However, there are also semi-diurnal components, which propagate counter-clockwise around two amphidromic points within The Gulf (Figs. 4.10 and 4.11), one in the northwest and the other in the southwest (Lehr 1984).

In the coastal waters off Kuwait and Bahrain, the generally irregular diurnal nature of the tides ameliorates the extreme temperature conditions for shallow and intertidal biota to some degree (Sheppard 1993). High tide covers the shallow areas in the daytime during the summer and exposes them at night, while in the winter, when night air temperatures are low, the high tide occurs at night providing protection. This advantage does not occur throughout the entire Gulf, however, and in parts of Qatar, Saudi Arabia, and the UAE the intertidal areas are often exposed during the day in summer (Sheppard 1993).

In Bahrain, a complex of coral buildups and sand shoals (Fasht Al-Adhm) stretches across the northern portion of The Gulf of Salwah. It is believed that these structures are partially responsible for restricting water exchange and significantly

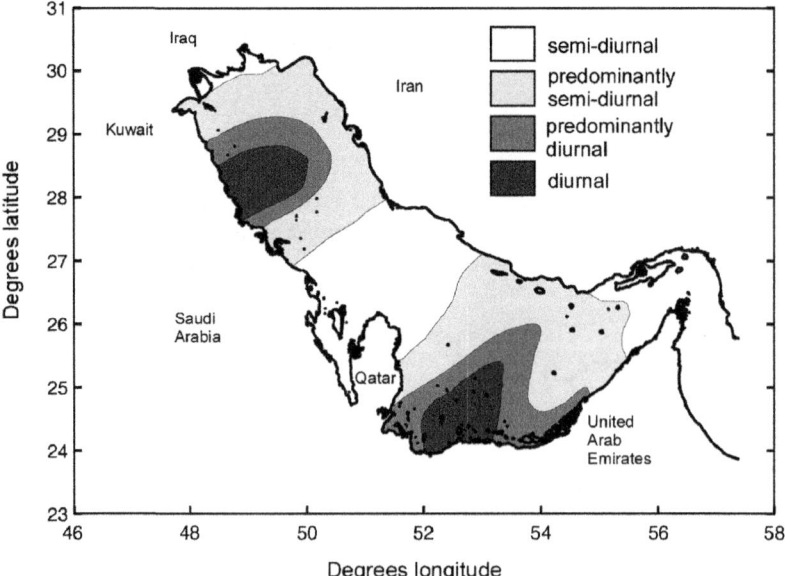

Fig. 4.10 Tidal classification in The Gulf (Modified from Jones 1986)

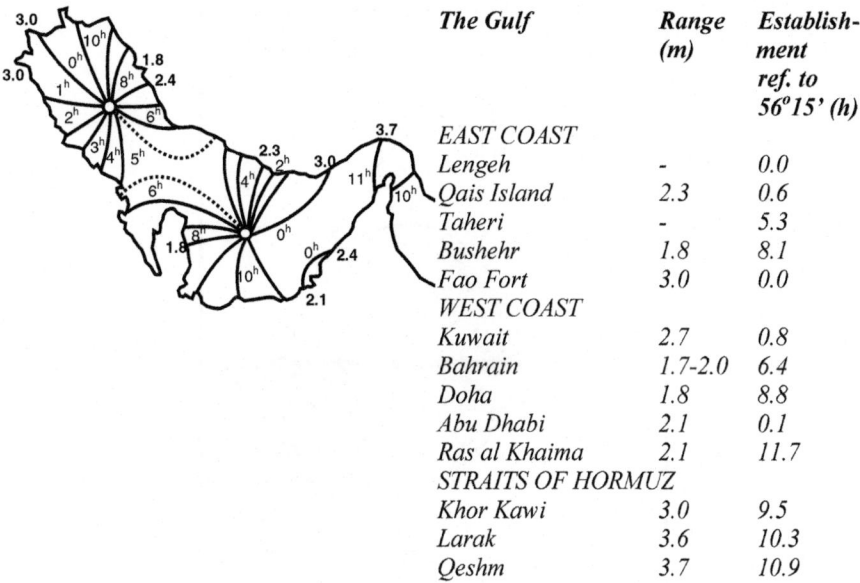

The Gulf	Range (m)	Establish-ment ref. to 56°15' (h)
EAST COAST		
Lengeh	-	0.0
Qais Island	2.3	0.6
Taheri	-	5.3
Bushehr	1.8	8.1
Fao Fort	3.0	0.0
WEST COAST		
Kuwait	2.7	0.8
Bahrain	1.7-2.0	6.4
Doha	1.8	8.8
Abu Dhabi	2.1	0.1
Ras al Khaima	2.1	11.7
STRAITS OF HORMUZ		
Khor Kawi	3.0	9.5
Larak	3.6	10.3
Qeshm	3.7	10.9

Fig. 4.11 Tidal ranges and amphidromic points in The Gulf (From Defant 1961; Reynolds 1993)

reducing the tidal range from 1.2 m just north of Bahrain to 0.5 m in the south of The Gulf of Salwah (Sheppard 1993). The significance of this restriction is profound when viewed in light of the fact that The Gulf of Salwah produces a portion of the dense saline bottom water that drives circulation in the Arabian Gulf (John et al. 1990, 1991). In this case a biological parameter (i.e. coral buildups and associated sedimentary bodies around the Fasht Al-Adhm) is a controlling factor not only upon the local tidal range, but also upon the overall circulation within The Gulf.

4.3.5   Currents and their Sedimentological Significance

The Gulf's current regime is a complex and variable combination of three components: tidal, wind-driven and density-driven. The tidal component consists of complex currents running mainly parallel to The Gulf's long axis, with velocities in excess of 50 cm/s when measured 0–4 m from the bottom (Seibold 1973). Tidal streams commonly exceed 1–2 m/s past islands and through constrictions (Sheppard 1993). John (1992b) presents records of tidal current observations from current meters located in the western Gulf. The local tidal type does not necessarily match the local tidal current regime (e.g. a diurnal tide may result in a semi-diurnal current regime) and flow reversals may be directionally bimodal or rotary (Figs. 4.12 and 4.13).

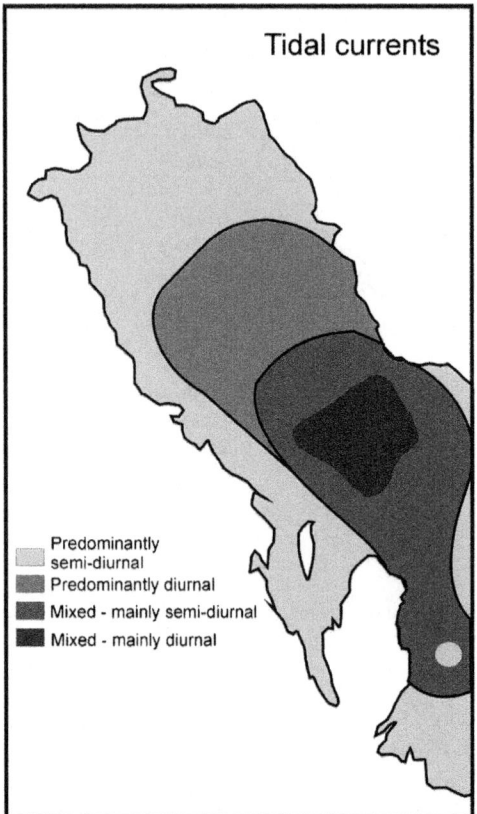

Fig. 4.12 Tidal current regime in the northern Gulf (From Lehr 1984. Permission by UNESCO)

Wind-driven currents are produced by local phenomena (e.g. land-sea breeze regimes) but are influenced on a Gulf-wide scale by the Shamal (see also Section 4.3.3).

Density driven currents result primarily from the high evaporation rate. Wind and density-driven currents are collectively called the residual current, which can be characterized as: a surface flow of Arabian Sea water entering The Gulf through the Strait of Hormuz, then flowing northwest along the Iranian coast toward the head of The Gulf where it follows the coastline around to the southeast. Upon reaching the northern reaches near the Shatt el Arab, water cools in winter and sinks. Also in The Gulf of Salwah and other portions of the southeast Gulf, it becomes dense enough to sink and flow northeast back toward the Strait of Hormuz where it exits as a bottom current with a velocity of about 10 cm/s (Koske 1972; Grasshoff 1976).

The kinetic energy of the water velocity associated with the three current-driving mechanisms can be partitioned among tidal, wind, and density at approximately 100, 10, and 1, respectively (Reynolds 1993).

Fresh water runoff has a significant impact on the residual currents in the northern Gulf (Galt et al. 1983). On the Iranian side, terrigenous material is introduced by

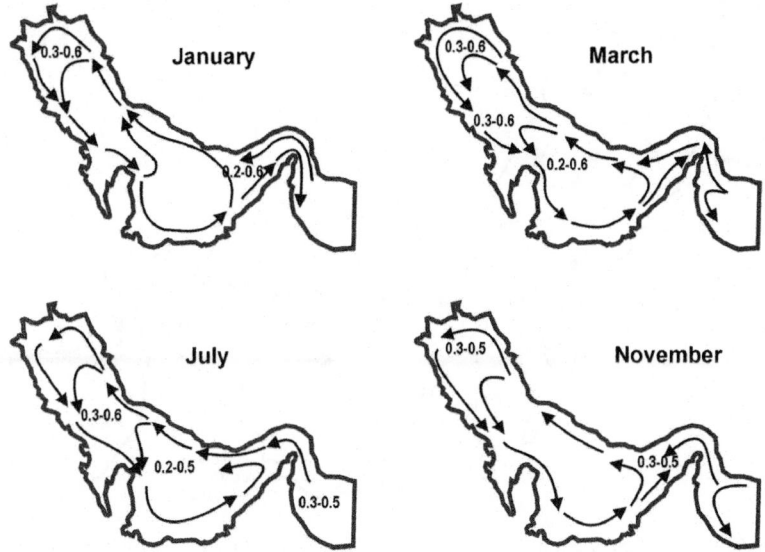

Fig. 4.13 General surface currents, average speed in knots (From U.S. Navy Hydrographic Office 1960)

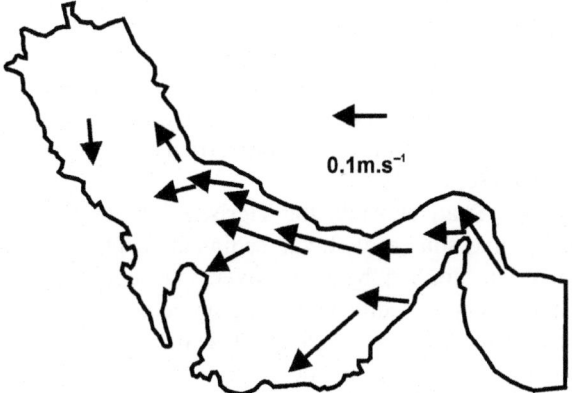

Fig. 4.14 Schematic of surface currents and circulation processes in The Gulf (From Reynolds 1993 and Lardner 1993. Permission by Elsevier)

intermittently flowing rivers from the Zagros Mountains. Upon entering The Gulf, this river input is deflected to the right by the Coriolis force and forms a 20 km wide plume continuing along the Iraqi coast towards Kuwait (Matthews et al. 1979; Hunter 1983a, b) (Fig. 4.14). This fluvially derived sediment has built up a young Holocene blanket deposit in the West Basin extending up to 30 km into The Gulf. The supply rate of fluvial source material is enough to allow sedimentation rates on the order of several m/ka, which renders the several cm/ka supply rate of the wind-transported fraction insignificant in comparison (Seibold 1973).

Fig. 4.15 Residual circulation (From Hunter 1983a and Lardner et al. 1993; ship drift data. Permission by Elsevier)

The tidal currents, in concert with wind-driven currents and waves, then transport the fluviatile material parallel to the coast towards the SE (Seibold 1973), which is opposite to the reported counter-clockwise flow of the residual current (Figs. 4.13 and 4.15). This dominant transport direction also helps explain why the Central Swell has prograded into The Gulf and built up to a height reaching 10–40 m. Further to the southeast, longshore transport is interrupted by islands and steeper bottom slopes. Here, the fluviatile material is carried directly into the deeper water of the Central Basin (Seibold 1973).

The rivers of the Shatt el Arab, the Tigris, Euphrates, and Karun are The Gulf's only year round source of fluvial sediments (Fig. 4.16) and as a consequence, the material supplied by these rivers has constructed a large delta. Complicating the history of sedimentation at the delta front is diastrophic emergence and submergence (Lees and Falcon 1952) due to compaction and eustasy. The net result is rapid growth of the delta front at a particular distributary delta arm that will then remain at that location for a long time whenever the course of that particular distributary changed. Not all of the sediment is deposited at the delta-front (Emery 1956). Wilson (1925) estimated that 90% of the sediment load carried by the rivers is deposited on the subaerial part of the delta. The remaining 10% of the Tigris and Euphrates river sediments and all of those carried by the Karun are either deposited at the delta-front or are carried far into The Gulf. Seasonal turbidity near the delta and along part of the Iranian coast is probably the reason that oyster beds are rare but abundant in the southeastern part of The Gulf (Emery 1956). The Coriolis effect deflects the Shatt el Arab river plume towards Kuwait and excludes most of the Iranian coast from a sedimentary influx from this main river system.

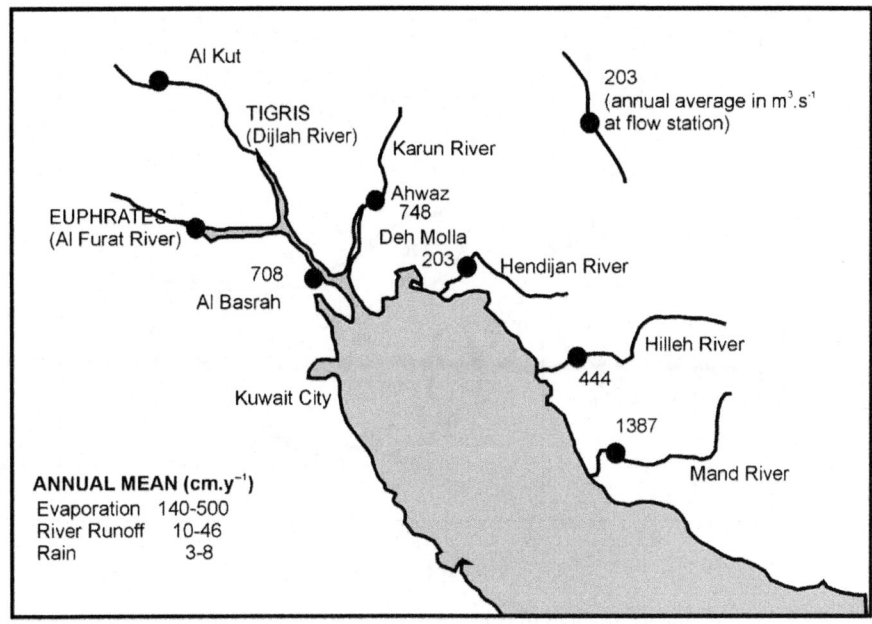

Fig. 4.16 River systems and fresh water import into The Gulf (Reynolds 1993). The river formed by the confluence of the Tigris and the Karun River is called Shatt el Arab, which is the major fresh water input into The Gulf. The Hilleh River is also referred to as Rud Hilla River (Permission by Elsevier)

Table 4.5 Location and sedimentary parameters of the delta region of the Rud Hilla (= Hilleh in Fig. 4.16) River in Iran (From Melguen 1973. Permission by Springer)

Distance from the estuary or the coast (km)	Water depth (m)	Current velocity (cm/s)	Sedimentation rate (approx.) (m/ka)
15	8	<50	4–5
15–17	8–15	<50	2–4
19	25	<50	1.5–2
19	25	<50	1
18	21	<50	0.8
80–120	21–46	<50	0.8–1
18	21	<50	<0.8

Current velocities were measured at various locations within the United Arab Emirates barrier island complex where tides range from a maximum of 2.5 m just seaward of the islands to 1 m along the southern shores of the lagoons (Evans et al. 1973). Current magnitudes varied from 25 cm/s outside of the islands to 65 cm/s at the lagoon inlets between the islands and 25 cm/s at the inner shores of the lagoons. Transport directions in the area are parallel to the coast offshore the islands resulting in longshore spit development and normal to the shore in the inter-island lagoons resulting in island tail accretion and development of tidal deltas (see also Section 4.5.5) (Table 4.5).

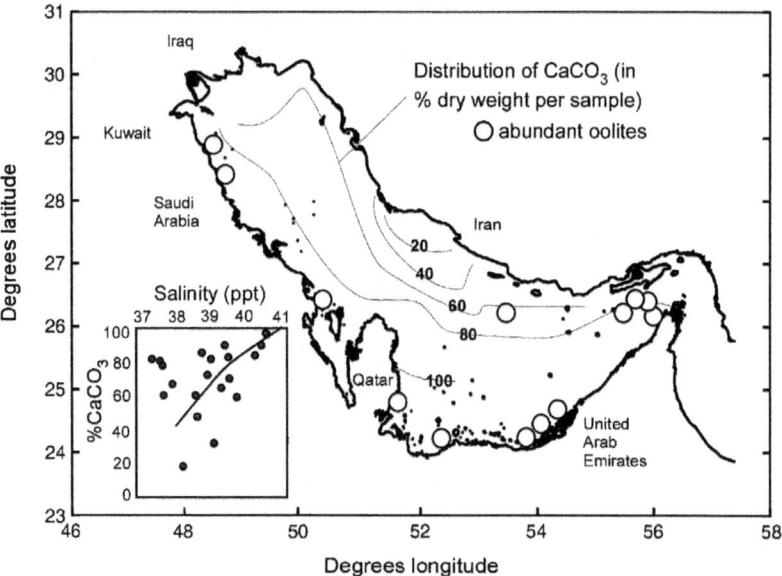

Fig. 4.17 Map illustrating locations of ooid sands, calcium carbonate distribution, and, inserted graph, relationship of carbonate distribution and salinity (Combined from Emery 1956; Loreau and Purser 1973; Gischler and Lomando 2005; own observations)

Also the distribution of ooids and erosion in the Straits of Hormuz are current-mediated sedimentary phenomena. Ooid sands mainly concentrated in the SE Gulf where heated and more saline waters in response to evaporation aid the carbonate precipitation on the ooids (Fig. 4.17; Emery 1956).

4.3.6 Temperature

The seasonal range of sea surface temperatures underlines the variability of the basic hydrographic parameters (Grasshoff 1976). Highs can exceed 36°C in summer and fall below 15°C in winter (John et al. 1990, Table 4.6). Temperatures along the northern coast (Iran) are documented by Grasshoff (1976) and are based on observations made by the R/V Meteor in 1965 at 133 stations. The most spatially continuous depictions of the sea temperature conditions of The Gulf, however, are those of Reynolds (1993).

Winter Shamals are often accompanied by low air temperatures that chill the seawater. A documented Shamal of 1964 (Shinn 1976) that lasted several days, brought an air temperature of 0.5°C to Qatar, a record low in that country. The air cooled the surface water to 4°C, while 80 km offshore at a depth of 17 m, the water temperature dropped to 14.1°C. Air temperatures approached the freezing point in 1982 off Kuwait, driving sea surface temperatures to below 7°C (Downing 1985).

Table 4.6 Temperature extremes and ranges of Arabian reef areas (Sheppard 1993)

Location	Latitude (°N)	Minimum (°C)	Maximum (°C)	Range
Saudi Gulf	27	11.4	36.2	24.8
Qatar Gulf	24	14.1	36.0	21.9
Abu Dhabi Gulf	25	16.0	36.0	20.0
Kuwait Gulf	29	13.2	31.5	18.3
Suez Red Sea	29.5	17.5	30.0	12.5
Aqaba Red Sea	29	20.0	28.0	8.0

Sheppard (1993) and Sheppard et al. (1992) state that off the Saudi Arabian coast, fish kills are reported almost annually following such cold weather episodes.

From a sedimentological perspective, the high summer temperatures are favorable for biogenic (Emery 1956; Seibold 1973) as well as chemogenic (Emery 1956) carbonate production, resulting in ooid production near the Strait of Hormuz (Fig. 4.17), where ooids account for over 80% by weight of the calcium carbonate in the sand size fractions. However, the high variability in temperature limits the biological productivity of certain skeletal carbonates (e.g. corals).

4.3.7   Light Penetration

In shallow seas like The Gulf, suspended sediment can lead to considerable reduction of light penetration (Hughes Clarke and Keij 1973). Based on the relative abundance of light-dependent benthic organisms as an indicator of light penetration, the boundary between the well-lit zone and deeper water is at 20 m in The Gulf's southern portion (Hughes Clarke and Keij 1973). This division is based on the presence of one or more of the following groups: blue-green algae, calcareous and other algae, hermatypic corals and large perforate and certain imperforate foraminifera. The euphotic zone extends to 30 m or more in the less turbid, axial parts of The Gulf (Purser and Seibold 1973).

4.4   Chemical Factors Influencing Carbonate Sedimentation

4.4.1   Salinity

Marked horizontal and vertical salinity gradients exist within The Gulf (Brewer and Dyrssen 1985). Inflowing surface waters from the Arabian Sea are at a salinity of about 36.5‰ and increase rapidly to about 40‰ roughly along The Gulf's E-W axis, splitting it into a northern, less saline, and southern, more saline, realm. Towards the Shatt el Arab, the only continuous freshwater input, salinities again

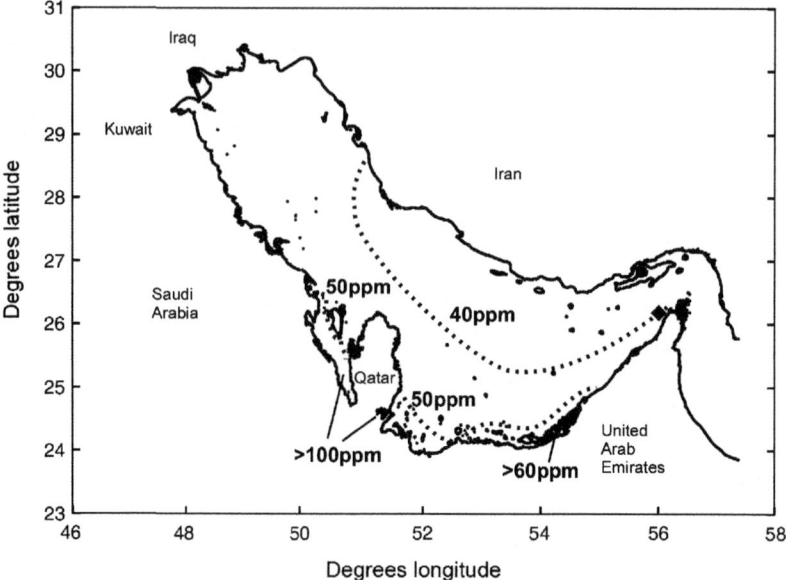

Fig. 4.18 Major salinity trend in The Gulf (From Purser 1973a. Permission by Springer)

decrease to about 36‰ (Fig. 4.18). Due to the higher density of hyperpycnal waters, these form at least part of the bottom waters. Routinely high salinities are found on the shallow Arabian side (salinities in excess of 43‰) (Azam et al. 2006). The densest waters are actually created in the northern Gulf in winter, which together with the highly saline waters from the Arabian side forms the deep water on the Iranian side, with salinities above 40‰ (Brewer and Dyrssen 1985; Reynolds 1993; Swift and Bower 2003).

4.4.2   Oxygen and Nutrients

Nutrient-rich waters enter The Gulf through the Straits of Hormuz from the Arabian Sea and Coriolis deflection then concentrates them on the Iranian side (Johns et al. 2003). Thus, the highest nutrient concentrations are found in shallow waters of the northern Gulf. These waters originate in the north-eastern Arabian Sea, which has a strong vertical oxygen gradient with a maximum at about 50 m depth (Brewer and Dyrssen 1985; Reynolds 1993). Oxygen levels can be so low that denitrification and secondary nitrite maxima are observed leading to high nutrient values at relatively shallow depths. Oxygen consumption and nutrient buildup within The Gulf is not high due to the short residence time and the deep outflow (Brewer and Dyrssen 1985).

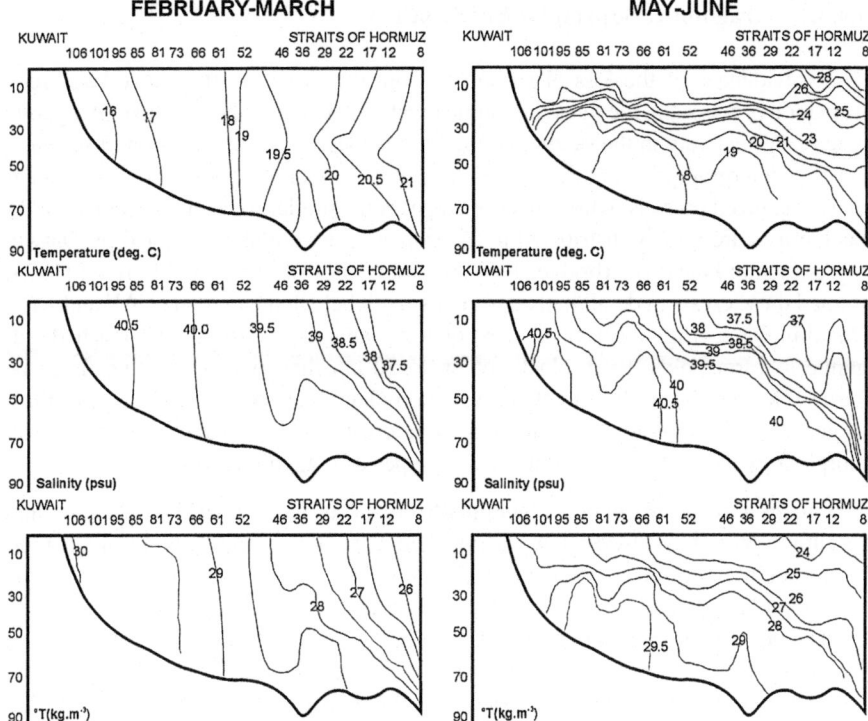

Fig. 4.19 Current-Temperature-depth profile of Gulf axis in summer and winter (From Reynolds 1993. Permission by Elsevier)

Silicate values are low, but rarely zero. The highest values (~6 mol/kg) were measured in the outflow of the Shatt el Arab (Brewer and Dyrssen 1985) (Fig. 4.19).

4.4.3   The Carbonate System

The Gulf is an area of active carbonate deposition, both biogenically and inorganically (or, as present thinking goes, organically mediated) as whitings (areas of milky water where aragonite precipitates in the water column). In most areas of The Gulf the $CaCO_3$ fraction of the sediments is >50%, except near the Iranian coast (Hartmann et al. 1971). This active carbonate depositional regime is also expressed in solution chemistry with clear alkalinity losses of waters after entering The Gulf (Brewer and Dyrssen 1985). The greatest carbonate losses occur in the shallow areas off the coast of the UAE, which correspondingly is an area with very high biogenic carbonate sedimentation, submarine lithification and formation of whitings. Low alkalinity with high salinity is also found in the saline Gulf bottom waters. Part of these depleted, highly saline waters derive from the area of alkalinity depletion off the coast of the UAE (Brewer and Dyrssen 1985; Swift and Bower 2003).

4.4.3.1   Aragonitic, Supratidal Encrustations

Direct evidences of the loss of carbonate from the water are the active beachrock formation process and the thick aragonitic crusts that occur in the intertidal. Aragonite precipitates actively in the sandy intertidal in response to the evaporation of seawater either as intergranular, pore-destroying cement in rock or as a glossy crust of tangentially organized crystals, which gives an appearance similar to lacquer (Evamy 1973). Such crusts are widely distributed in the tropics and were already noted by Darwin on Ascension Island (Fairbridge, 1957 in Evamy, 1973). Purser and Lorreau (1973) studied these crusts in detail in the area between Jebel Dhannah in the UAE and Khor Odaid in Qatar (Fig. 4.20). Sand-sized quartz grains were coated with polished aragonite to form ooids. Even blocks of up to 50 cm diameter, originating from the Tertiary formations at the shoreline, were completely coated thus forming pisoliths. In general, the morphology of the aragonitic crusts showed progressive changes in morphology along the beach profiles. Two types of aragonitic crust occur:

1. **Polished aragonite crusts** ("**pelagosite** crusts" of Evamy, 1973) in the inter and supratidal. They always have a polished appearance and only form on nonporous rock surfaces. They consist of laminae of tangentially oriented crystals with at least five different nanofabrics (Purser and Lorreau 1973).
2. **Irregular aragonite crusts** which form in the highest part of the "splash zone", i.e. the higher supratidal, where water, and thus chemicals, are only imported intermittently. Their morphologies resembled cave or spring tufas with crenulated surfaces. Also individual dripstones were found. The depressions in the crusts are believed to be formed and maintained in their complex morphology by biological attack, subsequent trapping of sediment and renewed encrustation.

Purser and Loreau (1973) believe that the crusts are formed by precipitation and lithification of aragonite and are later settled by micro floras, rather than the aragonite being precipitated directly into an algal matrix.

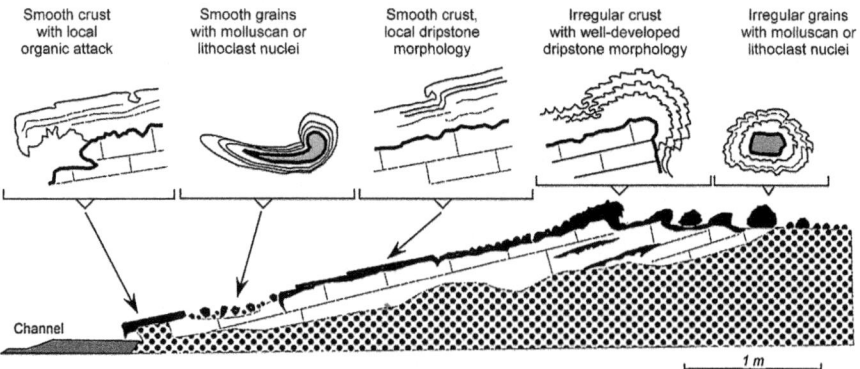

Fig. 4.20 Profile across the intertidal zone near Jebel Dhannah (Abu Dhabi) intertidal showing the morphology of aragonitic crusts (From Purser and Lorreau 1973. Permission by Springer)

4.4.3.2 Synsedimentary Submarine Lithification

Wide areas of The Gulf, in particular on the Arabian homocline, are characterized by marine hardgrounds, i.e. lithified Holocene sandy seafloors (Fig. 4.21). Shinn (1969) studied submarine lithification in The Gulf in detail (Fig. 4.22). He provided evidence (by using pottery and other artifacts in the lithified sediment) that the lithification process is indeed synsedimetary, taking place in depths ranging to 30 m. Shinn's criteria for recognizing recent cement of submarine origin included (1) acicular and fibrous aragonite, (2) micritic cement of high-magnesium calcite, (3) recrystallization of aragonite skeletal fragments, (4) fossils that are Recent and the same as those in overlying unconsolidated sediment (Shinn 1969). The principal types of cement are aragonite and microcrystalline magnesium calcite. The principal physical factors involved in this rapid lithification process are relatively low rates of sedimentation, sediment stability and a high initial permeability of the sediment.

Areas of rapid sediment accumulation, such as the leeward flanks of offshore highs and the nearshore belt of spits, bars, and tidal flats are areas with only slight submarine cementation (Shinn 1969). The lithification leads to the expansion of the layers and thus the generation of "submarine anticlines" and overthrust tepees. These areas are important for the settlement of calcareous fauna (Fig. 4.22).

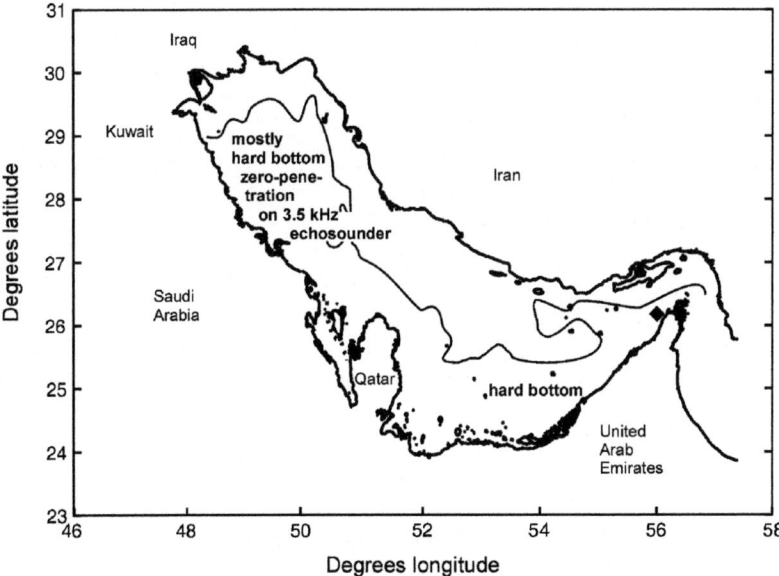

Fig. 4.21 The distribution of hard bottom in The Gulf (From Uchupi et al. 1996). The southern Gulf is mostly hardground (Permission by Elsevier)

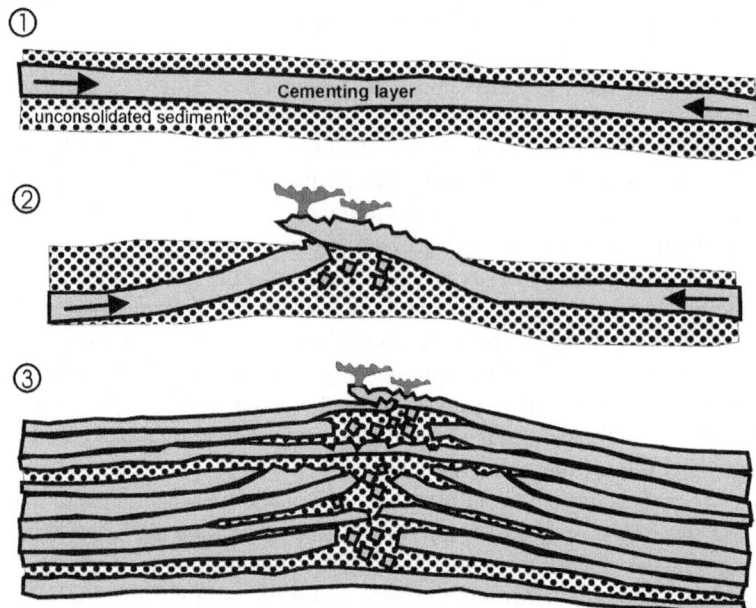

Fig. 4.22 Three stages in the evolution of a hardground ridge (From Shinn 1969). (1) a cementing layer forms inside the sediment which eventually buckles up and (2) cracks, exposing a hardground ridge which frequently serves as a focal point for coral framework development. (3) several generations of hardground ridges can be formed. This explanation for the formation of ridges in the shallow Gulf complements the structural control theory for many such features by Lomando (1999) (Permission by Springer). Figure 4.23: Facies distribution of the northern Gulf (From Seibold et al. 1973. Permission by Springer). *Arrows* indicate transport direction

4.5   Geometries of Facies Belts

4.5.1   Northern Gulf: Overview

The Iranian side of The Gulf is dominated by marly (consolidated or unconsolidated calcareous mud; Tucker 1996) sediments with an overall sand content of less than 50% (Seibold et al. 1973). Grain size and carbonate content both increase with depth, sediments closest to shore containing less than 50% sand size fraction and less than 30% carbonate content, sediments along The Gulf axis containing 50–90% sand size fraction and 30–40% carbonate content (Seibold et al. 1973).

Sand components include fecal pellets and lumps, non-carbonate minerals, Pleistocene reworked relict sediment, as well as benthic and planktic biogenic components. Seibold et al. (1973) divided the northern Gulf into five sedimentological facies (Tables 4.7 and 4.8; Fig. 4.23):

1. Clayey marl with few coarse grains
2. Calcareous marl rich in coarse material
3. Calcareous marl
4. Coarse calcarenite
5. Clay

The non-carbonate sand components originate from three sources: (1) from the Iranian rivers bringing in well-sorted sands, (2) from relict minerals made of poorly sorted medium sand, and (3) from reworked material pushed up by salt domes that

Table 4.7 Distribution and carbonate-CO_2% of five sedimentological facies of the Northern Gulf (From Seibold et al. 1973). Permission by Springer).

Sample type	Distribution	Carbonate-CO_2%
Clayey marl with few coarse grains	Nearshore (river mouths)	18.2
Calcareous marl rich in coarse material	Nearshore with limited terrigenous inflow and offshore in basins	32.3
Calcareous marl	Off river mouths (in shallow water)	23.4
Calcarenite (coarse)	Shallows and island slopes	41.7
Clay	Continental slope of Gulf of Oman	10.6

Table 4.8 Composition of the 0.06–2.0 mm fraction of five sedimentological facies of the northern Gulf, tr = trace (From Seibold et al. 1973. Permission by Springer)

Sample type	Clayey marl with few coarse grains (%)	Calcareous marl rich in coarse material (%)	Calcareous marl (%)	Calcarenite (%)	Clay (%)
Total 0.06–2.0 mm fraction	0.16	57.4	37.12	89.65	14.74
Foraminifera:					
Calcareous	41.3	11.25	1	6.9	1.06
Arenaceous	0.5	4.65	1.7	0.9	–
Planktonic	–	0.06		–	28.8
Sponges	tr	–		–	–
Corals	–	Tr		–	–
Bryozoa	–	0.16	tr	1.2	–
Mollusks:					
Benthonic	22.3	40.9	1.1	7.5	tr
Planktonic	5.4	3.05	tr	0.25	tr
Crustacea:					
Balanus	0.1	Tr	tr	0.1	–
Decapods	0.9	3.95	0.2	0.9	tr
Ostracods	5.45	1.1	0.5	tr	0.1
Echinoids	14.9	1.05	0.45	0.3	tr
Ophiurids	0.05	0.85	0.3	0.3	–
Fish remains	1.65	0.1	tr	0.12	0.55
Plant remains	1.15	Tr	0.3	–	0.1
Calcareous algae	–	–	–	tr	–
Biogenic relict	–	10.7	–	68.6	–
Miscellaneous	6.3	22.0	93.8	12.7	69.2

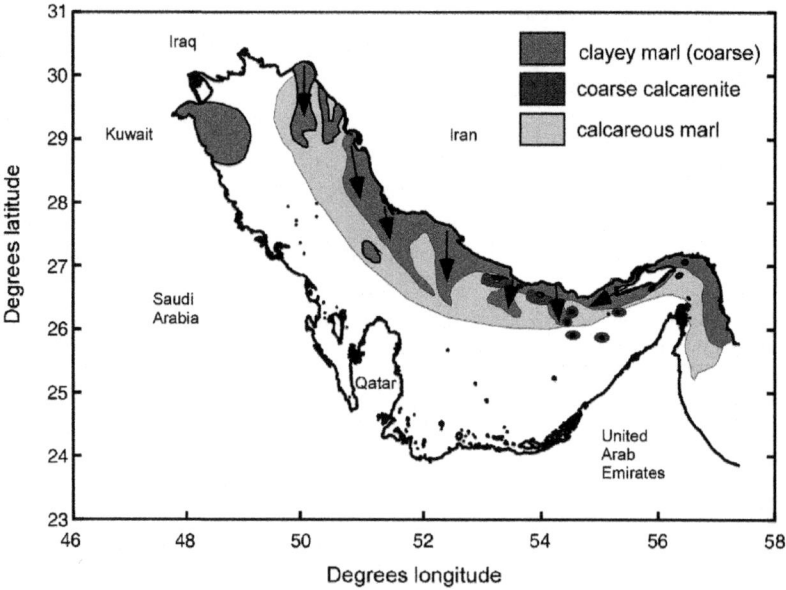

Fig. 4.23 Facies distribution of the northern Gulf (From Seibold et al. 1973). *Arrows* indicate transport direction (Permission by Springer)

is made of medium sand to gravel sized sediments (Seibold et al. 1973). Aeolian transport also contributes small amounts of very fine sand (Seibold et al. 1973).

The carbonate sediments are mostly of Holocene biogenic origin with admixed reworked Pleistocene sediment in many areas, especially those with low sedimentation rates. Late Pleistocene relict sediments are mostly aragonitic and show intricate facies patterns reflecting environments during the transgression (Seibold et al. 1973). Near the continental shelf margin (Gulf of Oman) biogenic sediments consist of roughly 50% benthic and 50% planktic organisms while in the Central Basin of The Gulf the benthic component makes up as little as 5% of the carbonate sediment (Seibold et al. 1973). Areas influenced by Iranian river inflow contain abundant ostracods, benthic foraminifera, and land-plant fragments up to 30 km offshore (Seibold et al. 1973). Sedimentation rates vary regionally in response to the locations of river mouths and variations in topography (Figs. 4.24 and 4.25; Seibold et al. 1973).

The northern Gulf is divided into seven depositional facies (Sarnthein and Walger 1973): **deep water, outer delta** (referring to the deltas of rivers draining into The Gulf basin from the Iranian highlands), **delta axis, inner delta, off-delta shelf plain, shelf plain, and bank/starved shelf** (Figs. 4.25 and 4.26).

Outside The Gulf, a **deep-water facies** is found at depths greater than 200 m and has an average of a 0.1–10% coarse-grained fraction (Sarnthein and Walger 1973). Sediment components include fecal pellets, planktic foraminifera, and fish remains. It is a benthos-poor, calcitic clay. Coarse sand is transported into the deep by the outflowing Gulf bottom current that can be traced up to 100 km into the Arabian Sea (Fig. 4.24; Sarnthein and Walger 1973).

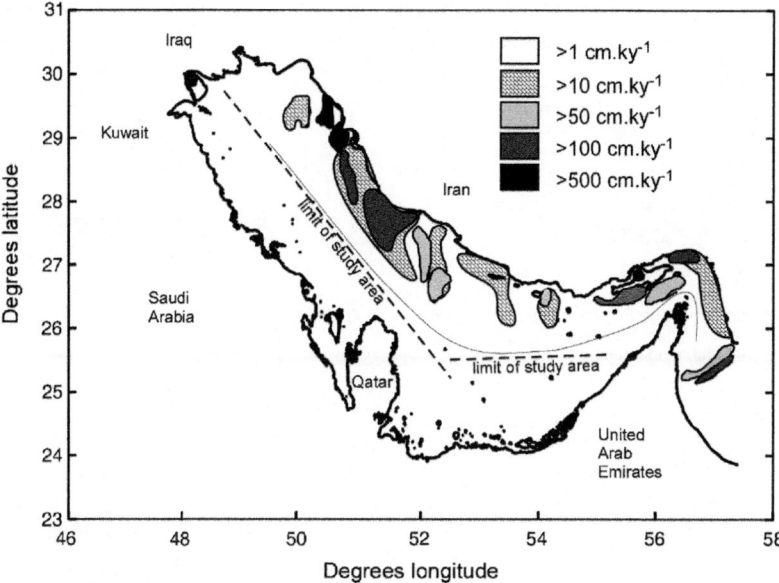

Fig. 4.24 Distribution of relative sedimentation rates in the northern Gulf. Values calculated using planktonic mollusk frequency (From Seibold et al. 1973. Permission by Springer)

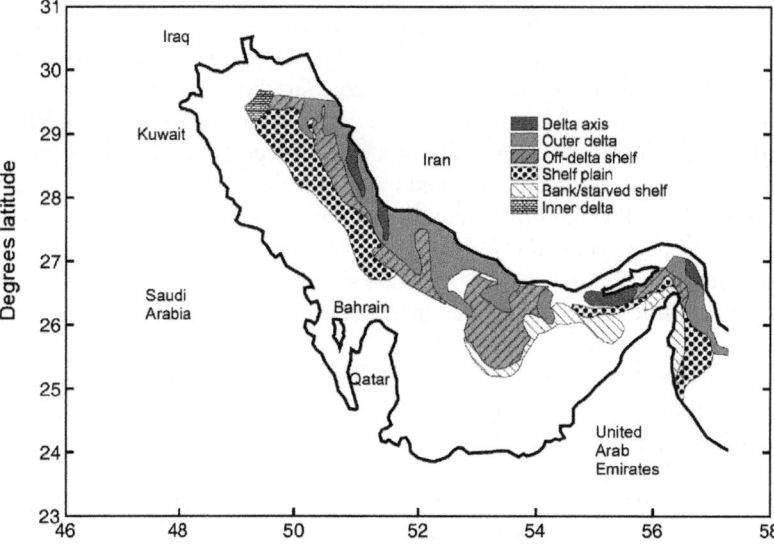

Fig. 4.25 Depositional facies of the northern Gulf (From Sarnthein and Walger 1973. Permission by Springer)

a Frequency of calcareous foraminifera and ostracods

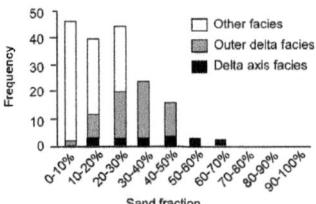

b Frequency of fecal pellets, planctonic foraminifera, and fish remains

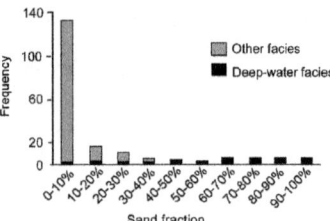

c Frequency of echinoderms, scaphopods, planctonic mollusks, plant fragments, terrigenous sand-
sized material, mud lumps

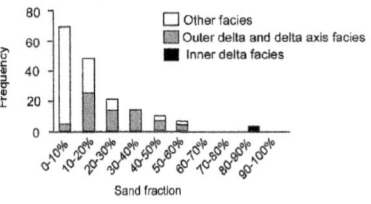

d Frequency of Relict Sediment

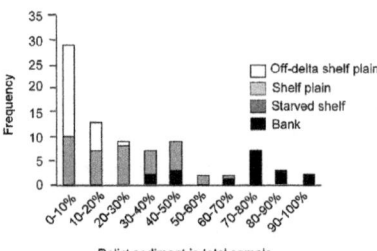

Fig. 4.26 (**a–c**) Sand fraction frequencies of selected depositional facies of the northern Gulf
(From Sarnthein and Walger 1973). (**d**) Relict sediment content in selected depositional facies
(Recalculated from Sarnthein and Walger 1973. Permission by Springer)

An **outer delta facies** is found at water depths <100 m and extends 24–50
km offshore. The coarse sediment fraction consists of foraminifera, ostracods,
echinoderms, plant fibers, terrigenous sand, and planktic mollusks and was called
a foram-mollusk-echinoderm marl (Fig. 4.25; Sarnthein and Walger 1973). The
sediments contain twice as much carbonate as the deep-water facies (40–60%).

A **delta axis facies** occurs in zones around the outer delta lobes of major rivers.
The coarse fraction is 0.1–1%. The sand fraction is a foram-ostracod marl including
echinoderms, planktic mollusks, plant fibers, terrigenous sand, ostracods, and foraminifera. Carbonate content is 40–60%.

An **inner delta facies** has a 35–70% fine sand component in an abundant coarse
fraction. This fine sandy calcareous marl contains 80–90% quartz and carbonate
rock detritus.

An **off-delta shelf plain facies** is found beyond the delta facies and is marked
by an increase in carbonate content to 50–70%, a lower sedimentation rate, epibenthic skeletal grains and relict grains. It is a mollusk-rich coarse-grained marl.

The **shelf plain** encompasses the area extending up to 100 km from The Gulf's
E-W axis and following the axis to the continental shelf breaks. Sediment components of this coquina marl include relict grains, fixed epibenthic skeletal grains,
tests from arenaceous foraminifera, decapods, balanids, and a 0.2–0.5% quartz sand
fraction likely due to eolian sedimentation.

The **bank or starved shelf** has little terrigenous input and comprises coarse grains and/or carbonate sediments that include relict grains, larger foraminifera, *Lithothamnium* algae, and gravel-sized terrigenous material. The shallow bank sediments are a large foraminiferal-coquina arenite and the starved shelf sediments a marly coquina calcarenite (Sarnthein and Walger 1973).

4.5.2   Southern Gulf: Overview

On The Gulf's southern, Arabian, side, the seafloor slopes more gently towards its center than on the Iranian side. Over much of the year, the Arabian side is alse the windward shoreline due to a predominantly northerly wind-direction (Wagner and van der Togt 1973). Whether it represents a classical carbonate ramp system (Wilson 1975; Wilson and Jordan 1983; Read 1985; Tucker and Wright 1990; Burchette and Wright 1992; Kirkham 1998; Gischler and Lomando 2005) or should better be referred to as a foreland basin with a mixed carbonate-clastic fill (Walkden and Williams 1998) is a matter of debate. Wave protection of the sediments increases with depth (Wagner and van der Togt 1973). Sediments grade from skeletal, oolitic, pelletoidal sand and fringing reefs near the margin, to compound grains, and skeletal muddy sands to muds in the basin (Wagner and van der Togt 1973) (Fig. 4.27).

Fourteen sedimentary facies exist on The Gulf's Arabian margin (Table 4.9; Wagner and van der Togt 1973). Sediment on the Arabian side trends from impure

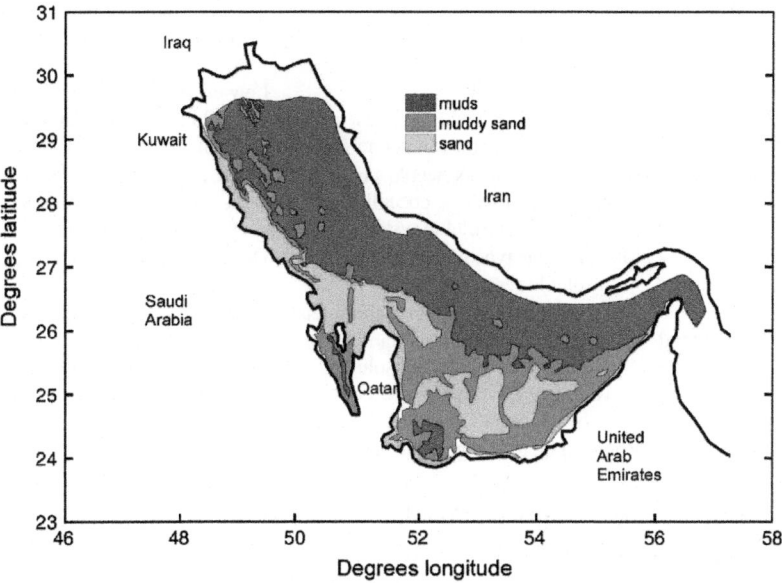

Fig. 4.27   Distribution of sediments in The Gulf (From Wagner and van der Togt 1973. Permission by Springer)

Table 4.9 Sedimentology and environmental setting of sedimentary facies in Te Gulf (Wagner and van der Togt 1973. Permission by Springer)

Facies	Sedimentology	Relation to environment
Pelecypod sand	Rounded to angular, moderately-sorted, medium sand-sized pelecypod fragments; 4% mud	Shallow marine, non-restricted, high-energy environments with absence of hard-rock substrate
Compound grain-pelecypod sand	Angular fragments of pelecypods, small gastropods, large (Operculina) and small foraminifera, echinoid debris, compound grains; 4% mud	Shallow-marine non-restricted conditions with energy slightly lower than that of pelecypod sands
Foraminiferal sand	Perforate large foraminifers (Heterostegina, Amphistegina), rhodoliths	Unrestricted offshore highs at depths between 15–35 m
Coral-algal sand	Coral and coralline algae fragments, mollusks, echinoid fragments, foraminifera; moderately to poorly sorted; 3% mud	Small patches in high-energy conditions, high sedimentation rates; form sand bars and tails downwind from reefs
Foraminifer-pelletoidal sands	Imperforate foraminifers, algally bored skeletal grains, pelletoids, compound grains from adjacent lagoons	Restricted, moderate to low energy environments, water depths <5 m, e.g. tidal flats
Gastropod sands	Poorly sorted, medium coarse grained, gastropods, pelecypods, fewer pelletoids, compound grains, imperforate foraminifers	In depths <5 m, restricted environments e.g. coastal barrier complexes, rock substrates
Oolitic sands	Ooids with some mollusk fragments	<3 m depth, active tidal bidirectional currents, close to shore
Pelecypod muddy sand facies	Angular fragments of infaunal pelecypods, foraminifers, echinoid grains, black compound grains; 25% mud	Shallow marine, unrestricted low energy environments between 15–45 m of water depth
Gastropod muddy sand facies	Poorly-sorted, medium to coarse grained gastropod fragments, foraminifera, and compound grains	Low energy, restricted, depths <10 m, protected embayments such as The Gulf of Salwah.
Pelecypod mud	Pelecypod fragments, whole small shells, fecal pellets, coccoliths; 80% matrix, 6% insoluble residue	Extensive patches in broad depressions, 15–35 m depth on the proximal homocline
Argillaceous pelecypod mud	Like previous type but with higher insoluble residue, 21%	Deep marine distal homocline and axial areas
Foraminifer/gastropod mud	Light-gray carbonate mud with skeletal grains of imperforate foraminifera; mud may be pelleted; insoluble residue <5% except when close to eolian source	Highly protected, shallow, <5 m, within coastal embayments and on lee sides of barrier complexes; salinities exceed 50 ppm frequently
Quartz sand, muddy sand	Rounded quartz, angular skeletal debris	Leeward coasts: SE coast of Qatar, becomes muddier seaward
Sedimentary gypsum	Fauna: Cyprideid ostracods	Highly restricted, depths <2 m, at ends of long embayments

carbonate muds in the center of the basin to bioclastic sands and ooids in high-energy areas nearshore and muddy sediment in coastal embayments (Fig. 4.27).

The low-energy center of the basin contains argillaceous pelecypod muds with both unbroken and angular grains. Grain size increases towards the Arabian shoreline as energy increases and carbonate skeletal production increases. Sediments change from mud supported sediment of the distal basin to grain-supported pelecypod muddy sand at the foot of a terrace at 36 m depth. In Kuwait, this transition occurs between 15–20 m depth (Gischler and Lomando 2005). At depths between 10–20 m, winnowed compound grains and pelecypod sands are deposited at moderate energy and slow sedimentation rates (Wagner and van der Togt 1973). High-energy conditions along the coastline produced by the north-westerly Shamal winds produce broken pelecypod sands. Also ooid sands are found at the seaward end of tidal channels, in particular near the Abu Dhabi barrier islands, but also near Jebel Dhannah and towards Musandam (Straits of Hormuz) and in Kuwait (Gischler and Lomando 2005). Fringing reefs and coral-algal sands replace the pelecypod sands in more windward areas (Wagner and van der Togt 1973).

Muds accumulate in low energy conditions such as in coastal embayments and lagoons that are protected by spits or barriers. If isolation is thorough enough, evaporative conditions may produce precipitated gypsum. Stromatolitic algal mats form on the flanks of protected embayments and lagoons where wide inter- and supratidal flats form (Wagner and van der Togt 1973; Duane and Al-Zamel 1999).

Sediments are not arranged as shore-parallel belts due to: (1) variation in the homocline's orientation to the NW winds, (2) the presence and absence of barriers, and (3) variation in the inclination of the homocline.

4.5.3 Tidal Flats and Sabkhas

On tidal flats on the northeast coast of the Qatar Peninsula or in Kuwait, where coastal accretion is occurring, southward longshore transport creates carbonate cheniers (narrow, vegetated beach ridges parallel to the prograding shoreline), bars and spits (Table 4.10; Shinn 1973; Lomando 1999). Chenier beach ridges are up to 50–100 m wide, 2 m or more thick, and 12 km long and contain abundant cross-bedding (Shinn 1973). They are found in intertidal and supratidal environments (Shinn 1973). In Qatar, cheniers migrate south and accretion occurs on their south-ern ends where spits are located (Shinn 1973).

The southern Gulf is particularly well known for its extensive tidal flats and sab-khas. "Sabkha" is Arabic for "salt flat" and according to this general definition sev-eral types of sabkha exist (such as inland sabkhas, the biggest and best known being the Sabkha Matti in Abu Dhabi), but the best studied and in the present context most important are the coastal sabkhas. Gulf tidal flats are characterized by sabkha devel-opment in the supratidal zone and in general relatively fewer tidal channels in the

Table 4.10 Tidal sediments at the northeast coast of the Qatar Peninsula (From Shinn 1973. Permission by Springer)

Environment	Sedimentology	Distribution
Subtidal	Soft-pelleted mud and silt-sized carbonate grains, some skeletal grains; fine-grained sediment is 30% windblown dolomite; sand found in channel lags and narrow belts of rippled sand	Muds and silts more than 8 m thick; underlie intertidal and supratidal sediments, cover lagoonal floors
Intertidal	Pelleted carbonate mud and silt, some skeletal material; light-tan in color; some stromatolitic algal laminations; fine and coarse sand found in channels, small eolian dunes, and beaches	Few meters to kilometers wide; wide belts laced with meandering tidal channels; sand drifts stabilized by vegetation found on interchannel areas
Supratidal	Pellet rich mud, some skeletal lags during storms	Up to 5 km in width, around 10 km long, thickness up to 50 cm, salt and gypsum crusts may form up to 2–3 cm thick

intertidal zone in comparison to tidal flats of the Bahamas. Extreme salinities and evaporite formation in the supra- and intertidal zones limits the presence of burrowing organisms. Tidal flats in the areas of the UAE and Qatar develop cemented winnowed grainstone layers in the upper subtidal and lower intertidal zones. These are efficient in retarding fluid flow due to their low permeablities while progradation of the flats continues (Shinn 1983). Other crusts form further offshore in 3–4 m of water and can be traced dozens of kilometers – they also impede fluid flow (Shinn 1983). Down-cutting by tidal channels is limited by submarine crusts. However, when channels erode through the crust they can reach up to 18 m in depth (Shinn 1983). Documented channel sediments consist of a ~10 cm thick basal, lithified lag deposit, overlain by 50 cm of graded, uncemented muddy channel lag, then by ~25 cm of light-gray mud, and 0.5–1 m of sorted, cross-bedded carbonate sand on top (Shinn 1973). When abandoned, these channels are first filled by layers of stromatolitic algae and finally by eolian sand. Therefore, their surfaces, even when completely filled, are visibly different from the surrounding sabkha surface (Shinn 1983).

Sabkhas are found in supratidal settings of the Arabian coast from Musandam to Kuwait (Schneider 1975). Sabkas in the region south of Abu Dhabi are among the best-studied of The Gulf and are part of a tidal flat/lagoonal complex protected by islands and shoals. In the southern Gulf, they are on average 2 m above sealevel and have a gradient of 1 m over a distance of 10 km (Sanford and Wood 2001). Salinities of nearby open water reach 38‰ and in the inner lagoon reach 60‰ (Schneider 1975).

The Abu Dhabi sabkha began to form about 4,000 year BP when the sea began to regress from a highstand of about 1 m above present sea level (McKenzie et al. 1980). During regression, the sediments of the tidal environment prograded and early diagenetic alteration of the carbonate sediments occured underneath the surface of the sabkha. Due to the arid environment, evaporitic minerals precipitated

Fig. 4.28 Aspects of the Abu Dhabi sabkha. (**a**) Crincle sabkha (**b**) Polygonal sabkha with halite crust (**c**) Gypsum crystals. These crystals can crop out at deflation surfaces (**d**) Nodular anhydrite overlying gypsum mush as is a typical sequence in the polygonal sabkha (Photos by Riegl)

including dolomite, gypsum, anhydrite, and halite (McKenzie et al. 1980). Gulf sabkhas all have their unique characteristics according to the availability of water, but by and large they can be characterized into the zonation pattern illustrated by Alsharhan and Kendall (1994). Vertical sequences in a sabkha are illustrated in Kirkham (1998). Figure 4.28 illustrates some key features.

The sabkha has played an important role for the development of dolomitization concepts. Schneider (1975) and McKenzie et al. (1980) described the Holocene dolomitization process occurring in the Abu Dhabi sabkha tied to a three stage hydrological cycle consisting of:

- Flood recharge
- Capillary evaporation
- Evaporative pumping

According to the classical view of evaporative pumping, this is induced by the aridity and warm climate. The primary physical parameters acting on/in these sabkha deposits are tides, wind (responsible for water recharge by causing evaporation which in turn causes the capillary rising of water within the sabkha) and evaporation (responsible for hydraulic pumping and enriching the reservoir water with heavy isotopes and cations). Fluctuations in sea level in the region of Abu Dhabi are mainly tidal with normal tidal ranges of greater than 1 m (Hsü and Schneider 1973). Studies of potential evaporation, conducted at the Abu Dhabi airport, indicate

Table 4.11 Diagenetic facies of the Abu Dhabi Sabkha (From Schneider 1975)

Diagenetic facies	Sedimentology
Upper intertidal	Gypsum mush, celestite, dolomite, calcite
Lower supratidal (lower sabkha)	Gypsum mush, bassanite, dolomite, calcite nesquehonite
Upper supratidal (intermediate sabkha)	Anhydrite nodules, large gypsum disks, dolomite, magnesite, huntite, halite, some polyhalite
Continental sabkha	Anhydrite layer, secondary gypsum, dolomite and anhydrite are secondary; only gypsum is a primary precipitate; anhydrite can rehydrate gypsum

an annual rate of 8 m water loss per year (McKenzie et al. 1980). Actual loss of groundwater due to evaporation was considered to be one order of magnitude less than the potential evaporation (McKenzie et al. 1980) (Table 4.11).

Sanford and Wood (2001) and Wood et al. (2002) showed that the predominant source of water to the sabkha is indeed recharge from rainfall with negligible lateral inflow and small groundwater inflow (in a rectilinear volume of sabkha 1 m wide, 10 km long and 10 m deep: 1 m^3/year of water enters and exits by lateral groundwater flow, 40–50 m^3/year enter by upward leakage and 640 m^3/year enter by recharge from rainfall). A solute budget found the dominant source of solutes to be the underlying Tertiary formations. Natural convection was found to occur due to evaporation with less-dense brine entering from below and being concentrated by a process of evaporation and re-solution of salts at the land surface. Based on their findings, they supplant the previously existing models (seawater flooding and evaporative pumping) (Hsü and Siegenthaler 1969; Hsü and Schneider 1973; McKenzie et al. 1980) with their "ascending-brine model" (Fig. 4.29; Wood et al. 2002).

Ascending brines are important for geochemical processes since they provide a mechanism for ionic transport. Calculations show that the lateral groundwater velocities in the range of 10^{-3}–10^2 cm/year are too low for effective ionic transport necessary for dolomitization within the time available (McKenzie et al. 1980). Vertical groundwater movement is much more efficient in this respect. In the sabkhas of Abu Dabhi, vertical velocities of approximately 1 cm/day are recorded in both upward and downward directions depending on whether the sabkhas are being recharged or undergoing evaporative loss (McKenzie et al. 1980). The hydrologic situation is sufficient to replenish the system with Mg for continuous dolomitization but recent studies indicate that dolomitization is probably mediated by microbial activity as demonstrated in other areas and in the laboratory (Vasconcelos et al. 1995; Vasconcelos and McKenzie 1997).

4.5.4 Sand Bodies

The coast of the UAE offshore Abu Dhabi contains linear tidal bar belts. Where tidal flow is strongest between islands and the mainland, ooid shoals develop (Halley et al. 1983; Kirkham 1998). Bars of skeletal sand also line the inner edge of the

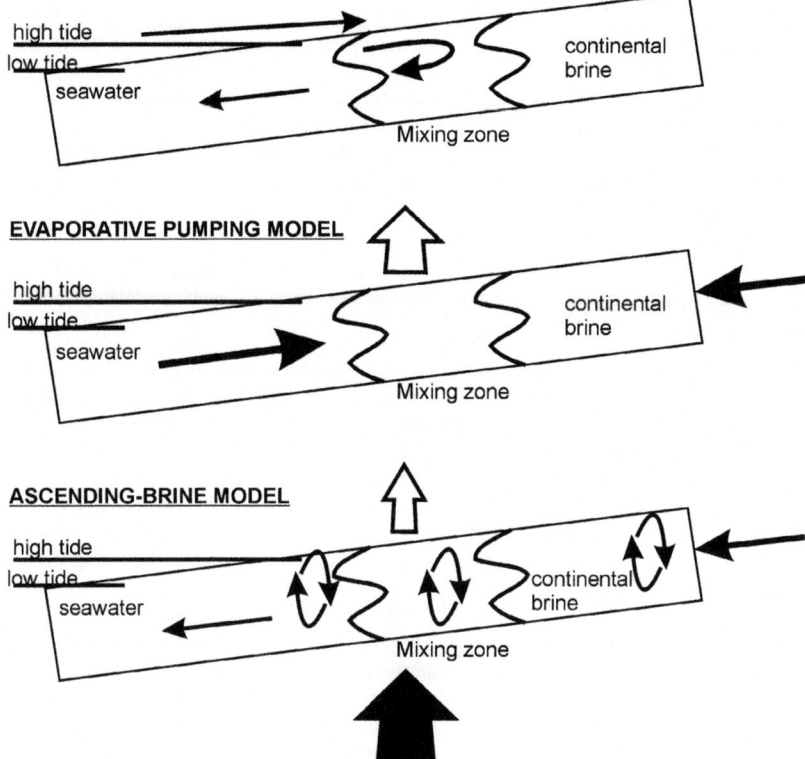

Fig. 4.29 Diagrams outlining the models developed to explain the source of solutes to the coastal sabkhas in The Gulf (From Wood et al. 2002). *Arrows* indicate water movement (Permission by Springer)

Great Pearl Bank (Table 4.12). Deltaic sand bodies form as tidal currents flow through inter-island channels of a barrier island complex that developed along the central part of the coast of the UAE (Purser and Evans 1973), Abu Dhabi being on one of these islands. Tidal deltas of the coast of the UAE can extend offshore 1–2 km with 5–6 km width (Purser and Evans 1973). Table 4.12 lists qualitative attributes of the Great Pearl Bank and shoreward lagoon created by the barrier island.

4.5.5   Lagoons

Extensive lagoonal systems exist in the SE Gulf, with distinct sedimentary and biotic environments. Most lagoons are formed behind offshore highs (for example the Khor al Bazam-Great Pearl Bank lagoon; Kendall et al. 1969) or nearshore barrier islands (the Abu Dhabi islands/lagoon complex). Each of these settings has a distinct facies.

Table 4.12 Characteristics of sand bodies (Purser and Evans 1973. Permission by Springer)

	Essentially submerged, west	Paritally emerged	Essentially emerged, east
Barrier	Between Ghasha and Bazm al Gharbi islands	Between Bazam al Gharbi and East end of Abu al Abyad islands	Between Abu al Abyad islands and Ras Ghanada
	Average depth 5–10 m; locally emergent islands (as small as 1 km in diameter) along lagoonal margin	Average depth on axis 2–5 m; large islands along lagoonal edge with intertidal flats	Depth in channels 2–10 m; 75% of barrier emergent as barrier islands with downwind tails and/ or lateral spits due to longshore transport
	Lamellibranch sand and small patch reefs along lagoonal margin	Well-developed reefs on seaward flank, lamellibranch and coral-algal sand on axis and as spill-over sheets along lagoonal edge; mud and evaporites on intertidal flats	Oolitic sands; very small reefs in channels and in front of barrier islands. Wide intertidal flats with algal mats and evaporites along lagoonal margins
Lagoon	Average depth 20 m	Average depth 2–5 m	Lagoon dissected by accretion tails into secondary lagoons (depth 2–3 m)
	Salinity 40°/$_{oo}$	Salinity 40–50°/$_{oo}$	Salinity 50–60°/$_{oo}$
	Lamellibranch muds in lee of barrier; coral-algal sands on lagoonal highs, molluskan sand and coral-algal debris nearer shore	Limited areas of imperforate foram, mud along lee of islands' pelletal and compound grain sands widespread; frequent submarine lithification	Imperforate foram. Muds in very sheltered areas behind barrier islands; hard pelletal sands dominant; lithified submarine crusts frequent
Continental shoreline	Rocky headlands with narrow tidal flats in embayments	Regular morphology consisting mainly of 1–2 km wide sabkhas	Maximum devlopment of sabkha plain (2–5 km) accreting over adjacent lagoon

The Khor al Bazam lagoon, a classic location of sedimentary studies, is protected by the Great Pearl Banks (and in particular the topographic highs of the Bu Tina and Mubarraz shoals and Murrawah/Bazam al Gharbi islands) and the string of nearshore island between Sir Bani Yas and Abu al Abyad. The oceanic environment becomes more restricted towards the east and correspondingly salinity increases from 40 to >50 ppm. From the offshore shoals to landward, and along the easterly salinity gradient, open water calcareous biota like corals, echinoids and algae decrease in importance to favor gastropods and foraminifera and eventually, in the supratidal, algal mats. The sediments within the lagoon are mostly skeletal pelletal sands (mainly made up by bivalves and, to a lesser extent, gastropods) (Wagner and van der Togt 1973), with relatively little lime mud (foramineran/gastropod muds with dasycladacean fragments in the most restricted areas).

A well defined complex of lagoons is developed behind and between the islands of the Abu Dhabi barrier islands system. These lagoons are highly restricted and water reaches salinities upwards of 60 ppm. The dominant sedimentary facies are skeletal-pelletal sands in areas with higher circulation and mud and pelletal mud in the most protected parts, containing imperforate foraminifera and gastropods. Around the edges of the lagoons, mangroves and sabkha with thick algal mats are developed. There is evidence that some of the lagoonal muds may be direct precipitates from the water, since whitings are known to occur in the area (Kinsman and Holland 1969). Strong tidal flows between the islands at the mixing zone with more oceanic waters create well-developed ooid tidal deltas.

Facies distribution in and around these lagoons shows a progression from the coarser more exposed sediments on the windward side of the islands and in the passes between the islands to the finer sediments in more sheltered areas close to the mainland. Also the growth pattern of the islands, which is essentially determined by the sedimentary "tail" developing in their lee, reflects this gradation of wave energy.

Khor Odaid in Qatar is a true lagoon, almost entirely enclosed by the mainland and only connected to the ocean by a relatively narrow channel. It is one of the areas with the highest recorded salinities in the entire Gulf. Biotically, especially in the inner reaches, it is essentially a desert. The typical sediments are gastropod/imperforate foraminifera carbonate muds.

Other lagoons, however of a less enclosed nature, exist in many places along The Gulf shoreline in the lee of coastal reefs (for example inshore the Umm Said-Masaeed reef complex in SE Qatar) or behind islands (such as between Abu Ali Island and the Saudi mainland coast) and inside deep embayments (Tarut Bay, Manifa Bay, Dauhat ad Dafi, Dauhat al Musallamiya and others in Saudi Arabia) (Fig. 4.30).

All these areas are primarily characterized by lamellibranch sand in the more open areas that move towards a foraminiferan/gastropod mud facies in the more sheltered reaches. A well-developed lagoon exists behind the winged spits of the Jebel Dhanna headland. This lagoon is characterized by pelletoidal gastropod sand in the interdital zone and muddy peletoidal gastropod sand in the central, infratidal part (Purser and Loreau 1973). Also extensive, synsedimentarily cemented hardgrounds exist in the shallow subtidal in the southern part of the bay. This lagoon, as well as Khor Odaid lagoon, is the location of aragonitic supratidal carbonate precipitation (see Chapter 3) (Fig. 4.31).

4.5.6   Islands

Islands in The Gulf are of variable origin. In the extreme northern Gulf, near the delta of the Shatt el Arab, they are depositional and consist largely of mud derived from the river. Also mainly depositional are the barrier islands of the Abu Dhabi nearshore region. These consist of an older core (usually of Pleistocene or older origin)

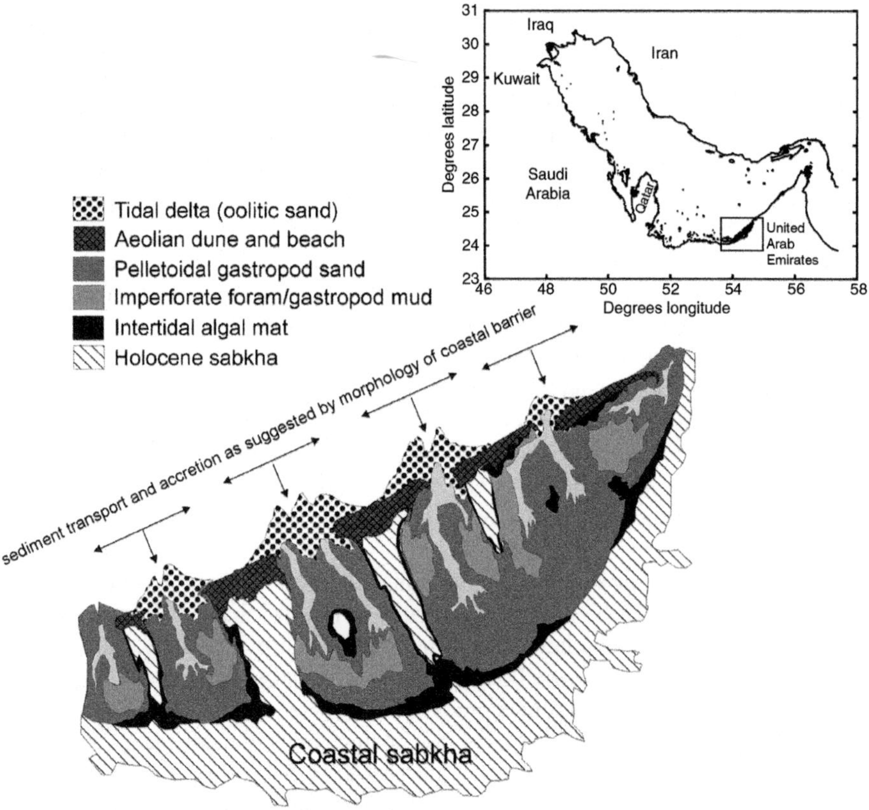

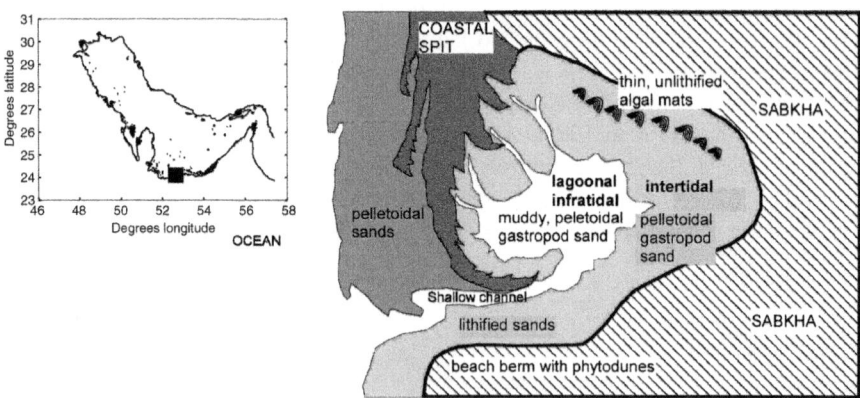

Fig. 4.30 Facies distribution of the Abu Dhabi coastal system. *Arrows* show dominant directions of water flow: wind-driven along-shore flow and, perpedicular, tide-driven flow in between the islands (From Purser and Evans 1973. Permission by Springer)

Fig. 4.31 The lagoon on the western side of the Jebel Dhannah headland (From Purser and Loreau 1973). The area is surrounded by polygonal sabkha, except on the beach berm, the coastal spit (which are composed of gastropod and bivalve sands) and the Tertiary ouitcrop that is Jebel Dhannah. The inner side of the coastal spit is the location of inter and supratidal aragonite precipitation (Permission by Springer)

and an accretionary tail of Holocene sands that accumulates in their lee (Section 4.5.5). These islands were fronted by reefs (or at least well-developed coral communities) until recently (some still are; in others the reefs were removed by dredge-and-fill projects) and have well-developed oolitic bars in the tidal deltas between adjacent islands (Purser and Evans 1973). The offshore islands in the southern Gulf, and most in the northern Gulf, are of structural origin. Bahrain is an anticline and the island chain off Iran is part of the Zagros fold belt (Ross et al. 1986). Most of the islands in the SE Gulf are salt diapers. Salt diapers cause continuous rise, and have created terraces of raised reef and shoreline deposits, for example at Kish, Qeshm, Larak and Hormoz (Pirazzoli et al. 2004; Bruthans et al. 2006).

4.6 Controls of Structure and Antecedent Topography on Facies Distribution

Lomando (1999) interpreted continuous facies distribution on the Arabian coast as controlled by strike of structural features, creating a trend of facies basinward from coastal lagoons and sabkhas to sand bodies to basin muds. Oolitic sand bodies extend along the headlands that are presumed to be located on fold axes. Facies between Ras Al-Qulay'ah and Ras Al-Zour trend parallel to the coast due to energetic constraints by the marine environment (Lomando 1999). These same facies are fragmented between Ras Al-Zour and Ras Bard Halq due to faults and their associated dips (Lomando 1999).

Topographical relicts from sea level highs and lows exert control in the submarine environment. During the last sea level lowstand, the Tigris-Euphrates river system cut valleys into the platform of The Gulf that subsequently drowned when sea level rose (Kassler 1973; Ross et al. 1986). Kassler (1973) identified six levels of submarine terraces off the Arabian coast speculated to have formed during sea level rises and are now located at depths of 120, 100, 66–80, 29–37, 18, and 9 m (Fig. 4.32). Drainage along the Arabian coast is directed to the northwest in the basin and into local depressions which over time has smoothed the topography (Kassler 1973; Ross et al. 1986).

4.7 Biotic Carbonate Sedimentation

The biotic component of carbonate deposition is strongly influenced by the ecology of organismal assemblages and their reaction to the environment. The Gulf is a system with extreme temperature and salinity fluctuations situated in a high-latitude area of marked attenuation of the tropical fauna.

In their original treatise of carbonate sediment-producing organisms in The Gulf, Hughes, Clarke and Keij (1973) differentiated three subdivisions of the marine environment with respect to vitality, and therefore diversity, abundance and sediment-production potential of calcareous biota:

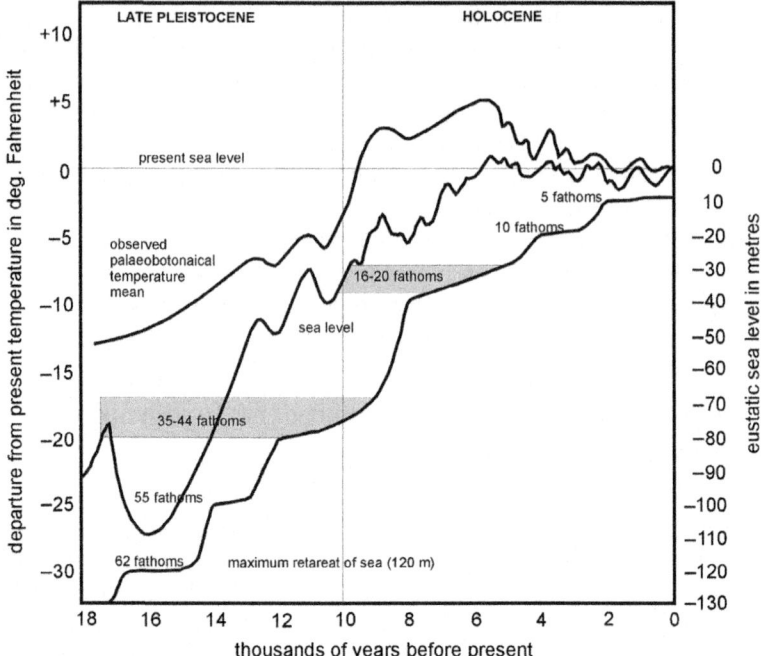

Fig. 4.32 Submarine platforms in The Gulf as response to sea level and climate history (From Kassler 1973. Permission by Springer)

1. The **normal marine environment** of salinities up to ~50‰ (Fig. 4.18) which is the most widely distributed environment, roughly comparable to the normal marine environment of the Indo-Pacific, however poorer in number of species. A further decline in species is observed upon entering areas with restricted flow and even higher salinities (such as The Gulf of Salwah, Fig. 4.3) where at about 45 ppm salinity some important fauna is lost (e.g. the perforate foraminifera *Operculina*, *Heterostegina*, *Amphistegina*; the gastropods *Strombus*, *Conus*, *Xenophora*, all pectinids and all echinoids except *Clypeaster*, as well as all extensive coral build-ups dominated by *Acropora* while *Porites* dominated, sparse assemblages remain).

2. The **restricted environment** of salinities between 50‰ and 70‰, which is found in many coastal areas and dominated by imperforate foraminifera and gastropods, in particular of the genus *Cerithium*. These organisms probably dominate because of widespread synsedimentary lithification that generates favourable substratum for macroalgal beds, favorite habitats for aforementioned foraminifera and gastropods.

3. The **highly restricted environment** of salinities exceeding 70‰, which is found only in some isolated lagoons, such as Khor al Odaid in Qatar and other lagoonal areas. These areas are largely devoid of fauna, except cyprideid ostracods.

A comparable, but less extreme, differentiation and gradation from normal marine to highly restricted is also observed in Belize, where conditions inside Chetumal Bay also become increasingly restricted, and some of the costal playa lakes in the Bahamas (e.g. on Long Island and Inagua).

4.7.1   Corals

4.7.1.1   Substratum for Coral Growth

Consolidated hard substratum favors the establishment of coral assemblages. In particular in the southern Gulf, marine chemistry favors the rapid precipitation of carbonate cements and this active, ongoing cementation of sandy seafloor creates abundant areas of hard substratum, often referred to as caprock since it often caps deeper sand reservoirs. These substrata can be flat or follow underlying topography, such as on the numerous offshore shoals, which are mostly due to salt diapirism (Fig. 4.33; Purser 1973a).

4.7.1.2   The Distribution of Corals in The Gulf

Records of coral growth in The Gulf are widely spread throughout the literature and few compilations regarding their specific occurrence in the entire region have been attempted except for Sheppard and Wells (1988) and Spalding et al. (2001) who

Fig. 4.33 The usual substratum for coral growth is synsedimentarily lithified hardground (Shinn 1969). In particular, ledges, such as this "overthrust teepee" (clearly visible in the background, see Section 4.4.3.2 for explanation). Corals preferentially recruit to such ledges, which can then serve as nucleus for the development of dense frameworks (Photo by Riegl)

give overview descriptions and rough site maps. Detailed records and maps can be found in Purser (1973a) for the southeastern Gulf (Abu Dhabi emirate and Qatar). Shinn (1976) provides a rough sketch of the distribution of *Acropora* communities in the southern Gulf. Downing (1985, 1988) describes coral community dynamics and distribution of reefs in Kuwait. Krupp et al. (1996) provide maps for the Jubail area and some of the offshore islands. Sheppard and Salm (1988) provide an overview for Musandam but sites are not specified, Sheppard and Sheppard (1991) provide descriptions of reefal areas mainly from Bahrain. Fadlallah et al. (1993) list sites in Qatar, Bahrain, Saudi Arabia and Kuwait and Fadlallah et al. (1995a, b) give details about the Tarut Bay reef. Shokri et al. (2000) list sites with coral growth and attempt a classification of type for Iran. George and John (1999, 2000a, b) and John and George (2003) describe coral growth from Abu Dhabi Emirate, Riegl (1999, 2001, 2002) and Riegl et al. (2001) and Purkis and Riegl (2005) and Purkis et al. (2005) provide detailed maps and sketches of the reefs in Dubai emirate. Lomando et al. (2003) map and discuss coral areas in Kuwait. Corals grow virtually throughout the entire Gulf (Fig. 4.34), with best development on offshore shoals but important fringing systems along the mainland shoreline (in particular UAE, Qatar, Saudi Arabia). Relatively few records exist of coral assemblages on the Iranian mainland coast (Shokri et al. 2000; Maghsoudlou et al. 2008), which can be due to runoff from the mountainous hinterland creating unfavorable conditions and insufficient

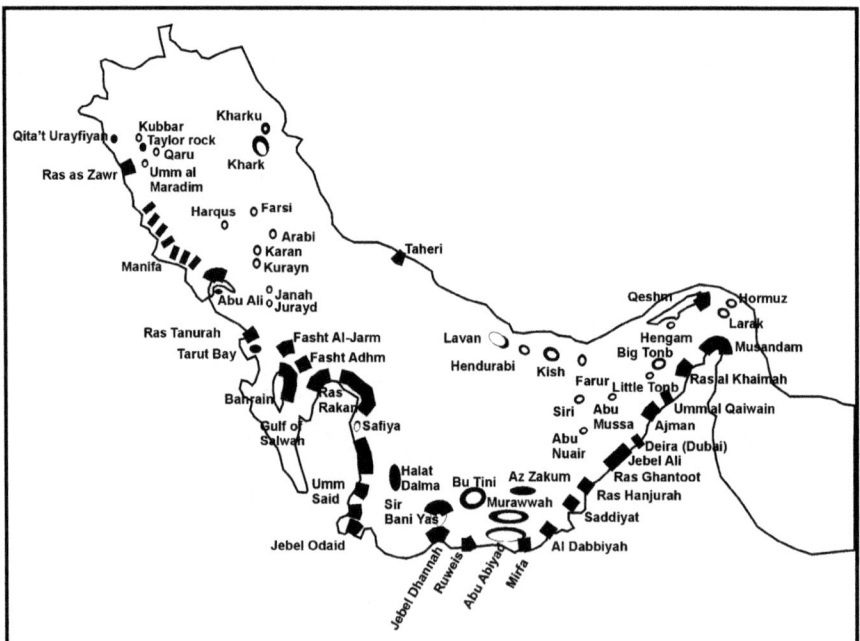

Fig. 4.34 The location of areas with extensive coral growth reported in the literature are indicated here by black lines

records in literature available in the West. Pleistocene (MIS 7 and 5e) reefs are known from several Iranian islands (Pirazzoli et al. 2004).

The total Gulf coral fauna is ~40 scleractinian (reef-building) species (Coles 2003) and 31 alcyonacean (soft coral) species (Samimi Namin and van Ofwegen 2009), but local diversity is usually lower. The richest local coral fauna is recorded from Saudi Arabian offshore islands (50 species, in Basson et al. 1977), which is maybe an over-estimation, Vogt 1996) and the UAE (34 scleractinian species, Riegl 1999). Iran should have a rich coral fauna, probably the richest in The Gulf, due to more benign oceanographic conditions. Local species richness in The Gulf is subject to temporal fluctuations caused by mass mortality events that preferentially affect the branching *Acropora* (Shinn 1976; Riegl 1999). Taxonomic composition of Gulf corals is typically Indo-Pacific. Two endemic *Acropora* species (*Acropora arabensis*, Hodgson and Carpenter 1995; *Acropora downingi*, Wallace 1999) and one endemic *Porites* (*P. harrisoni*, Veron 2000) are known. The closest faunistic proximity to other reefs of the Indo-Pacific is naturally to The Gulf of Oman and then the Red Sea (Sheppard and Sheppard 1991; Veron 2000) due to a shared paleo-oceanographic history of restriction during the last sea-level low stand and simultaneous flooding during the Holocene transgression (Sheppard and Sheppard 1991; Uchupi et al. 1996). While the Red Sea has marked endemism (18 species), The Gulf has only three endemic coral species (Hodgson and Carpenter 1995; Wallace 1999; Veron 2000). The coral fauna in the southern Gulf (Peninular Arabian coast) is characterized by rarity of alcyonacean soft corals and absence of reef building hydrozoa, which are common on other high latitude reefs in the Red Sea and the Indian Ocean.

4.7.1.3   Types of Frameworks

True coral reefs, consisting of in-situ, interconnected framework built by the corals themselves are small and relatively rare. The rocks of true coral reefs are solid and extend laterally over several tens to hundreds of meters, forming a clearly three-dimensional framework that rises steeply from the surrounding (usually sandy) seafloor and reaches the water surface. Such framework reefs (Fig. 4.35) can be subdivided into:

- *Fringing reefs*: such as developed around most offshore islands. These classify best as incipient, or at least very young, fringing reefs in comparison with other reef areas, such as the Great Barrier Reef (Hopley et al. 2007).
- *Stringer reefs*: elongated structures, often forming a maze of several parallel and inter-connecting stringers of several hundreds of meters (up to kilometers length, for example at Bu Tinah, Mubarraz).
- *Patch reefs*: small, usually round structures of sizes of meters to tens of meters. They are distinguished from blocks of other substratum by being made up entirely by coral skeletons.

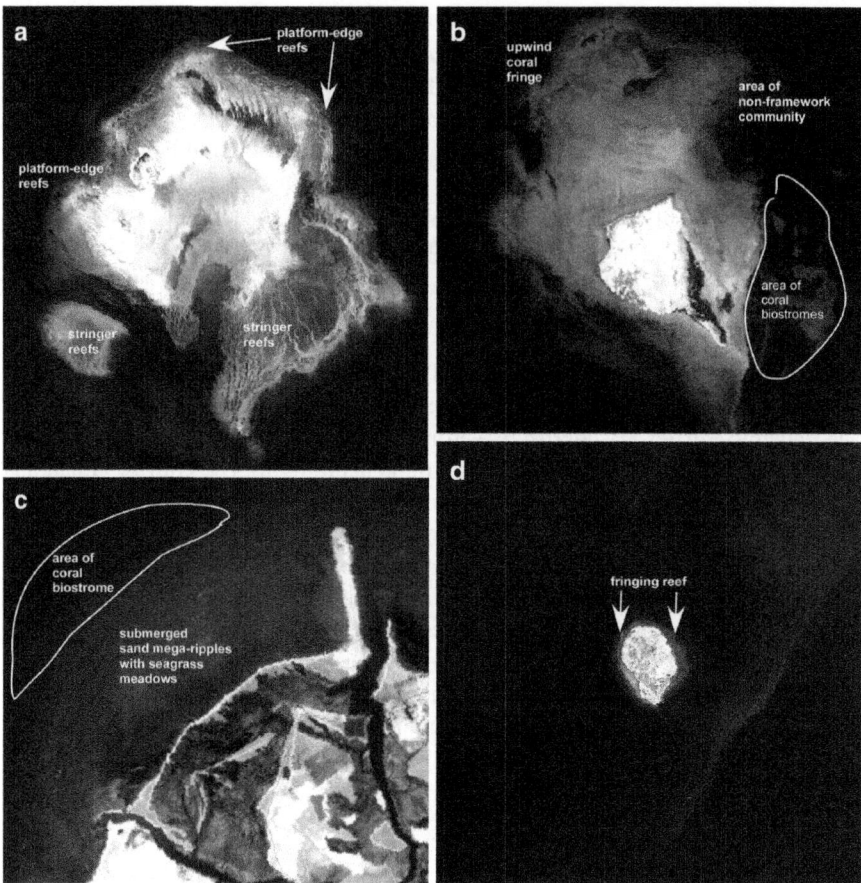

Fig. 4.35 Reef frameworks in The Gulf. (**a**) Bu Tinah is an emergent shoal with a sand island. .Stringer reefs and a thin upwind framework reef are developed. (**b**) Bazm-al Gharbi with thin upwind frameworks and an extensive semi-exposed coral framework. (**c**) The headland of Ras Ghanada in Abu Dhabi has one of the best-developed coral biostromes of the SE Gulf. (**d**) Zirku is also an offshore island (UAE) with narrow upwind coral fringe. Weakly-developed frameworks do not break the surface and do not create much topography. Previously better developed, their constituent corals suffered mass mortality in 1996 and most frameworks have since been bio-eroded

Coral carpets (or biostromes) consist of laterally thin, but continuous frameworks of coral that do not reach the water surface (Fig. 4.35). Coral carpets are extensive and do not develop into reefs because of (1) underlying flat topography, (2) disturbance history of the local coral assemblage.

Non-framework-building coral communities are sparse assemblages of few, widely-spaced corals that generally do not touch or interlock and therefore do not form a framework (i.e. coral rock). They are widely distributed throughout The Gulf in any suitable habitat between about 2 and 20 m depth. Generally, biodiversity is lower in these systems than in either coral carpet or coral reef.

4.7.2   *Foraminifera*

The Gulf has a relatively rich foraminiferan fauna of 98 species (Cherif et al. 1997). Foraminifera are either infaunal (in or on the sediment) such as *Operculina* and some *rotaliids,* or epifaunal (on fronds of weeds). Many of the important sediment producers such *peneroplids, milliolids,* and *Amphistegina* are epifaunal and later deposited in-situ or displaced by currents. This shows the importance of non-calcareous seaweed as a requirement for epifaunal development, and secondly, that the area of production of such epifauna may not coincide with the area of deposition.

Few planktonic foraminifera occur further north than the Straits of Hormuz, and due to the current pattern are mainly found along the Iranian side of The Gulf (Fig. 4.36; Seibold et al. 1973). Overall, more benthic than planktic form are found, with only *Globigerinoides* (*G. ruber, G. trilobus*) penetrating into The Gulf as far as Qatar, *Orbulina* (*O. universa*), *Globorotalia, Globigerina* (*G. calida*) and *Globigerinita* (*G. glutinata*) being restricted to waters near the Straits of Hormuz (Hughes Clarke and Keij 1973). The highest proportion of planktic foraminifera in the sediments is found in the deeper parts of the Central Basin, while the highest diversity in the overall foraminiferal assemblage was found by Cherif et al. (1997) near the Shatt el-Arab and in the Western Basin. The Central Trough and the near-shore areas have relatively low diversity.

Three regional assemblages exist (Fig. 4.36; Evans et al. 1973; Hughes Clarke and Keij 1973; Cherif et al. 1997).

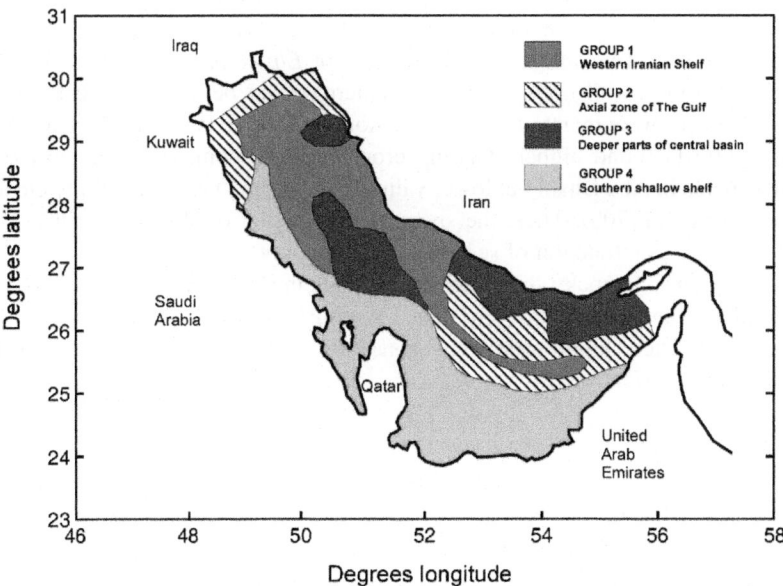

Fig. 4.36 The taxonomically based foraminiferal assemblages of Cherif et al. (1997) (Permission by Elsevier)

- The **imperforate assemblage** from the coastal zone and The Gulf of Salwah (equivalent with group 4 "**assemblage of the southern shallow shelf**" of Cherif et al. 1997). The further offshore, the more arenaceaous foraminifera occur so that in some areas (i.e. off the coast of the UAE) a **mixed imperforate-arenaceous assemblage** can be found.
- The **arenaceous assemblage**, from the Umm Shaif high, around the UAE offshore islands and the proximal parts of the Arabian homocline. With increasing depth and decreasing restriction this becomes a **mixed arenaceous-perforate assemblage** extending to a depth of about 55 m (equivalent to Cherif et al.'s (1997) "**assemblage of the axial zone**").
- The "**perforate assemblage**" occurs in the central basin (lamellibranch muds) and corresponds to Cherif et al.'s (1997) assemblage of "**the deeper parts of the Central Basin**".
- A "**foraminiferal association of the Western Iranian shallow shelf**" was observed by Cherif et al. (1997) in addition to the above three facies.

The distribution of larger benthic foraminifera on the Arabian homocline shows a clear gradation from a nearshore *Peneroplis/Sorites* assemblage, to an *Operculina/Pseudorotalia* assemblage in the distal reaches of the ramp and an *Ammonia* assemblage in the deeper areas (Fig. 4.36). A *Heterostegina/Amphistegina* assemblage is found around the offshore islands (Hughes Clarke and Keij 1973).

Foraminiferal live and dead assemblages can differ significantly and live assemblages dominated by infaunal taxa (*Ammonia, Elphidium, Brizalina*) can change to death assemblages composed largely of transported epiphytic taxa (Basson and Murray 1995). There is also high annual variability in the standing crop of foraminifera, ranging from 1–118 individuals per 10 cm^3. Over summer (starting July) standing crop in *Ammonia beccarii* doubled, in *Elphidium advenum* it increased four to five times, while in *Nonium* sp. the increase was between 30 and 100 times. These densities were maintained over a 7-month period (Basson and Murray 1995). Murray (1991) found higher standing crop values in comparable shallow water environments in Abu Dhabi but lower values (<1–47.cm^{-3}) in high-salinity environments (Murray 1970b). Thus, the species composition of above described facies may also vary as a function of season and taphonomy.

Locally, on some rocky shorelines for example in Iran or at off-shore highs, large foraminifers can account for more than 150 g/m^2/year sediment production (Lutze et al. 1971). These high-productivity zones are usually associated with red-algal oncoids (Fig. 4.37).

4.7.3 Mollusca

The molluscs are the most important sediment producers in The Gulf (Hughes Clarke and Keij 1973). The bivalves, due to their high population densities in subtidal and intertidal sand, contribute strongest to sediment-forming processes and

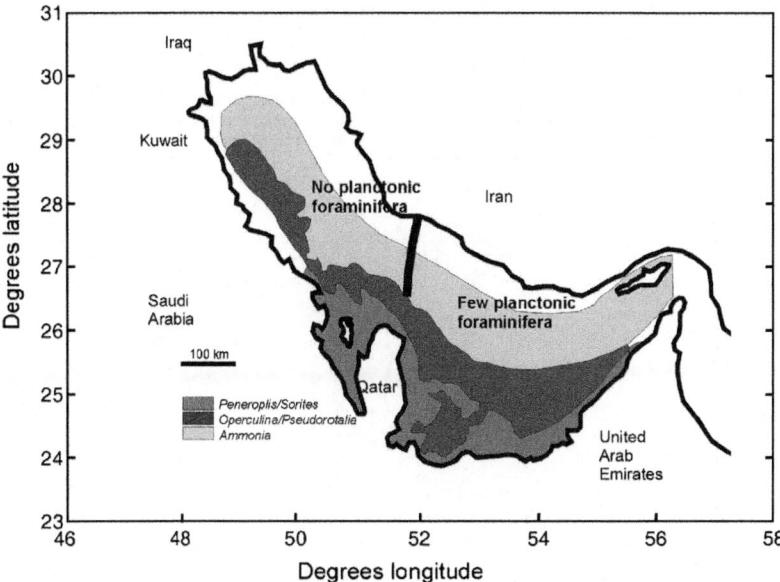

Fig. 4.37 Preferential distribution of foraminiferal genera in The Gulf (From Hughes Clarke and Keij 1973. Permission by Springer)

many of the shallow sands in The Gulf are "lamellibranch sands" (Fig. 4.38; Wagner and van der Togt 1973). The molluscan fauna of The Gulf is rich and varied and typically Indo-Pacific in character (Bosch et al. 1995). Gastropods are important both in the planktonic and the benthic realm. Pteropods and heteropods are sediment-forming in the deeper waters of The Gulf and towards the Iranian shore-line, where they form an important part of the sediments. There appear to be endemic forms of pteropods and heteropods in Gulf waters (Seibold et al. 1973). The proportion of planktonic versus benthic molluscs in the sediments decreases from the Straits of Hormuz towards the Shatt el-Arab (Seibold et al. 1973).

Oysterbeds are important to sedimentary processes since they can cover extensive areas. The main constituent fauna are *Pinctada radiata* oysters and beds can be dense and almost exclusively oyster-dominated, or can be mixed communities with abundant sponges, intermittent corals and mainly brown algae (*Hormophysa cuneiformis*). Sometimes they are intermixed with dense growth by the calcifying red algae *Jania rubens* and thus have high sediment-producing potential.

In seagrass meadows, particularly of *Halodule uninervis*, dense assemblages of the zebra oyster *Pterelectroma zebra* (a small bivalve up to 20 mm) occur. Fragments of their shells make up an important percentage of the sands in these seagrass beds.

The sand is populated by rich and diverse assemblages frequently dominated by carditids (in particular *Cardites bicolor*), cultellids (in particular *Siliqua polita*) and tellinids. In intertidal sands, mactrids are common. In muds, which are frequently hypoxic or dysoxic, specialized communities of lucinid bivalves (like *Bellucina*

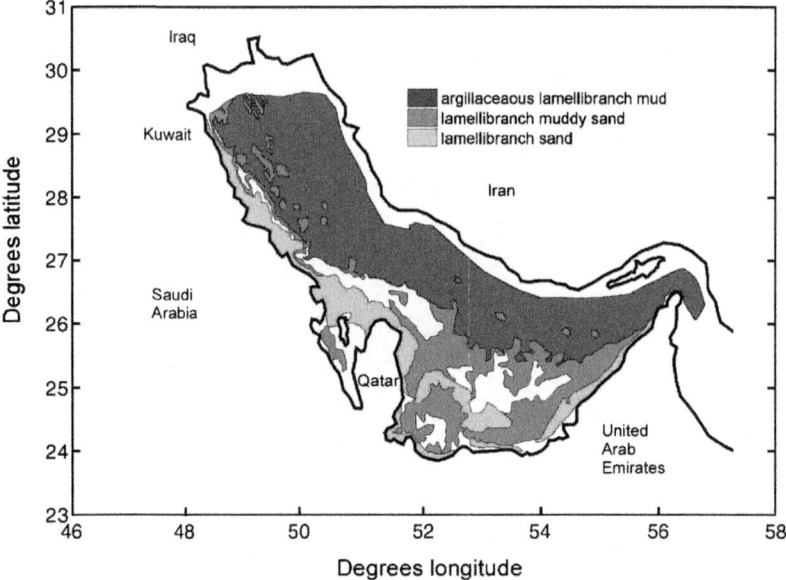

Fig. 4.38 Bivalves (lamellibranchs) are important sediment producers (Wagner and van der Togt 1973). White areas in southern Gulf are other sediment types (mainly foraminiferan sands and mixed sands) in northern and western Gulf unsurveyed areas. Coralgal sands overlap with lamellibranch sands and are not shown (Permission by Springer)

semperiana, one of the most common taxa in Saudi Arabia (Coles and McCain 1990) with chemosymbiontic bacteria that allow sulfate reduction, are found.

Infaunal bivalve assemblages in The Gulf vary with salinity, grain-size of the substratum (Coles and McCain 1990) and the availability of space, which is frequently limited by hardgrounds that are covered by only a thin layer of sand. Dominant taxa in coarse sediments are venerids, tellinids and cardiids. The sizes of the infauna depends on the availability of enough living space, bivalves in deep sand reservoirs are bigger than those living in shallow sand overlying hardgrounds. Particularly in shallow sediment bodies overlying hardgrounds, the small size of the infaunal bivalves and high mobility of the sediment with resultant abrasive action, leads to a rapid breakup of shells and rapid in situ production of sand.

Bivalves are also important as carbonate producers when they overgrow other substrates, like dead corals. Greiss and Riegl (2000) found that overgrowth of dead corals by first *Chama aspera* and the *Spondylus marisrubris* (among other, less important species) increased carbonate mass of the skeletons almost twofold. A clear succession in overgrowth was observed: within the first year, a dense cover by *Chama aspera* covered almost the entire skeleton. These original settlers then died in the second year and gave way to more encrustation by red algae, serpulids and maily *Spondylus marisrubris*. Hassan (1998) found a similar pattern of original bioaccretion, and only later bioerosion, in the Red Sea.

Gastropods are a common faunal component, however gastropod sands are overall far less important than lamellibranch (bivalve) sands or foraminiferal sands. Gastropods are among the few taxa that occur in high density in restricted and highly restricted marine environments (see Section 4.7; Hughes Clarke and Keij 1973) and cerithids are clear indicator species of such environments.

Many of the "bivalve sands" are in fact mixed bivalve/gastropod sands. Gastropods are usually epifaunal but many also use the upper sediment layers as refuge. Dominant taxa are Dialidae and Risoellidae (Coles and McCain 1990). On subtidal soft and hardgrounds locally dense accumulations of strombids (*Strombus persicus*) can occur, and dense patches of Cerithiidae are common. Cerithiids are also common in the intertidal and in some restricted lagoons where they share the environment with few species of large imperforate foraminifera.

Similar to the distribution of planktonic foraminifera, planktonic gastropods are mainly controlled by the inflowing Gulf surface current which concentrates them on the eastern side of The Gulf with a decreasing number of planktonic forms towards the northwest. A decline in pteropod fauna is observed at the entrance into The Gulf (Hughes Clarke and Keij 1973) but pteropods are common in basin center muds (in particular *Cavolinia longirostris*).

4.7.4 Echinoderms

Due to its shallow nature, The Gulf provides abundant habitat for echinoderms. Echinoids, asteroids, ophiuroids and holothuroids are common, although limited in species richness. The most common echinoids are *Echinometra mathai* and *Diadema setosum*, which occur mainly in areas with coral growth but also in oyster beds.

Echinoids in particular are important due to their activities as bioeroders, which makes them a major controlling factor for reef-framework production and its breakdown to fine sediment. Hughes Clarke and Keij (1973) quote Shinn (personal communication) as having measured a sand production rate of 0.5 g medium grained sand per echinoid per day when browsing on dead *Acropora* coral skeletons.

4.7.5 Plants

4.7.5.1 Calcareous Algae

The Gulf lacks many of the algal sediment producers that are found in the Atlantic and Indo-Pacific (like *Halimeda*, *Penicillus*, etc.). Red algae can occur as encrusters primarily associated with other calcareous substrata, such as reef corals, or they can occur in more or less dense growths by themselves. In both instances, they have carbonate production potential.

In The Gulf, the articulated geniculate red alga *Jania rubens* occurs frequently together with brown algae (*Hormophysa cuneiformis* and *Sargassum* spp.).

These assemblages are found on hardgrounds. Red algae can also be associated with coral reefs where they are important as encrusters of dead substrata. Red algae can enhance the settlement of larvae of other calcareous organisms, such as corals. They are also common in the subtidal of rocky shores where conditions are not suitable for corals. On offshore high-energy shoals, they can form oncoids. *Acanthophora spicifera* is an important, non-calcifying red algae species that can be a dominant alga in pearl-oyster (*Pinctada radiata*) beds and sandy areas, where it colonizes, and thus stabilizes large rubble.

Filamentous red algae like *Spyridia* associated with large brown algae cover shallow subtidal rocks (Basson et al. 1977). Filamentous red algae flourish in salinities of up to 100 ppt, while most other algae are not reported in water of salinities above 50 ppt. There is no carbonate production potential associated to these assemblages. The only calcareous algae among the rhodophyta are *Peysonellia simulans* and four corallinacean genera (*Hydrolithon farinosum, Jania rubens, Lithophyllum kotschyanum, Sporolithum molle*), among the chlorophyta, *Acetabularia calyculus*. *L. kotschyanum* is common, particularly on reefs, and is a major producer of carbonate sediments.

Both *L. kotshyanum* and *S. molle* occasionally form rhodoliths. However, not with the same frequency as rhodolith formation is observed in the Bahamas or in Belize. Hughes Clarke and Keij (1973) report characteristic algal oncoids from offshore shoals.

4.7.5.2 Other Green and Brown Algae

Green algae do not play an as active role in sediment production in The Gulf as in the Bahamas or Belize. However, green and brown algal weeds are widespread throughout the shallow water environment and host important epifaunal biota (Hughes Clarke and Keij 1973). Brown algal beds are common and widely distributed. They occur both in the shallow subtidal, frequently forming dense cover on rocky reefs in less than 2 m depth, or can also form dense assemblages on deeper subtidal areas (up to 10 m depth), particularly on hardgrounds. Dominant species include *Hormophysa cuneiformis* and several species of *Sargassum* and *Cystoseira*. Stands can be dense, monospecific or in association with other species. While many algae show a marked seasonality and are less common in summer, *H. cuneiformis* remains present year-round. Its thick, triangular fronds form the substratum of locally dense calcareous assemblages, consisting of red algae, serpulid polychaetes, small bivalves and even coral recruits. Subtidal *Hormophysa* assemblages frequently are in association with the calcareous red algae *Jania rubens* and therefore have carbonate production potential.

On shallow subtidal rocky reefs, dense assemblages of several species of *Sargassum* (*S. binderi, S. boveanum, S. angustifolium, S. decurrens, S. latifolium*) and *Cystoseira* (*C. myrica, C. trinodis*) can be found (De Clerck and Coppejans 1996). These assemblages provide shelter to a number of calcareous fauna but they have no carbonate production potential. *Sargassum* sometimes actively competes for space with reef-building corals and can therefore have deleterious effects on carbonate production. Particularly when corals die, due to temperature (and other) stresses, the area can become overgrown by dense *Sargassum* beds, which then makes re-settlement by coral larvae difficult to impossible.

Green algae with carbonate production potential are *Acetabularia calyculus*, which can occur in similarly dense growths as in the Caribbean. While *Avrainvillea* has a definite carbonate production potential in the Caribbean, The Gulf *Avrainvillea amadelpha* has soft fronds and does not precipitate calcium-carbonate. *Halimeda* species do not occur in The Gulf (De Clerck and Coppejans 1996). Hughes Clarke and Keij (1973) state that *Acetabularia* is characteristic for sediments in highly saline (>50 ppm) areas. However, the algae themselves can be common in normal marine conditions as well.

4.7.5.3 Blue-Green Algae

Although they do not produce a carbonate skeleton, blue green algae can aid in the formation of carbonate hardgrounds and the precipitation of carbonates and evaporites. Blue-green algae in form of algal mats are a key feature of inter- to supratidal environments in The Gulf. Their binding effect is important, as is their role as barriers to interstitial water circulation. Based on morphology, Kendall and Skipwith (1968) defined four algal mat zones in the supra/intertidal area, from the high water mark to the lagoon water: flat zone, the crinkle zone, the polygonal zone and the cinder zone (Fig. 4.28). These zones are defined on the morphology of the algal mat that, in turn, is controlled by wave energy and salinity of the water as well as frequency and duration of subaerial exposure. There is an increase of salinity of the superficial water and ground water as well as an increase in desiccation (due to prolonged exposure) from the lagoon towards the high water mark.

To develop algal mats, a combination of low water energy, a maximum amount of sunlight associated with moisture and contact with water are needed. Algal mats can withstand high salinities and high temperatures and do not need hard substrates to grow on.

The relative absence of tidal creeks and channels (Kirkham 1998) has contributed to the preservation of relatively continuous microbial mats, especially in the southern Arabian Gulf, by avoiding erosion due to channel migration. Thickly developed microbial mats are primarily found in highly protected, saline, intertidal environments. They do not occur along exposed peritidal flats. No Pleistocene microbial mats have been found (Kirkham 1998).

Modern stromatolitic environments are known from the United Arab Emirates (Kendall and Skipwith 1968; Alsharhan and Kendall 1994; Kirkham 1998) and from Kuwait (Duane and Al-Zamel 1999). In this tidal creek stromatolitic setting, the algal mat area is protected from direct wave impact by 1–3 km wide consolidated oolitic beach ridges. Stromatolites are being formed by *Microcoleus chthonoplastes*, *Lyngbya estuarii*, *Lyngbya limnetica*, and *Oscillatoria* sp. Six stages of stromatolite growth and burial were proposed which provide an alternate explanation (to p.ex. Hoffmann 1969) to the origin of domal stromatolites (1) growth of flat-topped stromatolites with laminated growth until a bioherm is formed (2) spalling of the bioherm during weathering (3) development of fractures which are invaded by crinkle stromatolites which build microcolumns in a vertical framework (4) erosion and exposure of microcolumns (5) flat-topped stromatolites grow and regeneration of columnar bioherm structure (6) drowning by risen sea-level and preservation of fossilized bioherm.

4.7.5.4 Seagrass

Seagrass meadows are an important component of carbonate sedimentary systems both in Atlantic and Indo-Pacific due to calcareous epifauna and – flora on their leaves and the sediment-stabilizing effect of their rhizome mats. The Gulf has a relatively poor development of seagrass meadows. Four species occur: *Halodule uninervis*, which can form dense meadows, *Halophila stipulacea*, *Halophila ovalis* and *Syringodium isoetifolium*, which is the rarest. Basson et al. (1977) estimated total productivity of seagrass blades in The Gulf to be about 100 g/cm²/year and the annual energy equivalent of seagrass beds to be 1.4×100^{10} kCal/year, which is equivalent to 95,000 barrels of oil (Rao and Al-Yamani 2000). Primarily in *Halodule uninervis* meadows, calcareous epiphytes (like the pelecypod *Pterelectroma zebra*) have carbonate production potential.

4.7.5.5 Mangrove

Biologically important throughout the subtropics, the mangal eco- and sedimentary system is also widely distributed in The Gulf. The sedimentary importance is by providing attachment space for a rich and typical calcareous fauna in an otherwise usually muddy environment without footholds and the production of mangrove peat, which readily fossilizes. Holocene mangrove paleosols are reported by Plaziat (1995) from Abu Dhabi (Umm al Nar). Fossil magrove roots (rhizoliths) are region-ally known from the Miocene eastern Arabia (Whybrow and McClure 1981) and the Oligocene of Egypt (Bown 1982).

The mangrove flora of The Gulf is depauperate with *Avicennia marina* as the only species. Typical calcareous fauna indicating these environments are the oyster *Saccostrea cucullata* growing on the pneumatophores (breathing knees of the roots) *Telescopium telescopium* and *Terbralia palustris* gastropods.

4.8 Regional Comparison of Biogenic Carbonate Sedimentation

The distribution of calcareous faunal elements on The Gulf carbonate ramp shows clear differences to those observed on the Bahamas oceanic rimmed shelf and the attached rimmed shelf of Belize. This is partly a result of different morphology (ramp versus shelf), but also due to The Gulf's more northernly situation and more extreme environment.

In the Caribbean rimmed shelves most coral frameworks are concentrated on the platform edges, where they form barrier or shelf-edge reefs. On the platform interior, numerous patch-reefs and, in the case of Belize, rhomboid shoals are formed. In any case, reef development preferentially takes place away from the shoreline of the mainland, with a marked preference for the distal areas of the shallow platform. In The Gulf, however, much framework development takes place in the most proximal reaches of the ramp, very close to the mainland shore where extensive coast-fringing coral frameworks are developed. In the more distal reaches of the Arabian homocline, coral frameworks are concentrated around topographic highs, where corals grow best on windward sides. This is different on the top of the platform in Belize and the Bahamas, where well-developed back-reef and coral frameworks also develop in the lee of the barrier reef, the shelf-edge reef or the islands. However, also in the Bahamas, the lee of Grand Bahama Bank is largely devoid of reefs, a situation that is mirrored in much smaller scale on The Gulf offshore banks.

The Gulf coral fauna, with about 40 species is poorer than that of the Caribbean (about 60 species; Chiappone et al. 1996). In all three areas (Belize, Bahamas and Gulf), the scleractinian genus *Acropora* is the most important framebuilder and dominates in the shallowest areas. Massive faviids usually occur in slightly deeper water below the zone of *Acropora* dominance. In the Caribbean, two species of *Acropora* dominate the framebuilding process – the palmate *Acropora palmata*, which forms large, primarily upward oriented, interlocking, rigid frameworks (Geister 1983; i.e. forming positive topography above the surrounding sedimentary strata) and the open arborescent (i.e. with sprawling, thin branches) *Acropora cervicornis*, which tends to form non-rigid frameworks (Geister 1983; i.e. dense layers of branch fragments filled in with sand that can either stand above or be at same level with the surrounding sediment). In The Gulf, only the tabular *Acropora clathrata/A. downingi* group of species forms rigid frameworks. No open arborescent species that can be compared to the Caribbean *A. cervivornis* either in aspect or in function are found. In all three study areas, recent framework production has been severely hampered by mass mortalities that killed large parts of the framebuilder populations. In the Caribbean, a disease pandemic killed most *A. cervicornis* and brought the production of non-rigid frameworks by this species largely to a halt. The standing dead skeletons then rapidly broke down and were covered by the surrounding sediment. At least in Belize, the production of non-rigid frameworks was then taken over by *Agaricia tenuifolis* (Aronson and Precht 1997). Also the main producer of rigid frameworks, *Acropora palmata*, has been the victim of extensive die-back throughout the Caribbean. Therefore, also the construction of rigid frameworks is greatly decreased or has completely ceased in many parts of the Caribbean. However, due to the sturdiness of the skeletons, and strong binding by red algae, these frameworks remain standing and erosion is slow. No other species have taken the place of *A. palmata*. In The Gulf, the cause of *Acropora* mass mortality is temperature anomalies, either negative or positive (Coles 1988;

Coles and Fadlallah 1991; Fadlallah et al. 1995a, b; Riegl 2001, 2003). They lead repeatedly to a removal of most or all *Acropora* in the system, which then leads to slow breakdown of the framework. The framework is maintained when new corals, in particular *Acropora*, recruit into the same area.

The most important non-*Acropora* framebuilders in all three areas belong to the genus *Porites*. In The Gulf, *Porites harrisoni* forms coherent, rigid frameworks. Especially in areas of more extreme salinities and temperature variations, these take the place of the *Acropora* frameworks. *Porites* frameworks are usually smaller than *Acropora* frameworks. The same is true in the Caribbean, where the *Porites porites* group of species (*P. porites, P. furcata, P divaricata*) can form rigid (*P. porites*) or non-rigid frameworks (all three species). In some instances, *Porites* non-rigid frameworks can replace *A. cervicornis*.

Overall, framebuilding in The Gulf happens on a much smaller scale than in Belize or the Bahamas. Frameworks in The Gulf are thin, if present at all, while they are widespread in Belize and the Bahamas. This difference can be attributed to the harsher climate and the observed and postulated (Riegl 2001; Riegl and Purkis 2009) repeated mass mortality events that interrupt the reef-building process. The Bahama Banks are also subject to temperature variability caused by North American cold-air outbreaks (Roberts et al. 1992) or El-Nino influenced extreme heating (Lang et al. 1988), that make this area unsuitable for the development of extensive coral framework. Coral reefs are largely absent from the platform interior (with the exception of small patch reefs) and are concentrated around tidal passes where they are under the influence of open oceanic water. Even the shelf-edge reefs suffer from unusually heated or cooled platform water, which is delivered in the form of density currents, and can cause mortality of deeper reef biota that would normally be outside the reach of extreme temperature oscillations (Lang et al. 1988; Wilson and Roberts 1992).

Due to the lower wave-energies in The Gulf, the coral frameworks (if developed) show more characteristics of the moderate energy frameworks (Bosence 1985). The coral frameworks in Belize and the Bahamas all show gradation from high-energy frameworks on the shelf-edge to moderate- to low-energy frameworks on the platform in the lee of the barrier reef or the islands. Following the classification of Wilson (1974), the modern Gulf can serve as a facies model for a sea of moderate to low energy, while Belize and the Bahamas have all the characteristics of high-energy seas, at least along the margins.

With regard to the reef associated flora, the important Caribbean sediment-producing calcareous green algal genus *Halimeda* is absent in The Gulf. Also calcareous red algae have less taxonomic diversity in The Gulf than the Caribbean. While in the Caribbean red algae are being considered important and efficient binders, either of rubble or reef framework (Blanchon et al. 1997; Perry 1999) the same is not true in The Gulf, where red algae encrust and bind dead reef frameworks, but cannot prevent their break-down and do not appear to efficiently cement rubble (Riegl 2001; Rasser and Riegl 2002).

Many echinoderms are reef-associated, such as the echinoids (sea urchins), which fulfil in both oceans an important role as bioeroders. In Belize and the Bahamas, the previously common long-spined urchin *Diadema antillarum* died during a mass mortality event in the early 1980s (Lessios 1988). The absence of this species, as one of the key herbivores on the reefs, appears to have facilitated an increasing algal dominance on Caribbean coral reefs, to the detriment of carbonate production by corals. This phenomenon is absent from The Gulf, where *Diadema setosum* and other urchins are still common and play a very active role in the breakdown of dead coral frameworks as well as the grazing of macro-algae.

The distribution of molluscan and foraminiferal facies grades from coarser to finer along The Gulf homocline. Parallels can be found in the distribution of these facies from more open to the most restricted conditions between The Gulf and, for example, Chetumal Bay in Belize.

4.9   Summary

1. The Gulf consists of a southern carbonate domain and a northern mixed domain, strongly influenced by fluviatile, siliciclastic input.
2. It is an area of active carbonate deposition, by planktic and benthic organisms, and as ooids, whitings, aragonitic encrustations and subtidal hardground formation.
3. Non-carbonate input is due to aeolian and fluviatile transport.
4. Carbonate sediments vary from argillaceous peleypod or foraminiferan muds in the distal part of the Arabian homocline to skeletal sand in the proximal parts and around topographic highs.
5. According to its pattern of facies distribution, at least the southern side of The Gulf can indeed be treated as a carbonate ramp setting and thus used in comparison with fossil ramp settings.
6. Extensive tidal flats with evaporite formation (sabkhas) are a distinctive feature.
7. Barrier islands and lagoons are formed primarily in the southeastern Gulf.
8. Strong structural control of facies belts along dip and along strike of tectonic structures as well as by antecedent morphology is observed.
9. Wave energy, driven largely by wind, is the main factor determining facies patterns in areas of comparable bathymetry.
10. Wind, tides and salinity changes due to evaporation are the main driving factors of ocean currents, which also influence the distribution of facies belts.
11. Tides vary from diurnal to semi-diurnal.
12. Ocean salinity varies from >100 ppt (Gulf of Salwah, Qatar and Abu Dhabi lagoons) to around 40 ppt near the Strait of Hormuz.
13. The most significant freshwater input is in the northwest.

A wide variety of sedimentary components compose sandy sediments. Mollusks, foraminifers, ooids, and peloids are typically dominant, but a certain percentage of quartz is always admixed, which reflects the importance of the Mesopotamian fluviodeltaic and the Arabian continental eolian depositional systems (Walkden and

Williams 1998). Because of the shallow water depth of large parts of The Gulf, an extensive area of the sea-floor is in the photic zone. This results in high productivity of carbonate secreting organisms in all depositional sub-environments.

Hydrology and composition of the coastal sabkhas (intertidal flats composed mainly of fine, windblown material within which evaporites form) is a direct result of the ramp morphology and the regional climate. Most important is the creation of a vertical hydraulic gradient that for the most part is directed upward by capillary evaporation. This vertical water movement is responsible for sufficient Mg transport for pene-contemporaneous dolomitization, and the formation of evaporites.

In the basinal areas, the ramp morphology, the antecedent topography of the basin, and especially the tidal and wind-induced currents exert a major control on the distribution of facies belts. Despite the relatively smooth gradient of the gently dipping sea floor, facies belts are not arranged in a parallel fashion because, (1) the northwestern winds influence the orientation of facies on the homocline, (2) off-shore barriers (many of these highs are determined by salt diapirism) and reefs modify energy flux, and (3) slight changes in inclination of the homocline result in large changes of facies associations. Antecedent topography is created by recent sea-level changes, tectonic movements and locally extensive salt tectonics (Purser 1973a; Pirazzoli et al. 2004; Bruthans et al. 2006). Ancient river valleys were created during sea-level lowstands. Submerged beach ridges and shorelines were formed during the Holocene sea-level rise. These bathymetric highs range from tens of meters to kilometers in diameter, are generally elongated and, at least in the northern and western parts of The Gulf, trend along The Gulf's axis. Close to the shoreline, reefs can be found on some of these highs, while foraminiferal sands accumulate on offshore highs.

Wind energy is an important factor in The Gulf's oceanic and sedimentary regimes. One of the most important winds is the Shamal (northerly winds blowing from the Iranian highlands into The Gulf) besides a very active landbreeze-sea-breeze regime. The winds directly influence the distribution of sediment. A key process is wave-induced long-shore transport of sands. In addition, the strong winds generate high waves and deepen the wave base razor. Wave induced long-shore currents deposit arrays of extensive carbonate cheniers, bars and spits. The Shamal and the seabreezes are also responsible for eolian transport either of air-borne particles or as dunes along the coast.

The currents in The Gulf are controlled by tides, wind, and seawater density. The kinetic energy of the water velocity associated with the three current driving mechanisms can be partitioned at approximately 100, 10, and 1. Wind and tidal driven currents affect sediment transport direction, and have formed series of long-shore drifts consisting of carbonates and a considerable amount of fluviatile derived clastic material in the northern Gulf.

High seasonal temperature variation, high salinity, high turbidity, and nutrient poor water characterize The Gulf aquatic environment. Salinity is a major control on the biota. Corals have partly adjusted to this high variability and can tolerate higher salinity and temperature changes than their Indo-Pacific or Caribbean counterparts. Areas above 50 ppt salinity are generally no longer able to sustain corals, calcareous algae, echinoderms or most of the foraminifera. In the high salinity waters, changes of pCO_2 can trigger whitings. In addition, a direct correlation exists between salinity and carbonate mineralogy. In areas with fresh water input, carbonate mud consists mostly of Mg-calcite, while in higher saline areas the mud is mostly aragonite. This relationship is also observed in the Belize lagoon.

The Gulf exhibits virtually all components of a typical (sub)tropical carbonate to mixed carbonate/siliciclastic sedimentary system, allows a relatively clear delineation of the controlling factors, and has thereby served to greatly advance understanding of carbonate depositional processes.

References

Abuzinada AH, Krupp F (1994) The status of coastal and marine habitats two years after the Gulf War oil spill. Cour Forsch Inst Senckenberg 166, p 80

Alsharhan AS, CGStC K (1994) Depositional setting of the Upper Jurassic Hith Anhydrite of the Arabian Gulf: an analog to Holocene evaporates of the United Arab Emirates and Lake McLeod of Western Australia. AAPG Bull 78(7):1075–1096

Alsharhan AS, CGStC K (2003) Holocene coastal carbonates and evaporites of the southern Arabian Gulf and their ancient analogues. Earth Sci Rev 61:191–243

Alsharhan AS, Nairn AEM (1997) Sedimentary basins and petroleum geology of the Middle East. Elsevier, The Netherlands, 843

Aronson RB, Precht WF (1997) Stasis, biological disturbance, and community structure of a Holocene coral reef. Palaeobiology 23(3):326–346

Azam MH, Elshorbagy W, Ichikawa T, Terasawa T, Taguchi K (2006) 3D model application to study residual flow in the Arabian Gulf. J. Watrway, Port. Coas Ocean Eng Sept/Oct 2006:388–400

Barth H-J (2001) Characteristics of the wind regime north of Jubail, Saudi Arabia, based on high resolution wind data. J Arid Environ 47:387–402

Basson PW, Murray JW (1995) Temporal variations in four species of intertidal foraminifera, Bahrain, Persian Gulf. Micropaleontology 41(1):69–76

Basson P, Burchard JH, Hardy JT, Price A (1977) Biotopes of the western Arabian Gulf. Aramco Department of Loss Prevention and Environmental Affairs, Dharhan, Saudi Arabia, p 284

Behairy AKA, El-Sayed MK, Durgaprasda Rao NVN (1985) Eolian dust in the coastal area north of Jeddah, Saudi Arabia. J Arid Environ 8:89–98

Biggs HEJ (1973) The marine mollusca of the Trucial coast, Persian Gulf. Bull Brit Mus Nat Hist Zoology 24:343–421

Blanchon P, Jones B, Kalbfleisch W (1997) Anatomy of a fringing reef around Grand Cayman: storm rubble, not coral framework. J Sed Res 67:1–16

Borgensen F (1939) Marine algae from the Iranian Gulf. In: Jessen K, Sparck R (eds) Danish scientific investigations in Iran I. Copenhagen, Einar Munksgaard

Bosch DT, Dance SP, Moolenbeek RG, Oliver PG (1995) Seashells of eastern Arabia. Motivate Publishing, Dubai, p 296

Bosence DWJ (1985) Preservation of coralline algal frameworks. Proceedings of the 5th International Coral Reef Congress, Tahiti, vol 2, pp 39–45

Bown TM (1982) Ichnofossils and rhizoliths of the nearshore fluviatil Jebel Qatrani formation (Oligocene) Fayum province, Egypt. Palaeogeogr, Palaeoclim, Palaeoecol 40:255–309

Brewer PG, Dyrssen D (1985) Chemical oceanography of the Persian Gulf. Prog Oceanogr 14:41–55

Bruthans J, Filippi M, Gersl M, Zare M, Melkova J, Pazdur A, Bosak P (2006) Holocene marine terraces on two salt diapers in the Persian Gulf, Iran: age, depositional history and uplift rates. J Quat Sci 21:843–857

Burchard JE (1979) Coral fauna of the western Persian Gulf. ARAMCO, Dahran, Saudi Arabia, p 129

Burchette TP, Wright VP (1992) Carbonate ramp depositional systems. Sed Geol 79:3–57

Butler GP (1969) Modern evaporite deposition and geochemistry of coexisting brines, the sabkha, Trucial Coast, Arabian Gulf. J Sed Petrol 39:70–89

Burt J, Bartholomew A, Usseglio P (2008) Recovery of corals a decade after a bleaching event in Dubai, United Arab Emirates. Mar Biol 154:27–36

Cherif OH, Al-Ghabdan A-N, Al-Rifaihy A (1997) Distribution of foraminifera in the Persian Gulf. Micropaleontology 43(3):253–280

Chiappone M, Sullivan KM, Lott C (1996) Hermatypic scleractinian corals of the southeastern Bahamas: a comparison to western Atlantic reef systems. Caribb J Sci 32:1–13

Coles SL (1988) Limitations of reef coral development in the Persian Gulf: temperature or algal competition? Proceedings of the 6th International Coral Coral Reef Symposium 3, pp 211–216

Coles SL (2003) Coral species diversity and environmental factors in the Arabian Gulf and the Gulf of Oman: a comparison to the Indo-Pacific region. Atoll Res Bull 507:1–19

Coles SL, McCain JC (1990) Environmental factors affecting benthic infaunal communities of the Western Arabian Gulf. Mar Environ Res 29:289–315

Coles SL, Fadlallah YH (1991) Reef coral survival and mortality at low temperature in the Arabian Gulf: new species-specific lower temperature limits. Coral Reefs 9:231–237

De Clerck O, Coppejans E (1996) Marine algae of the Jubail marine wildlife sanctuary, Saudi Arabia. In: Krupp F, Abuzinada AH, Nader IA (eds) A marine wildlife sanctuary for the Arabian Gulf. Environmental research and conservation following the 1991 Gulf War oil spill. NCWCD, Riyadh and Senckenberg Research Institute, Frankfurt a.M., pp 199–286

Defant A (1961) Physical oceanography, vol 2. Pergamon, London, 598

Diester-Haas L (1973) Holocene climate in the Persian Gulf as deduced from grain-size and pteropod distribution. Mar Geol 14:207–223

Downing N (1985) Coral reef communities in an extreme environment: The Northwest Arabian Gulf. Proceedings of the 5th International Coral Reef Congress, Tahiti, vol 6, pp 343–348

Downing N (1988) The coral reefs and coral islands of Kuwait. Proc ROPME Workshop on Coastal Area Development. UNEP Regional Seas Reports and Studies No 90, ROPME Publ No GC-5/006

Downing N, Roberts C (1993) Has the Gulf War affected coral reefs of the northwestern Gulf? Mar Poll Bull 27:149–156

Duane MJ, Al-Zamel AZ (1999) Syngenetic textural evolution of modern sabkha stromatolites (Kuwait). Sediment Geol 127:237–245

Emery KO (1956) Sediments and water of the Persian Gulf. Bull AAPG 40:2354–2383

Endlicher SL, Diesing CM (1845) Enumeratio algarum, quas ad oram Karek, sinus Persici, legit Theodorus Kotschy. Botanische Zeitung 3:268–269

Enos P (1983) Shelf Environment. In: Scholle PA, Bebout DG, Moore CH (eds) Carbonate depositional environments. AAPG Memoir 33:267–295

Evamy BD (1973) The precipitation of aragonite and its alteration to calcite on the Trucial Coast of the Persian Gulf. In: Purse BH (ed) The Persian Gulf: Holocene carbonate sedimentation and diagenesis in a shallow epicontinental sea. Springer, Berlin, pp 329–342

Evans G (1966) The recent sedimentary facies of the Persian Gulf region. Phil Trans R Soc Lond, Ser A 259(1099):291–298

Evans G (1970) Coastal and nearshore sedimentation: a comparison of clastic and carbonate deposition. Proc Geol Ass Lond 81:493–508

Evans G (1995) The Arabian Gulf: a modern carbonate-evaporitic factory; a review. Cuad Geol Iber 19:61–96

Evans G, Shearman DJ (1964) Recent celestite from sediments of the Trucial coast of the Persian Gulf. Nature 202(4930):385–386

Evans G, Kendall CG, Skipwith PA (1964) Origin of the coastal flats, the sabkha, of the Trucial coast, Persian Gulf. Nature 202(4934):579–600

Evans G, Schmidt V, Bush P, Nelson H (1969) Stratigraphy and geologic history of the sabkha, Abu Dhabi, Persian Gulf. Sedimentology 12:145–159

Evans G, Murray JW, Biggs HEJ, Bate R, Bush PR (1973) The oceanography, ecology, sedimentology and geomorphology of parts of the Trucial Coast barrier island complex. In: Purser BH (ed) The Persian Gulf: Holocene carbonate sedimentation and diagenesis in a shallow epicontinental sea. Springer, New York, pp 233–278

Fadlallah YH (1996) Synchronous spawning of *Acropora clathrata* coral colonies from the Western Arabian Gulf (Saudi Arabia). Bull Mar Sci 59:209–216

Fadlallah YH, Lindo RH, Lennon DJ (1992) Annual synchronous spawning event in *Acropora* sp. from the Persian Gulf. Proceedings of the7th International Coral Reef Symposium, Abstracts: 29

Fadlallah YH, Eakin CM, Allen KW, Estudillo RA, Rahim SA, Reaka-Kudla M, Earle SA (1993) Reef coral distribution and reproduction, community structure and reef health (Qatar, Bahrain, Saudi Arabia, Kuwait): Results of Mt. Mitchell Cruise, May 1992. Final Rep Workshop Sci Workshop Results RV Mt Mitchell Cruise in ROPME Sea Area, Kuwait. Annex III, pp 1–27

Fadlallah YH, Allen KW, Estudillo RA (1995a) Damage to shallow reef corals in the Gulf caused by periodic exposures to air during extreme low tides and low water temperatures (Tarut Bay, Eastern Saudi Arabia). In: Ginsburg RN (ed) Global aspects of coral reefs: health, hazards and history, Rosenstiel School of Marine and Atmospheric Science, University of Miami, pp 371–377

Fadlallah YH, Allen KW, Estudillo RA (1995b) Mortality of shallow reef corals in the western Persian Gulf following aerial exposure in winter. Coral Reefs 14:99–107

Fairbridge RW (1957) The dolomite question. SEPM Spec Pub 5:125–178

Fischer P (1891) Liste des coquilles recueillis par F. Houssay dans le Golfe Persique. J Conch Paris 31:220–30

Galt JA, Payton DL, Torgrimson GM, Watabayashi G (1983) Applications of trajectory analysis for the Nowruz oil spill. In: El-Sabh MI (ed) Oceanographic modeling of the Kuwait Action Plan (KAP) region. UNESCO Rep Mar Sci 28, Paris

Geister J (1983) Holozaene westindische Korallenriffe: Geomorphologie, Oekologie und Fazies. Facies 9:173–284

George JD, John DM (1999) High sea temperatures along the coast of Abu Dhabi (UAE), Persian Gulf – their impact upon corals and macroalgae. Reef Encounter 25:21–23

George JD, John DM (2000a) The effects of recent prolonged high seawater temperatures on the coral reefs of Abu Dhabi (UAE). In proceedings of international symposium on Extent of coral bleaching, pp 28–29

George JD, John DM (2000b) The coral reefs of Abu Dhabi, United Arab Emirates: past, present and future. In proceedings of the 2nd Arab international conference and exhibition on Environmental biotechnology (Coastal Habitats), Abu Dhabi, p 33

George JD, John DM (2002) Is it curtains for coral reefs in the southern Arabian Gulf? Abstracts volume: international society for reef studies European meeting, Cambridge, Sept 2002, p 36

Gischler E, Lomando AJ (2005) Offshore sedimentary facies of a modern carbonate ramp, Kuwait, northwestern Arabian-Persian Gulf. Facies 50:443–462

Glennie KW (1996) Geology of Abu Dhabi. In: Osborne PE (ed) Desert ecology of Abu Dhabi, Pisces Publications, Newbury, UK, pp 16–35

Grasshoff K (1976) Review of hydrographical and productivity conditions in the Arabian Gulf region. Report of consultation meeting on Marine science research for Gulf area. UNESCO technical papers in marine science, Paris, vol 26, pp 39–62

Greiss M, Riegl B (2000) Taphonomic alteration of coral skeletons following a mass mortality in the Persian Gulf. Proceedings of the 9th international coral reef symposium, Bali, p 254

Haas F (1952) Shells collected by the Peabody expedition to the Near East I. Molluscs from the Persian Gulf. Nautilus 65:114–118

Halley RB, Harris PM, Hine AC (1983) Bank margin. In: Scholle PA, Bebout DG, Moore CH (eds) Carbonate depositional environments. AAPG Memoir 33:463–506

Hartmann M, Lange H, Seibold E, Walger E (1971) Oberflaechensedimente im Persischen Golf und Golf von Oman. I. Geologisch-hydrologischer Rahmen und erste sedimentologische Ergebnisse. "Meteor" Forschungsergebnisse C, No. 4, Stuttgart, Berlin, pp 1–76

Hassan M (1998) Modification of carbonate substrata by bioerosion and bioaccretion on coral reefs of the Red Sea. Shaker Verlag, Aachen, p 126

Hodgson G, Carpenter K (1995) Scleractinian corals of Kuwait with description of a new species. Pac Sci 49:227–246

Hoffmann HJ (1969) Attributes of stromatolies. Geol Surv Can Pap 69(39):1–58

Hopley D, Smithers SG, Parnell KE (2007) The geomorphology of the Great Barrier Reef: development, diversity, and change. Cambridge, p 520

Houbolt JJHC (1957) Surface sediments of the Persian Gulf near the Qatar peninsula. Doctoral thesis, University of Utrecht, Den Haag, Mouton and Co

Hsü KJ, Schneider JF (1973) Progress report on dolomitization, Abu Dhabi Sabkas, Arabian Gulf. In: Purser BH (ed) The Persian Gulf: Holocene carbonate sedimentation and diagenesis in a shallow continental sea. Springer, New York, pp 409–422

Hsü KJ, Siegenthaler C (1969) Preliminary experiments on hydrodynamic movement induced by evaporation and their bearing on the dolomite problem. Sedimentology 12:11–15

Hughes P, Hunter J (1979) A proposal for a physical oceanography program and numerical modeling of the KAP region. Project for KAP 2/2, UNESCO, Paris

Hughes Clarke MW, Keij AJ (1973) Organisms as producers of carbonate sediment and indicators of environment in the southern Persian Gulf. In: Purser BH (ed) The Persian Gulf: Holocene carbonate sedimentation and diagenesis in a shallow epicontinental sea. Springer, Berlin, Heidelberg, pp 33–56

Hunter JR (1983a) Aspects of the dynamics of the residual circulation of the Arabian Gulf. In: Gade MG, Edward A, Svenson H (eds) Coastal Oceanography. Plenum, New York, pp 31–42

Hunter JR (1983b) A review of the residual circulation and mixing processes in the KAP region with reference to applicable modelling techniques. Symposium on Oceanographic modelling of the Kuwait action plan region, Dhahran, Saudi Arabia

Illing LV, Wells AJ, Taylor JMC (1965) Penecontemporaneous dolomite in the Persian Gulf. In: Pray LC, Murray RC (eds) Dolomitization and limestone diagenesis: S.E.P.M. Special Publication 13:89–113

John DM, George JD (2003) Coral death and seasonal seawater temperature regime: their influence on the marine algae of Abu Dhabi (UAE) in the Arabian Gulf. In: Chapman RO, Anderson RJ, Vreeland VJ, Davison IR (eds) 17th int Seaweed symposium, Cape Town (2001), pp 341–348

John VC (1992a) Circulation and mixing processes and their effect on pollutant distribution in the western Arabian Gulf. Appl Ocean Res 14:59–64

John VC (1992b) Harmonic tidal current constituents of the western Arabian Gulf from moored current measurements. Coast Eng 17:145–151

John VC, Coles SL, Abozed AI (1990) Seasonal cycles of temperature, salinity and water masses of the western Arabian Gulf. Oceanologica Acta 13(3):273–281

John VC, Kruss PK, Fadlallah YH (1991) Oceanographic monitoring program of the western Arabian Gulf. Mar Technol Soc J 25(2):22–28

Johns WE, Yao F, Olson DB (2003) Observations of seasonal exchange through the Straits of Hormuz and the inferred heat and freshwater budgets of the Persian Gulf. J Geophys Res 108 C12, 3391, doi: 10.1029/2003JC001881, 21-1-21-18

Jones DA (1986) A field guide to the seashores of Kuwait and the Arabian Gulf. University of Kuwait and Blanford press, Poole

Kassler P (1973) The structural and geomorphic evolution of the Persian Gulf. In: Purser BH (ed) The Persian Gulf: Holocene carbonate sedimentation and diagenesis in a shallow epicontinental sea. Springer, New York, pp 11–32

Kendall CG, St C, Skipwith PA (1968) Recent algal mats of a Persian Gulf lagoon. J Sed Pet 38:1040–1058

Kendall C.G.St.C., Sir Patrick A. d'E Skipwith, Bt. (1969) Geomorphology of a Recent shallow-water carbonate province: Khor al Bazam, Trucial Coast, South-West Persian Gulf. Geol Soc Amer Bull 80:865–891

Kendall, C.G.St.C., Lakshmi V, Althausen J, Alsharhan AS (2003) Changes in microclimate tracked by the evolving vegetation cover of the Holocene beach ridges of the United Arab Emirates. In: Sharhan AS, Wood WW, Goudie AS, Fowler A, Abdellatif EM (eds) The desertification in the third millennium. Swets & Zeitlinger Publishers (Balkema), Lisse, The Netherlands, pp 91–98. ISBN: 90 5809 5711

Kinsman DJJ (1964a) The recent carbonate sediments near Halat el Bahrani, Trucial Coast, Persian Gulf. Deltaic and shallow marine deposits. Developments in Sedimentology, Elsevier, vol 1, pp 185–192

Kinsman DJJ (1964b) Reef coral tolerance of high temperatures and salinities. Nature 202:1280–1282

Kinsman DJJ, Holland HD (1969) The co-precipitation of cations with CaCO3 IV. The co-precipitation of Sr2+ with aragonite between 16°and 96°. Geochim Cosmochim Acta 33:1–17

Kirkham A (1998) A Quaternary proximal foreland ramp and its continental fringe, Arabian Gulf, UAE. In: Wright VP, Burchette TP (eds) Carbonate ramps. Geo Soc Spec Publ 149:15–42

Koske P (1972) Hydrographische Verhältnisse im Persischen Golf. Beobachtungen von F.S. Meteor in Frühjahr 1965. Meteor Forsch Ergnbn Gebrüder Borntraeger, Berlin, pp 58–73

Krupp F, Abuzinada AH, Nader IA (1996) A marine wildlife sanctuary for the Persian Gulf. Environmental research and conservation following the 1991 Gulf war oil spill. NCWCD Riyadh and Senckenbergische Naturforschende Gesellschaft, Frankfurt a.M., p 511

Kukal Z, Saadallah A (1973) Aeolian admixtures in the sediments of the northern Persian Gulf. In: Purser BH (ed) The Persian Gulf: Holocene carbonate sedimentation and diagenesis in a shallow epicontinental sea. Springer, New York, pp 115–122

Lang JC, Wicklund RI, Dill RF (1988) Depth- and habitat-related bleaching of zooxanthellate reef organisms near Lee Stocking Island, Exuma Cays, Bahamas. Proceedings of the 6th international coral reef symposium, pp 269–274

Lardner RW, Al-Rabeh AH, Gunay N, Hossain M, Reynolds RM, Lehr WJ (1993) Computation of the residual flow in the Gulf using the Mt. Mitchell data and the KFUPM/RI hydrodynamical models. Mar Poll Bull 27:61–70

Lees GM (1948) The physical geography of southeast Arabia. Geogr J 121:441

Lees GM, Falcon NL (1952) The geographical history of the Mesopotamian plains. Geogr J 118:24–39

Lehr WJ (1984) A brief survey of oceanographic modeling and oil spill studies in the KAP region. UNESCO Reports in Marine Sciences 28:4–11

Lessios HA (1988) Mass mortality of *Diadema antillarum* in the Caribbean: what have we learned. Ann Rev Ecol Syst 19:371–393

Lomando AJ (1999) Structural influences on facies trends of carbonate inner ramp systems, examples from the Kuwait-Saudi Arabian coast of the Persian Gulf and northern Yucatan, Mexico. GeoArabia 4(3)

Lomando AJ, Gischler E, Al-Hazeem SH, Ameen M, Sajer AA, Al-Doheim A, Al-Wadi M (2003) Field seminar to the Holocene carbonate ramp of southern Kuwait. ChevronTexaco Geo2002 Seminar Notes

Loreau JP, Purser BH (1973) Distribution and ultrastructure of Holocene ooids in the Persian Gulf. In: Purser BH (ed) The Persian Gulf: Holocene carbonate sedimentation and diagenesis in a shallow epicontinental sea. Springer, New York, pp 279–328

Lutze GF, Grabert B, Seibold E (1971) Lebendbeobachtungen an Grosz-Foraminiferen (Heterostegina) aus dem Persischen Golf. "Meteor" Forsch. Ergebnisse, Reihe C, 6:21–40

Maghsoudlou A, Araghi PE, Wilson S, Taylor O, Medio D (2008) Status of coral reefs in the ROPME sea area (The Persian Gulf, Gulf of Oman and Arabian Sea. In: Wilkinson C (ed) Status of coral reefs of the world: 2008. Global Coral Reef Monitoring Network, Townsville, pp 79–90

Martens E von (1874) Ueber vorderasiatische Conchylien nach den Sammlungen des Prof. Haussknecht. Cassel, p 129

Matthews CP, Samuel M, Al-Attar MH (1979) The oceanography of Kuwait waters: some effects of fish population and on the environment. Annual Research Report 1979, Kuwait Institute for Scientific Research, Kuwait

McKenzie JA, Hsü KJ, Schneider J (1980) Movement of subsurface waters under the sabkha, Abu Dhabi, UAE, and its relation to evaporative dolomite genesis. In: Zenger DH, Dunham JB, Ethington RL (eds) Concepts and models of dolomitization. Soc Econ Paleontol Mineral Spec Publ 28:11–30

Melguen M (1973) Correspondence analysis for recognition of facies in homogeneous sediments off an Iranian river mouth. In: Purser BH (ed) The Persian Gulf: Holocene carbonate sedimentation and diagenesis in a shallow epicontinental sea. Springer, New York, pp 99–114

Melvill JC (1897) Description of thirty-four species of marine mollusca from the Arabian Sea, Persian Gulf, and Gulf of Oman (mostly collected by F.W. Townsend Esq.). Mem Proc Manchr Lit Phil Soc 41 (7):1–26

Melvill JC (1898) Further investigations into the molluscan fauna of the Arabian Sea, Persian Gulf and Gulf of Oman, with a description of fourty species (Mostly dredged by F.W. Townsend Esq.). Mem Proc Manchr Lit Phil Soc 42 (4):1–36

Melvill JC (1899) Notes on the Mollusca of the Arabian Sea, Persian Gulf and Gulf of Oman, mostly dredged by Mr. F.W. Townsend, with descriptions of twentyseven species. Ann Mag Nat Hist (Series 7) 4:81–101

Melvill JC (1904) Descriptions of twenty-three species of gastropoda from the Persian Gulf, Gulf of Oman and Arabian Sea, dredged by Mr. F.W. Townsend of the Indo-European Telegraph Service in 1903. Proceedings of Malacological Society of London, vol 6, pp 51–60

Melvill JC (1917) A revision of the Turridae (Pleurotomidae) occurring in the Persian Gulf, Gulf of Oman and North Arabian Sea as evidenced mostly through the results of dredging carried out by Mr. F.W. Townsend, 1893–1914. Proceedings of Malacological Society of London, vol 12, pp 140–201

Melvill JC (1928) The marine mollusca of the Persian Gulf, Gulf of Oman and North Arabian Sea as evidenced mainly through the collections of Captain F.W. Townsend, 1893–1914. – Addenda, corrigenda and emanenda. Proceedings of Malacological Society of London, vol 18, pp 93–117

Murray JW (1965a) The foraminiferida of the Persian Gulf. Part 1. Rosalina adhaerens sp. Nov. Ann Mag Nat Hist 8(Ser 13):77–79

Murray JW (1965b) The foraminiferida of the Persian Gulf. 2. The Abu Dhabi region. Paeogeogr Palaeoclim Palaeoecol 1:307–332

Murray JW (1966a) The foraminiferida of the Persian Gulf. 3. The Halat al Bahrani region. Paeogeogr Palaeoclim Palaeoecol 2:59–68

Murray JW (1966b) The foraminiferida of the Persian Gulf. 4. Khor al Bazam. Paeogeogr, Palaeoclim, Palaeoecol 2:153–169

Murray JW (1966c) The foraminiferida of the Persian Gulf. 5. The shelf of the Trucial Coast. Paeogeogr Palaeoclim Palaeoecol 2:267–278

Murray JW (1970a) The foraminiferida of the Persian Gulf. 6. Living forams in the Abu Dhabi region. J Nat Hist 4:55–67

Murray JW (1970b) The foraminifera of the hypersaline Abu Dhabi lagoon, Persian Gulf. Lethaia 3:51–68

Murray JW (1991) Ecology and palaeoecology of benthic foraminifera. Longman, Harlow, p 397

Murty TS, El-Sabh MI (1984) Storm tracks, storm surges and sea state in the Arabian Gulf, Strait of Hormuz and the Gulf of Oman. UNESCO Report Mar Sci 28:12–24

Newton L (1955a) The marine algae of Kuwait. In: Dickson V (ed) The wild flowers of Kuwait and Bahrain. Allan & Unwin, London, pp 100–102

Newton L (1955b) The marine algae of Bahrain. In: Dickson V (ed) The wild flowers of Kuwait and Bahrain. Allan & Unwin, London, pp 141–144

Patterson RJ, Kinsman DJJ (1977) Marine and continental sources in a Persian Gulf sabkha. In: Frost SH, Weiss MP, Saunders JB (eds) Reefs and related carbonates: ecology and sedimentaology. Am Assoc Pet Geol Stud Geol 4:381–397

Patterson RJ, Kinsman DJJ (1981) Hydrologic framework of a sabkha along the Arabian Gulf. Bull Am Assoc Petrol Geol 65:1457–1475

Perrone TJ (1981) Winter Shamal in the Persian Gulf, naval environmental prediction research facility, Monterrey, California. Technical Report I.R. 79-06, August, 1979

Perry CT (1999) Reef framework preservation in four contrasting modern reef environments, Discovery Bay, Jamaica. J Coastal Res 15:796–812

Pirazzoli PA, Reyss J-L, Funtugne M, Haghipour A, Hilgers A, Kasper HU, Nazari H, Preusser F, Radtke U (2004) Quaternary coral-reef terraces from Kish and Qeshm Islands, Perisan Gulf: new radiometric ages and tectonic implications. Quat Int 120:15–27

Plaziat J-C (1995) Modern and fossil mangroves and mangals: their climatic and biogeographic variability. In: Bosence DWJ, Allison PA (eds) Marine palaeoenvironmental analysis from fossils. Geol Soc Spec Pub 83:73–96

Purkis SJ, Riegl B (2005) Spatial and temporal dynamics of Arabian Gulf coral assemblages quantified from remote-sensing and in situ monitoring data (Jebel Ali, Dubai, U.A.E.). Mar Ecol Progr Ser 287:99–113

Purkis SJ, Riegl B, Andréfouët S (2005) Remote sensing of geomorphology and facies patterns on a modern carbonate ramp (Arabian Gulf U.A.E.) J Sedim Res 75:861–876

Purser BH (1973a) The Persian Gulf: Holocene carbonate sedimentation and diagenesis in a shallow epicontinental sea. Springer, New York, p 471

Purser BH (1973b) Sedimentation around bathymetric highs in the Southern Persian Gulf. In: Purser BH (ed) The Persian Gulf: Holocene carbonate sedimentation and diagenesis in a shallow epicontinental sea. Springer, New York, pp 157–178

Purser BH, Evans G (1973) Regional Sedimentation along the Trucial Coast, SE Persian Gulf. In: Purser BH (ed) The Persian Gulf: Holocene carbonate sedimentation and diagenesis in a shallow epicontinental sea. Springer, New York, pp 211–232

Purser BH, Lorreau J-P (1973) Aragonitic, supratidal encrustations of the Trucial Coast, Persian Gulf. In: Purser BH (ed) The Persian Gulf: Holocene carbonate sedimentation and diagenesis in a shallow epicontinental sea. Springer, New York, pp 343–376

Purser BH, Seibold E (1973) The principle environmental factors influencing Holocene sedimentation and diagenesis in the Persian Gulf. In: Purser BH (ed) The Persian Gulf: Holocene carbonate sedimentation and diagenesis in a shallow epicontinental sea. Springer, New York, pp 1–10

Rao DV, Al-Yamani F (2000) The Arabian Gulf. In: Sheppard CRC (ed) Seas at the Millenium: An environmental evaluation. Chapter 53. Elsevier, pp 1–16

Rasser MW, Riegl B (2002) Holocene coral reef rubble and its binding agents. Coral Reefs 21:57–72

Read JF (1985) Carbonate platform facies models. Bull AAPG 69:1–21

Reynolds RM (1993) Physical oceanography of the Gulf, Strait of Hormuz, and the Gulf of Oman – Results from the Mt Mitchell expedition. Mar Poll Bull 27:35–59

Riegl B (1999) Corals in a non-reef setting in the southern Persian Gulf (Dubai, UAE): fauna and community structure in response to recurring mass mortality. Coral Reefs 18:63–73

Riegl B (2001) Inhibition of reef framework by frequent disturbance: examples from the Persian Gulf, South Africa, and the Cayman Islands. Palaeogeogr, Palaeoclim, Palaeoecol 175(1–1): 79–102

Riegl B (2002) Effects of the 1996 and 1998 sea surface temperature anomalies on corals, coral diseases, and fish in the Arabian Gulf (Dubai, UAE). Mar Biol 140:29–40

Riegl B (2003) Global climate change and coral reefs: different effects in two high latitude settings (Arabian Gulf, South Africa). Coral Reefs 22:433–446

Riegl B, Korrubel JL, Martin C (2001) Mapping and monitoring of coral communities and their spatial patterns using a surface-based video method from a vessel. Bull Mar Sci 69(2):869–880

Riegl B, Purkis SJ (2009) Model of coral population response to accelerated bleaching and mass mortality in a changing climate. Eco Mod 220:192–208

Roberts HH, Wilson PA, Lugo-Fernandez A (1992) Biologic and geologic responses to physical processes: examples from modern reef systems of the Caribbean-Atlantic region. Cont Shelf Res 12:809–834

ROPME (1987) Dust fallout in the northern part of the ROPME sea area. Regional Organization for the Protection of the Marine Environment, Kuwait. GC-5/005

ROPME (1993) Final report of the scientific workshop on results of the R/V Mt. Mitchell cruise in the ROPME sea area. Kuwait, 24–28 January 1993

Ross DA, Uchupi E, White RS (1986) The geology of the Persian-Gulf-Gulf of Oman region: a synthesis. Rev Geophys 24(3):537–556

Samimi Namin K, van Ofwegen LD (2009) Some shallow water octocorals (Coelenterata: Anthozoa) of the Persian Gulf. Zootaxa 2058:1–52

Sanford WE, Wood WW (2001) Hydrology of the coastal sabkhas of Abu Dhabi, United Arab Emirates. Hydrogeo J 9:358–366

Sarnthein M (1970) Sedimentologische Merkmale für die Untergrenze der Wellenwirkung im Persischen Golf. Geol Rundsch 59(2):649–666

Sarnthein M, Walger E (1973) Classification of modern marl sediments in the Persian Gulf by factor analysis. In: Purser BH (ed) The Persian Gulf: Holocene carbonate sedimentation and diagenesis in a shallow epicontinental sea. Springer, New York, pp 81–98

Schlager W (2005) Carbonate sedimentaology and sequence stratigraphy. SEPM Concepts in Sedimentology and Paleontology 8, p 200

Schneider JF (1975) Recent tidal deposits, Abu Dhabi, United Arab Emirates, Arabian Gulf. In: Ginsburg RN (ed) Tidal deposits, a casebook of recent examples and fossil counterparts. New York, Springer, pp 209–214

Seibold E (1973) Sedimentation in the Persian Gulf. J Mar Biol Ass India 15(2):621–624

Seibold E, Vollbrecht K (1969) Die Bodengestalt des Persischen Golfs. "Meteor" Forschungs-Ergebnisse, Reihe C, No. 2:29–56

Seibold E, Diester L, Fütterer D, Lange H, Muller P, Werner F (1973) Holocene sediments and sedimentary processes in the Iranian part of the Persian Gulf. In: Purser BH (ed) The Persian Gulf: Holocene carbonate sedimentation and diagenesis in a shallow epicontinental sea. Springer, New York, pp 57–80

Shearman DJ (1963) Recent anhydrite, gypsum, dolomite and halite from the coastal flats of the Arabian shore of the Persian Gulf. Proc Geol Soc Lond 1607:63

Sheppard CRC (1993) Physical environment of the Gulf relevant to marine pollution: an overview. Mar Poll Bull 27:3–8

Sheppard CRC, Salm RV (1988) Reef and coral communities of Oman, with a description of a new coral species (Order Scleractinia, genus *Acanthastrea*). J Nat Hist 22:263–279

Sheppard CRC, Sheppard ALS (1991) Corals and coral communities of Arabia. Fauna of Saudi Arabia 12, p 170

Sheppard CRC, Wells SM (1988) Coral reefs of the world. Volume 2: Indian Ocean, Red Sea and Gulf. UNEP/IUCN, Cambridge, p 389

Sheppard CRC, Price ARG, Roberts CM (1992) Marine ecology of the Arabian region: patterns and processes in extreme tropical environments. Academic Press, London, p 359

Shinn EA (1969) Submarine lithification of Holocene carbonate sediments in the Persian Gulf. Sedimentology 12:109–144

Shinn EA (1973) Carbonate coastal accretion in an areas of longshore transport, NE Qatar, Persian Gulf. In: Purser BH (ed) The Persian Gulf: Holocene carbonate sedimentation and diagenesis in a shallow epicontinental sea. Springer, New York, pp 199–210

Shinn EA (1976) Coral reef recovery in Florida and the Persian Gulf. Environ Geol 1:241–254

Shinn EA (1983) Tidal Flat. In: Scholle PA, Bebout DG, Moore CH (eds) Carbonate depositional environments. AAPG Memoir, Tulsa, pp 345–440

Shokri MR, Haeri-Ardakani O, Abdollahi P (2000) Coral reef resources of the Islamic Republic of Iran. Reef Encounter 27:26–29

Smith EA (1872) Remarks on several species of Bullidae, with descriptions of some hitherto undescribed forms, and of a new species of *Planaxis*. Ann Mag Nat Hist 9(Series 4):344–355

Spalding MD, Ravilious C, Green EP (2001) World atlas of coral reefs. University of California Press, p 424

Swift SA, Bower AS (2003) Formation and circulation of dense water in the Persian/Arabian Gulf. J Geophys Res 108C1, doi: 10.1029/2002JC001360, 4-1-4-21

Taylor JCM, Illing LV (1969) Holocene intertidal calcium carbonate cementation, Qatar, Persian Gulf. Sedimentology 12:69–107

Tucker ME (1996) Sedimentary rocks in the field. Wiley, Chichester, p 153

Tucker ME, Wright VP (1990) Carbonate sedimentology. Blackwell, Oxford, p 482

Uchupi E, Swift SA, Ross DA (1996) Gas venting and late Quarternary sedimentation in the Persian (Arabian) Gulf. Mar Geol 129:237–269

Uchupi E, Swift SA, Ross DA (1999) Later Quaternary stratigraphy, paleoclimate and neotectonism of the Persian (Arabian) Gulf region. Mar Geol 160:1–23

UNESCO (1976) Marine sciences in the Gulf area. UNESCO Technical papers in marine science, vol 26, p 66

UNESCO (1984) Oceanographic modelling of the Kuwait Action Plan (KAP) region. UNESCO reports in marine science, vol 28, p 79

U.S. Navy Hydrographic Office (1960) Summary of Oceanographic conditions in the Indian Ocean. SP-53, Oceanographic Analysis Division Marine Sciences Division, p 144

Vasconcelos C, McKenzie JA (1997) Microbial mediation of modern dolomite precipitation and diagenesis under anoxic conditions (Lago Vermelha, Rio de Janeiro, Brazil). J Sediment Res 67:378–390

Vasconcelos C, McKenzie JA, Bernasconi S, Grujic D, Tien AJ (1995) Microbial mediation as a possible mechanism for natural dolomite formation at low temperatures. Nature 377:220–222

Veron JEN (2000) Corals of the world, 3 volumes. Austral. Inst. Mar. Sci, p. 463, 429, 490

Vogt H (1996) Investigations on coral reefs in the Jubail wildlife sanctuary using under water video recordings and digital image analysis. In: Krupp F, Abuzinada AH, Nader IA (eds) A marine wildlife sanctuary for the Persian Gulf. NCWCD Riyadh and Senckenbergische Naturforschende Gesellschaft, Frankfurt a.M., pp 302–327

Wagner CW, van der Togt C (1973) Holocene sediment types and their distribution in the Southern Persian Gulf. In: Purser BH (ed) The Persian Gulf: Holocene carbonate sedimentation and diagenesis in a shallow epicontinental sea. Springer, New York, pp 123–156

Walkden G, Williams A (1998) Carbonate ramps and the Pleistocene-Recent depositional systems of the Arabian Gulf. In: Wright VP, Burchette TP (eds) Carbonate ramps. Geol Soc (Lond) Spec Pub 149:43–53

Wallace CC (1999) Staghorn corals of the world. A revision of the coral genus Acropora (Scleractinia; Astrocoeniina; Acroporida) worldwide, with emphasis on morphology, phylogeny and biogeography. CSIRO Publishing, p 421

Wells AJ (1962) Recent dolomite in the Persian Gulf. Nature 194(4825):274–275

Whybrow PJ, McClure HA (1981) Fossil mangrove roots and palaeoenvironments of the Miocene of the eastern Arabian peninsula. Palaeogeogr, Palaeoclimatol, Palaeoecol 32:213–225

Wilson AT (1925) The delta of the Shatt-al-Arab, and proposals for dredging the bar. Geogr J 65:225–239

Wilson JL (1974) Characteristics of carbonate platform margins. Bull Am Ass Petrol Geol 58:810–824

Wilson JL (1975) Carbonate facies in geologic history. Springer, Berlin, p 266

Wilson JL, Jordan C (1983) Middle shelf environment. In: Scholle PA, Bebout DG, Moore CH (eds) Carbonate depositional environments. AAPG Memoir 33:297–345

Wilson PA, Roberts HH (1992) Carbonate-periplatform sedimentation by density flows: a mechanism for rapid off-bank and vertical transport of shallow-water fines. Geology 20:713–716

Wood WW, Sanford WE, Al Habshi AR (2002) Source of solutes to the coastal sabkha of Abu Dhabi. GSA Bulletin 114(3):259–268

Chapter 5
Summary: The Depositional Systems of the Bahamas, Belize Lagoon and The Gulf Compared

Gregor P. Eberli and Hildegard Westphal

5.1 Modern Analogues as Key to the Past

It is well-known that the present might not be a straight-forward key to the past, especially where biological evolution is involved, but also because of large-scale fluctuations in atmospheric CO_2 levels, and other factors that can produce anactualistic conditions. Nevertheless, an understanding of modern analogs is crucial for an understanding of past sedimentary systems of the rock record. It is only the modern environment that allows for examination of both, the sedimentary processes and their products that can thus be tied together. General rules extracted from observations of modern analogues help reconstructing past depositional processes and understanding the facies arrangement preserved in sedimentary rocks.

Facies architecture is the response to complex and interacting parameters including climate, biology, water composition, and water energy distribution. In contrast to opinions stated by others, facies belts are spatially organized in a non-random manner (e.g., Wilkinson et al. 1998). In many cases facies associations are always arranged in the same typical manner, but are modified with respect to, e.g., lateral extension by factors such as underlying topography. What in ancient examples typically prevents straight-forward predictions of facies distribution is the complex and non-linear nature of interaction, and an incomplete preservation. Nevertheless, general rules extracted from modern depositional settings help improve predictions for ancient carbonate provinces.

The three areas examined here, the Bahamas, Belize, and The Gulf, with their distinct differences in tectonic setting, climate, biota, etc., were chosen to enable the comparison and assessment of the controlling parameters tied to (sub-)tropical

G.P. Eberli (✉)
Rosenstiel School for Marine and Atmospheric Sciences, University of Miami,
Miami, Florida, USA
e-mail: geberli@rsmas.miami.edu

H. Westphal
MARUM and Department of Geosciences, Universität Bremen, Germany
e-mail: hildegard.westphal@uni-bremen.de

H. Westphal et al. (eds.), *Carbonate Depositional Systems: Assessing Dimensions and Controlling Parameters*, DOI 10.1007/978-90-481-9364-6_5,
© Springer Science+Business Media B.V. 2010

carbonate deposition and the resulting sedimentary assemblages. The Bahamas represent a sedimentation model for undisturbed, open marine, isolated carbonate platforms within the subtropical climatic zone. Belize is chosen as model of an attached lagoon with mixed sedimentation within the humid, tropical climatic zone. The Gulf can be regarded not only as a sedimentation model for a marginal sea within the arid climatic zone, but the Iranian side of The Gulf also provides a model area for the formation of a fully marine, marly, fine-grained molasse (Seibold et al. 1973). A comparison of the major differences and similarities of these three carbonate producing regions facilitates the utilization of data contained herein e.g. to basin reconstructions or stratigraphic models, by providing the context necessary for a greater understanding of the fundamental controls of these dynamic systems.

5.2 Setting and Geometries

5.2.1 Tectonic Setting

The contrasting architectural styles of the three study areas are not only related to different tectonic styles (passive margin setting for the Bahamas, strike-slip setting for Belize, and convergent margin setting for The Gulf), but also to the timing of tectonism.

The Bahamian archipelago has undergone a series of tectonic deformation episodes in the Mesozoic when the continental margin formed and rift morphology was created. In the Tertiary, the Cuban collision created structural deformation, but since the Miocene the archipelago sits on a slowly subsiding margin (Masaferro and Eberli 1999; Masaferro et al. 1999). This tectonic quiet time allowed the platform to heal the tectonic scars and to expand laterally (Eberli and Ginsburg 1988, 1989; Masaferro and Eberli 1999). As a result, Bahamian morphology is essentially decoupled from, or only mildly influenced by underlying basement tectonics. Its present isolated carbonate platform setting and shelf morphology mask the underlying tectonic structures (Purser 1973; Eberli and Ginsburg 1988). The present-day platforms are horizontal and extremely shallow (2–10 m) with very sharply defined shelf edges surrounded by deep oceanic waters (Purser 1973; Schlager and Ginsburg 1981).

In contrast, The Gulf is influenced by younger, Plio-Pleistocene tectonism resulting in the formation of a shallow depression (Kassler 1973) with a conduit to the Arabian Sea. The floor of The Gulf is gently inclined from its continental shore-lines to its linear bathymetric axis. The pronounced asymmetry in slope across the basin's bathymetric axis (175 cm/km on the north, Iranian, side versus 35 cm/km to the south, Arabian, side, offsetting the basin axis northward) is evidence of a transition between two contrasting tectonic provinces: The Arabian side of the basin constitutes part of the relatively stable Arabian foreland flanking the Pre-Cambrian Arabian shield, while the unstable Iranian side constitutes a major Tertiary fold belt (Lees 1948; Lees and Falcon 1952; Purser and Seibold 1973).

This morphological asymmetry results in different sedimentation patterns and different influence of terrigenous influx.

Belize lies in a geologically young and still active strike-slip region (Lara 1993). Tilted Pleistocene stalactites in karst holes document the ongoing deformation. Consequently, the architecture and extent of the shallow-water carbonate settings are reflecting the tectonic faults. This is true for the offshore atolls as well as for the lagoon with the barrier reef complex that delineates the seaward limits of its semi-restricted lagoon (Lara 1993; Gischler and Hudson 1997). The structural control is also visible in the distribution of reefal belts, which follow highs, and channels, which overlie structural lows. The restriction of the Belize lagoon is akin to The Gulf except with salinities below normal marine waters rather than above.

These differences in tectonic setting described above allow for comparison between a depositional setting with a level-bottom surface, a carbonate ramp, and a setting with a pronounced preexisting topography.

5.2.2 Geometries and Facies Relationships

The tectonic setting of each locality is a first-order determiner of morphology type and activity, i.e., attached or detached, active or inactive. The type of setting (isolated platform like the Bahamas, attached system like the Belize Lagoon and The Gulf) is important to the terrigenous input into the carbonate depositional system and for the facies arrangement. Accordingly, the three study areas are distinct in the geometries and lateral extend of the sedimentary facies. Where the setting is detached from a continent and tectonically quiet, such as the Bahamas, its facies pattern is a result of the wind regime. Where the setting is attached to a continent and tectonically active, as in The Gulf, no obvious facies pattern is apparent. If the environment is between such end members and is in a relatively inactive tectonic setting and is attached to the continent, such as Belize, the resultant facies belts occur parallel to the shore.

The facies arrangement of the present-day Bahamas with their stable tectonic setting (Hine et al. 1981a, b) is directly related to the energy received from currents, tides and winds. In contrast to the two other study areas, the isolated platforms in the Bahamas are a pure carbonate system, where the only source of siliciclastic material is airborne Saharan dust, and terrigenous influx does not play a role in the facies relationships. The facies geometry of the interior is characterized by relatively simple level-bottom relationships (Enos 1974, 1983). For this type of sheet-like, monotonous level-bottom facies arrangement that also typifies epeiric sea deposits the term "facies prosaic" was proposed (Cloud and Barnes 1967). In such a setting, the laterally extensive units (typically mud – to wackestones) show little interior facies differentiation; however, the energy breaks of the isolated Bahama platforms are characterized by high-energy facies, whereas protected areas are typified by more fine-grained material (Enos 1983). Generally, the platform margins are dominated by marine sand belts consisting of skeletal, pelletal, and oolitic grains (Hine et al. 1981b). Coral-algal reefs and sand bodies consisting of skeletal and ooid grains are found on the high energy windward sides of the Bahamian platforms (Harris 1979; Enos 1983). Raised platform margins with islands are more pronounced on the windward sides of the platforms, giving rise to asymmetric facies distribution in the platform interior (Smith 1940; Enos 1983). The low-energy leeward sides of the islands accumulate mud resulting in pelletoidal and skeletal packstones and wackestones and muddy tidal flats (Purdy 1963). The Bahamian archipelago has distinct windward and leeward sedimentary facies resulting from the depositional response to the prevailing easterly and northeasterly winds of the region. Because sediment transport is toward the west, windward eastern slopes of the platform are steeper than the western leeward sides and show much lower sedimentation and progradation rates (Grammer et al. 1993).

On a large scale, the Belize lagoon enclosed between the Yucatán peninsula and a barrier reef exhibits a parallel belt facies geometry with decreasing terrigenous influence in a seaward direction. Rivers import sediment from the Maya Mountains, resulting in a strongly siliciclastic facies belt nearest to the shoreline. It grades to a marl belt in the axial center and then a carbonate sediment belt featuring a 217 km long barrier reef that runs along the platform edge (James and Ginsburg 1979). On the leeward side of the barrier reef, extensive *Halimeda* sands are found. On a smaller scale, the depositional system is controlled by a complex of compartmentalized settings that result from the pronounced depositional relief (Wantland and Pusey 1975; Enos 1983). This depositional relief is inherited from young small-scale tectonic movements and is exaggerated by reef growth on tectonic highs (Purdy 1974; Lara 1993). Such compartmentalized settings are typified by rapid lateral and vertical facies transitions that might appear random ("facies mosaic" of Laporte 1967).

As for the Belize lagoon, the shallow-water carbonates of The Gulf have siliciclastics admixed. The amount of admixture is largely determined by the distance from shore and the position within the basin. The different tectonic activity is reflected in the runoff: Rivers emptying from the Tertiary fold belt of the Iranian coast bring in a significant amount of terrestrial material leading to marl sedimentation with carbonate contents of less than 10% to more than 40% on the Iranian side of The Gulf (Sarnthein and Walger 1973). Variations in clay and sand content of sediments at the Iranian coast are determined by river mouths and delta locations. In contrast, the Arabian side of The Gulf exhibits a stable tectonic regime and features a carbonate ramp. Sand bodies made of skeletal, oolitic, and pelletoidal grains are found nearest the coastline. Carbonate mud content tends to increase basinward but is present where low energy conditions such as coastal embayments, lagoons, and locally protected areas occur (Wagner and van der Togt 1973). Carbonate content on the Arabian side of The Gulf ranges from 40% at the basin axis to 100% in coastal regions (Emery 1956). The distribution and shape of the sand belts along the Arabian coast is determined by wind and tidally induced currents. Strong winds induce a deeper wave base that keeps the crest of the shoals in deeper water than those of the sand waves of the Bahamas.

5.3 Controls on Sediment Distribution

5.3.1 Climate

Climate is a factor that strongly influences the depositional systems. Temperature as such is not the only climatic factor that decides between warm versus cool to cold-water biotic communities. Notably aridity versus humidity, and evaporation rates are other important parameters. Although positioned in a similar latitude range, the climates of the Bahamas (22–28°N) and The Gulf (24–30°N) are remarkably different. This is mainly because temperatures in the Bahamas are kept stable by oceanic buffering, its islands being entirely surrounded by warm Atlantic Ocean water, whereas the continental setting of The Gulf promotes aridity and large seasonal fluctuations in temperature. Belize, in its more tropical latitude (roughly 16–19°N), remains warm throughout the year. However, its attached geographic setting allows for short-term continental influence and moderate temperature fluctuations that accompany northerly winds.

Evaporation in the arid Middle Eastern climate and gently sloping ramp setting of The Gulf is manifested in the formation of coastal sabkha deposits. Among the three study areas, this coastal facies is unique to The Gulf. The gently inclined coastal regions of the southern Gulf allow tide waters and occasionally storm surges to intermittently recharge these settings with seawater which typically becomes hypersaline in response to subsequent evaporation. Groundwater beneath these sabkhas range in salinities between 70 and 365 ppt. The aridity of this region also creates a difference in diagenetic patterns when compared to other regions examined. The climate of The Gulf stimulates the formation of evaporitic minerals, including widespread dolomite (Purser 1973; McKenzie et al. 1980).

The less arid climate of the Bahamas does not favor the formation evaporites, but evaporation on the Bahamas nevertheless results in the formation of hypersaline waters. The reflux of these denser waters through the platform results in dolomitization of the carbonate sediments, even though depositional and earliest diagenetic dolomite are less widespread, occurring only in association with algal mats in interlevee areas at the Andros tidal flats (Hardie 1977).

5.3.2 Wind

Within the framework of tectonic, morphologic (ramp versus steep-sided platform), and climatic setting, physical parameters influence the distribution of facies in a predictable manner as they respond to the resulting water energy patterns. The wind regime is a primary determinant of the development and distribution of sediment facies and grain size trends. In The Gulf with its ramp morphology a windward setting typically has a basinward increase in mud. At the same time wind influence increases with decreasing depth resulting in sand bodies nearshore, exceptions occur along the Northern coast of Saudi Arabia and the United Arab Emirates where these sand bodies protect restricted bays that become the site of mud accumulation. A windward setting on the steep-sided edge of the Belize platform is characterized by an increase in mud shoreward.

Easterly winds continually influence both the Bahamas and Belize. The Gulf, on the other hand, is buffeted by winds of a more sporadic nature, experiencing one or two major extra-tropical cyclones or NW winter shamals per year and a more continuous and milder summer shamal (Murty and El-Sabh 1984). Belize and the Bahamas can both experience hurricane force winds which produce waves that are responsible for transport of large debris up to boulder-size across their barrier reef crests. These hurricanes can be particularly devastating to the subaerially exposed cays and islands, but also have an impact on shallow reef structures. However, sediment redistribution by hurricanes appears to be minor when matched to its preservation potential in the stratigraphic record (Perkins and Enos 1968; Major et al. 1996).

For the isolated setting of the Bahamas, a distinct asymmetry between windward and leeward facies is obvious on the platform top. Sand bodies and islands occur preferentially on the shelf edge of the windward sides of the platforms. The carbonate mud accumulation is highest to the lee of these islands. Muddy sand is found on the leeward edges of the platforms where no islands are present.

The wind regime also strongly influences reef distribution in general. Reefs are best developed on windward, shelf edge margins, as seen in the Bahamas as well as in Belize with its large barrier reef. Wave energy is absorbed along the windward flanks of the carbonate platforms and barrier reefs of the Bahamas and Belize, whereas the absence of any true shelf edges within The Gulf permits a high degree of wave agitation along its coastlines (Purser and Seibold 1973). No barrier reefs or shelf edge sand bodies are known from The Gulf (Purser and Seibold 1973).

5.3.3   Tides, Currents, and Waves

Belize exhibits the smallest tidal ranges (15–40 cm), The Gulf having the largest ranges (100–400 cm), and the Bahamas the intermediate ranges (80–134 cm) of the three regions. The strength of resulting tidal currents corresponds to this order with the exception of heads of embayments in the Bahamas, which are subject to resonance and thus experience larger tidal ranges and increased current velocities.

In strongly tidally influenced environments of the study areas, sandbars develop. Depending on the velocity of the tidal currents, sandbars orient parallel or perpendicular to the platform margin, the latter being typical for the strong currents at the Tongue of the Ocean of the Bahamas.

In The Gulf, the largest tidal deltas are situated on the western side of The Gulf where tidal exchange (but not tidal range) is highest. The coastal morphology is important in focusing tidal currents. Smaller deltas are found in the Bahamas (e.g., the Exumas) where Pleistocene eolian dunes have created a complicated coastal morphology.

Prevailing broad-scale current systems continuously supply the Bahamian platforms and offshore atolls of Belize with normal open marine waters (except in the interiors of large Bahamian platforms). In contrast, The Gulf and Belize shelf lagoon are bathed in waters that reflect a more restricted character. The Belize Lagoon exhibits low salinities that are a response to high levels of rainfall, whereas The Gulf usually has high salinities associated with evaporation.

Waves and currents play a direct role in biological organization. Water energy levels are responsible for removing sedimentary fines, for mixing of waters and salinity, for turbidity and so light penetration. Water energy influences the distribution of high-energy species including coral, red algae, some mollusks and larger benthic foraminifera. There is a direct relationship between water flux and biotic diversity. The highly agitated waters of the reefal belt are habitat for a wide variety of species. In contrast, the slow moving waters of the bank interior of the Bahamas show a low diversity in its fauna and flora. Water motion as key role in the distribution of salinity and sediment transport directly affects the biological distribution. This means that high salinity areas tend to be monospecific while diversity is high in normal marine seawater. For planktonic species, current and nutrient patterns are the main factor in determining their distribution. Modern stromatolites as described from the Bahamas (Reid et al. 1995, 2000) are another example for the influence of water energy levels. These stromatolites occur in normal marine waters where high-energy settings are hostile to coral reefs because of the high sediment flux (ooids).

The relationship between wave-motion and reef development is well known (Roberts 1974; Hopley 1982). Where low wave motion takes place in shallow water depth as for example in more protected parts of a reef, delicate branching corals flourish. With increasing wave motion, massive branching corals dominate. Moderate wave motion in moderate water depths favors head corals, whereas platy corals dominate in deeper parts of a reef where light levels are low. Wave motion levels in the study areas show that maximum coral growth rate occurs in wave motion range of 80–380 cm/s. Above and below this range coral growth decreases until coral growth ceases at values below 50 cm/s and above 410 cm/s.

5.3.4 Light

The nature of organic communities is strongly influenced by the intensity of available light (e.g., Hughes Clarke and Keij 1973). The lower limit of euphotic and oligophotic zones is dependent upon turbidity, but also upon nutrient levels that influence plankton concentration, and upon clay influx (Purser and Seibold 1973; Hallock and Schlager 1986). Wave activity in The Gulf results in the presence of considerable quantities of suspended sediments in the water column and correspondingly reduces light penetration that is considerably lower than that in oceanic provinces like the Bahamas (Purser and Seibold 1973). While luxuriant growth of light-dependent biota reaches barely 20 m in The Gulf (Hughes Clarke and Keij 1973), comparable communities occur down to more than 40 m in Caribbean waters and in open marine settings even permit reef growth down to 119 m (Reed 1985) (Tables 5.1 and 5.2).

Table 5.1 Comparison of qualitative physical parameters of the three study areas (Compiled from other chapters in this book – for references see there)

Attribute/parameter	Bahamas	Belize	Persian Gulf
Tectonic setting	Passive continental margin	Strike-slip setting	Convergent basin
Morphology	Flat platforms with sharp shelf edges	Lagoon bound on W by continent and on E by shelf edge, barrier reef and offshore atolls	Restricted epi-continental sea within foreland basin ramp setting
Climate	Subtropical	Tropical	Subtropical
Marine influence	Fully marine	Marine and continental	Continental
Humidity (rain:evaporation ratio)	Humid	Humid (1.0)	Arid (0.1)
Wave energy distribution	Windward shelf edge reefs	Barrier reef and windward atolls	Along most coasts, highest in southeast
Winds	Easterly	Easterly	NW shamal
Temperature	Warm and stable	Warm and mostly stable	Hot in summer; highly variable with large range
Currents and sedimentary effect	Locally strong producing shelf edge sand bodies	Weak – little effect	Strong – scour bottom and reworking of sediments
Salinity	Normal to locally high	Low to normal	Very high to approaching normal at entrance (Strait)
Restriction effects	Locally high salinity in large platform interior	Lowered salinity in shelf lagoon	Very high salinities and large temperature fluctuations
Reefs	Barrier and patch reefs	Extensive barrier (shelf edge) reef and lagoonal patch reefs, pinnacles	Patch reefs, no shelf edge = no barrier reefs
Terrigenous input	No	Yes	Yes
Evaporite/dolomite precipitation	No	No	Yes

Table 5.2 Comparison of quantitative physical parameters of the three study areas (Compiled from other chapters in this book – for references see there)

Location/parameter	Bahamas	Belize	Persian Gulf
Wave height	1.4 m (storm: 15 m)	1–1.5 m (strom: 10 m)	(No continuous data) (storm: 6 m)
Wind speed	6–7 m/s (hurricanes: exceeding 250 km/h = 69 m/s)	4–5 m/s (storm: 259 km/h = 72 m/s)	(No continuous data) (storm: 100 km/h = 28 m/s)
Tidal range	80–134 cm	15–40 cm	100–400 cm
Light penetration in water column	Winter: 90 m, summer: 70 m	(No data)	20 m
Air temperature	25 °C (average) 23°C (winter) 27.5°C (summer)	24°C (winter) 27°C (summer) Range: 10–36°C	Range: 0.5–36°C
Sea surface temperature	27 °C (average)	25.5–30°C	Range: 4–35.5°C
Currents:			
– Tidal	20-200 cm/s	up to 60 cm/s	Up to 200 cm/s
– Prevailing	25-65 cm/s	(No data)	10–45 cm/s
– Storm		600 cm/s (bottom water velocity at fore-reef during hurricane)	(No data)

5.3.5   Sea Water Chemistry

Water composition including salinity, suspended sediment and nutrient levels influences the formation and accumulation of sediment in carbonate provinces in two ways: (1) by determining clay contents and influencing diagenesis etc., and (2) by controlling the carbonate producing biota. In particular for attached platforms, the hydrologic organization of the emerged hinterland, that is a product of climate and orography, controls the amount, the composition and the frequency of fresh water discharged into the marine system. Climate plays a direct role in determining the water composition that, in turn, has a direct impact upon the biological organization within a given region.

In the three regions studied, the Bahamas are unique in the isolation from land influence, whereas Belize is strongly influenced by water and sediment influx from the mountainous hinterland with a humid, tropical climate. A strong influx of fresh water to the system via perennial streams is typical especially for the southern Belize lagoon. On the other hand, influx of fresh water in The Gulf is localized in time and space. It can be torrential but has a low frequency except in the northwest.

Water exchange, evaporation rate, and input of fresh water control the salinity distribution. Salinity distribution in Belize is a function of river input and thus is related to the humid, tropical climate and the drainage of the Maya Mountains, especially during the rainy season, as well as to the shelf physiography and currents.

In The Gulf, the salinity distribution is the result of input of fresh water, residence time and a high evaporation rate associated with the dry and warm climate. In both, the Belize Lagoon and The Gulf, a distinct gradient is observed from the mouth of the rivers towards the basin axis. In an offshore archipelago like the Bahamas rainfall is the only source of fresh water. In the Bahamas, water chemistry depends mainly on residence times and evaporation on the banks.

Salinity is important for determining species distribution and biogenic carbonate production. Any deviation from normal seawater salinity produces specific biotic associations. Evans et al. (1973) have shown that in the Persian Gulf a salinity value of 50 ppm is the limit for most of the sediment producers, whereas certain foraminifers flourish in a hypersaline setting. High-salinity water associated with restricted environments is a habitat for biota dominated by cerithid gastropods and large imperforate foraminifers. Bioerosion is strongly diminished in high-salinity environments. A significant difference between the Bahamas on one side and the restricted northern Belize Lagoon and The Gulf on the other side, is the abundance of the large benthic peneroplid foraminifers in the latter two that flourish in high salinity water. In highest salinity waters as in the shallowest parts of the southern coast of The Gulf (UAE), algal mats develop.

Nutrient levels are another important determiner of biological system (e.g., Wood 1993; Mutti and Hallock 2003; Pomar et al. 2004). The concentration of nutrients in the water depends on multiple factors including water circulation and terrestrial influx and is time-variant. For coral reef systems the dependence on low nutrient environments and their high sensibility to changes in nutrient levels was recognized some 30 years ago (Margalef 1968). Nutrient overload leads to abundant growth of soft corals and algae and the demise of hard corals (Hallock and Schlager 1986; Hallock 1988). This mechanism is at least partly responsible for the absence of coral reefs in the northern Belize Lagoon. It might also influence the diminished reef growth in the vicinity of rivers in the southern Belize Lagoon and the northern Gulf. At its extreme the negative influence of nutrients on the coral reef system is observed presently on a global scale where coral reefs are deteriorating due to anthropogenic over-nutrification (Risk 1999; Hallock 2001).

In all three study areas, the waters are super-saturated with respect to both, aragonite and calcite. However, if purely "inorganic precipitation" exists, then the saturation state of the water is not a controlling factor ("inorganic precipitation" refers to precipitation that is not initiated by biologic activity or biologically triggered, e.g., by algal blooms or algal coatings that create the chemical and physical conditions that permit direct precipitation from sea-water). In the Bahamas, predominantly aragonite and to a minor extend high-Mg calcite are "inorganically precipitated." In Belize, besides aragonite and high-Mg calcite also low-Mg calcite is precipitated. In The Gulf, the precipitation of high-Mg calcite occurs only in the north where a perennial source of fresh water exists. What appears likely is that water composition affects the micro-biotic assemblage that, in turn, influences the type of $CaCO_3$ that precipitates. This hypothesis, however, remains to be proven.

5.4 Outlook

This book is an attempt to provide a comprehensive database and to extract general patterns from modern carbonate depositional settings. It aims at serving as a guide for interpretation of ancient carbonate rocks. Clearly, omissions were unavoidable. Questions that will need to be addressed in the future include: (1) verification of the observations for the ancient, that is, for the effect of biological evolution and other changes with time, (2) extending the compilation for the extratropical and the deep-water realms, (3) the effect of observation time.

5.4.1 The Effect of Biologic Evolution and Other Changes with Time

The present data compilation is based on the assumption that the modern is a valid analogue for the fossil record. However, carbonate systems are strongly influenced by the types of organic sediment producers present at a given geological time. Especially in carbonate producing systems, the geological record is strongly influenced by biologic evolution. For example, reef-building organisms considerably change through time, the modern coral-red algal reef being the exception rather than the rule (e.g., Kiessling et al. 2002). Additionally, carbonate depositional systems at times with reef development differ in many respects from times without reef growth. One effect of changes in biotic communities is variations in productivity and sedimentation rates through time.

To add to complexity, variations in water chemistry and atmospheric CO_2 levels might have profound effects on the carbonate system and the preservation of its products (e.g., Kleypas et al. 1999, 2001). To account for these variations in time, the parameters for reconstructing the fossil record need to be as scrupulously constrained as possible. Even though data on the fossil record in many cases are difficult to obtain, any information available from the ancient period studied has to be taken into account.

5.4.2 Extratropical Realm and Deep-Water

Although a thorough literature search was performed, the data presented here are limited with respect to areal coverage and completeness. The study areas presented here are restricted to subtropical and tropical shallow-water carbonate depositional environments. However, during the past years the importance of environments outside these classical settings with respect to complexity and productivity was acknowledged. These other settings include cool to cold-water realms (e.g., Nelson et al. 1988; James 1997, Freiwald 1998) as well as the deep water (e.g., Freiwald et al. 1997; Lazier et al. 1999; Rogers 1999, Freiwald and Roberts 2005), but also tropical settings with too high nutrient levels to allow for the formation of classical assemblages (e.g., Pomar 2001; Mutti and Hallock 2003; Pomar et al. 2004: Westphal et al. in press). Clearly the compilation presented here is only a starting point for further studies that include these other carbonate settings.

5.4.3 Effect of Observation Time

The quantitative data used in this report were assembled by scientists for various objectives. The time frame of the observation has been adjusted to their respective needs, and most data we compiled were measured in the modern system. However it has been shown that geological parameters like sedimentation rates are not time-invariant to the observation interval applied. For carbonate systems, Bosscher (1992) has demonstrated that the average accumulation rate decreases with increasing time increment. This is most obvious for the decreasing growth time of reefs with increasing observation time: whereas a present-day *Acropora cervicornis* might show growth rates of up to 20 cm per year, the rates of fossil reefs are orders of magnitude lower. Thus a major difficulty in reconstructing ancient deposition is that sedimentary systems have a non-linear behavior and many of the parameters involved are fractal (Plotnick 1986, 1988, 1995; Plotnick and Prestegaard 1993). This accounts not only for sedimentation rates, but also for the size and distribution of hiati and short-term events such as hurricanes. A fractal distribution means that the probability of including a hiatus increases with decreasing time coverage of the gap.

For reconstructing ancient depositional systems the scale and duration of the process needs to be adjusted to obtain geological relevance. For example, biological parameters discussed in this book typically are given for days and years, and a spatial distribution in the order of tens of meters. Geological time scales are an order of magnitudes larger and require an investigation through time, i.e. within a stratigraphic framework.

Chemical processes are likewise difficult to assess. The kinetic of reaction is generally slow, but some processes, like the precipitation in whitings, are of short duration. Burial diagenesis on the other hand, takes place over millions of years. Sedimentological processes can be similarly sporadic or long-lived. Torrential flooding in a Wadi as on the Iranian coast of The Gulf is a short time event but the settling of pelagic ooze is perennial. Therefore, one needs to consider a balance between high-impact short-lived events and low-impact long-term events.

References

Bosscher H (1992) Growth potential of coral reefs and carbonate platforms. Ph.D. thesis, Vrije Univeriteit, Amsterdam, p 160

Cloud PE, Barnes VE (1967) Early Ordovician sea in central Texas. In: Ladd HS (ed) Treatise on marine ecology and paleoecology. GSA Mem 67(2):163–214

Eberli GP, Ginsburg RN (1988) Aggrading and prograding infill of buried Cenozoic seaways, Northwestern Great Bahama Bank. In: Bally AW (ed) Atlas of Seismic Stratigraphy. AAPG Studies in Geology 27(2):97–103

Eberli GP, Ginsburg RN (1989) Cenozoic progradation of NW Great Bahama Bank – A record of lateral platform growth and sea-level fluctuations. In: Crevello PD et al. (eds) Controls on carbonate platform and basin evolution. SEPM Spec Publ 44:339–351

Emery KO (1956) Sediments and water of the Persian Gulf. AAPG Bull 40:2354–2383

Enos P (1974) Map of surface sediment facies of the Florida-Bahama Plateau. Map series MC-5, Boulder, CO. GSA: 4

Enos P (1983) Shelf environment. In: Scholle PA, Bebout DG, Moore CH (eds) Carbonate depositional environments. AAPG Mem 33:267–295

Evans G, Murray JW, Biggs HEJ, Bate R, Bush PR (1973) The oceanography, ecology, sedimentology and geomorphology of parts of the Trucial Coast barrier island complex. In: Purser BH (ed) The Persian Gulf: Holocene carbonate sedimentation in a shallow epicontinental sea. Springer-Verlag, New York, pp 233–278

Freiwald A (1998) Modern nearshore cold-temperate calcareous sediments in the Troms district, northern Norway. J Sed Res 68:763–776

Freiwald A, Henrich R, Pätzold J (1997) Anatomy of a deep-water reef mound from Stjernsund, West Finnmark, Northern Norway. In: James NP, Clarke JAD (eds) Cool-water Carbonates. SEPM Spec Publ 56:141–162

Freiwald A, Roberts JM (2005) Cold-water corals and ecosystems. Springer Verlag, Heidelberg, p 1243

Gischler E, Hudson JH (1997) Holocene development of three isolated carbonate platforms, Belize, Central America. Mar Geol 144:333–347

Grammer GM, Ginsburg RN, Harris PM (1993) Timing of deposition, diagenesis, and failure of steep carbonate slopes in response to a high-amplitude/high-frequency fluctuation in sea level, Tongue of the Ocean, Bahamas. In: Loucks RG, Sarg JF (eds) Carbonate sequence stratigraphy. AAPG Mem 57:107–131

Hallock P (1988) The role of nutrient availability in bioerosion: consequences to carbonate buildups. Palaeogeogr, Palaeoclim, Palaeoecol 63:275–291

Hallock P (2001) Coral Reefs in the 21st century: is the past the key to the future? In: Greenstein BJ, Carney CK (eds) Proceedings of the 10th symposium on the Geology of the Bahamas and other Carbonate regions, pp 8–23

Hallock P, Schlager W (1986) Nutrient excess and the demise of coral reefs and carbonate platforms. Palaios 1:389–398

Hardie LA (1977) Sedimentation on the modern carbonate tidal flats of North-west Andros Island, Bahamas: Baltimore. The Johns Hopkins University Press, Studies in Geology, No. 22, p 202

Harris PM (1979) Facies anatomy and diagenesis of a Bahamian ooid shoal. Sedimenta VII, University of Miami, Florida, p 163

Hine AC, Wilber RJ, Bane JM, Neumann AC, Lorenson KR (1981a) Offbank transport of carbonate sands along open, leeward bank margins: northern Bahamas. Mar Geol 42:327–348

Hine AC, Wilber RJ, Neumann AC (1981b) Carbonate sand bodies along contrasting shallow bank margins facing open seaways in northern Bahamas. AAPG Bull 65:261–290

Hopley D (1982) The geomorphology of the Great Barrier Reef. Quaternary development of coral reefs. Wiley Interscience, New York, p 453

Hughes Clarke MW, Keij AJ (1973) Organisms as producers of carbonate sediment and indicators of environment in the southern Persian Gulf. In: Purser (ed) The Persian Gulf: Holocene carbonate sedimentation in a shallow continental sea. Springer-Verlag, New York, pp 33–56

James NP (1997) The cool-water carbonate depositional realm. In: James NP, Clarke J (eds) Cool-water Carbonates. SEPM Spec Publ 56:1–20

James NP, Ginsburg RN (1979) The seaward margin of Belize barrier and atoll reefs. Int Ass Sedimentol Spec Publ 3, p 191

Kassler P (1973) The structural and geomorphic evolution of the Persian Gulf. In: Purser BH (ed) The Persian Gulf: Holocene Carbonate Sedimentation and Diagenesis in a Shallow Epicontinental Sea, pp 11–32

Kiessling W, Flügel E, Golonka J (2002) Phanerozoic reef patterns. SEPM Spec Publ 72, p 790

Kleypas JA, Buddemeier RW, Archer D, Gattuso J-P, Langdon C, Opdyke BN (1999) Geochemical consequences of increased atmospheric carbon dioxide on coral reefs. Science 284:118–120

Kleypas JA, Buddemeier RW, Gattuso J-P (2001) The future of coral reefs in an age of global change. Int J Earth Sci 90:426–437

Laporte LF (1967) Carbonate deposition near mean sea-level and resultant facies mosaic; Manlius Formation (Lower Devonian) of New York State. AAPG Bull 51:73–101

Lara ME (1993) Divergent wrench faulting in the Belize Southern Lagoon; implications for Tertiary Caribbean plate movements and quaternary reef distribution. AAPG Bull 77:1041–1063

Lazier AV, Smith JE, Risk MJ, Schwarcz HP (1999) The skeletal structure of *Desmophyllum cristagalli*: the use of deep-water corals in sclerochronology. Lethaia 32:119–130

Lees GM (1948) The physical geography of southeast Arabia. Geogr J 121:441

Lees GM, Falcon NL (1952) The geographical history of the Mesopotamian plains. Geogr J 118:24–39

Major RP, Bebout DG, Harris PM (1996) Facies heterogeneity in a modern ooid sand shoal – an analog for hydrocarbon reservoirs. Bureau of Econ Geology, Geological Circular 96-1, University of Texas, p 30

Margalef R (1969) The pelagic ecosystem of the Caribbean Sea. Symp in Investig and Resour of the Caribbean Sea and Adjacent Regions, Paris, France, UNESCO, pp 484–489

Masaferro J, Eberli GP (1999) Jurassic-Cenozoic structural evolution of southern Great Bahama Bank, southern Bahamas. In: Mann P (ed) Caribbean basins, sedimentary basins of the World, 4. Elsevier, Amsterdam, pp 167–193

Masaferro JL, Poblet J, Bulnes M, Eberli GP, Dixon TH, McClay K (1999) Paleogene – Neogene/present day(?) growth detachment folding in the Bahamian foreland in the Cuban fold and thrust belt. J Geol Soc 156:617–631

McKenzie JA, Hsu KJ, Schneider JF (1980) Movement of subsurface waters under the sabkha, Abu Dhabi, UAE, and its relation to evaporative dolomite genesis. In: Zenger DH (ed) Concepts and models of dolomitization. SEPM Spec Publ 28:11–30

Murty TS, El-Sabh MI (1984) Storm tracks, storm surges and sea state in the Arabian Gulf, Strait of Hormuz and the Gulf of Oman. UNESCO Report Marine Sci 28:12–24

Mutti M, Hallock P (2003) Carbonate systems along nutrient and temperature gradients: some sedimentological and geochemical constraints. Int J Earth Sci 92:465–475

Nelson CS, Keane SL, Head PS (1988) Non-tropical carbonate deposits on the modern New Zealand shelf. Sediment Geol 60:71–94

Perkins RD, Enos P (1968) Hurricane betsy in the Florida-Bahamas area; geologic effects and comparison with Hurricane Donna. J Geol 76:710–717

Plotnick RE (1986) A fractal model for the distribution of stratigraphic hiatuses. J Geol 94:885–890

Plotnick RE (1988) Fractal, random fractal, and random models for depositional hiatuses. GSA Abs Progs: A403

Plotnick RE (1995) Introduction to fractals. In: Middleton GV, Plotnick RE, Rubin DM (eds) Nonlinear dynamics and fractals – new numerical techniques for sedimentary data. SEPM Short Course 36:1–28

Plotnick RE, Prestegaard K (1993) Fractal analysis of geologic time series. In: Lam N, DeCola L (eds) Fractals in geography. Prentice-Hall, Englewood Cliffs, NJ, pp 45–54

Pomar L (2001) Types of carbonate platforms: a genetic approach. Basin Res 13:313–334

Pomar L, Brandano M, Westphal H (2004) Miocene tropical foramol-rhodalgal-bryomol associations of the western Mediterranean – a critical review. Sedimentology 51:627–651

Purdy EG (1963) Recent calcium carbonate facies of the Great Bahama Bank. 2. Sedimentary facies. J Geol 71:472–497

Purdy EG (1974) Karst-determined facies patterns in British Honduras: Holocene carbonate sedimentation model. AAPG, Bull 58:825–855

Purser BH (1973) Sedimentation around bathymetric highs in the Southern Persian Gulf. In: Purser BH (ed) The Persian Gulf Holocene carbonate sedimentation and diagenesis in a shallow epicontinental sea, p 471

Purser BH, Seibold E (1973) The principle environmental factors influencing Holocene sedimentation and diagenesis in the Persian Gulf. In: Purser BH (ed) The Persian Gulf: Holocene carbonate sedimentation in a shallow epicontinental sea. Springer-Verlag, New York, pp 1–10

Reed JK (1985) Deepest distribution of Atlantic hermatypic corals discovered in the Bahamas. Proc 5th Int Coral Reef Symp 6:249–254

Reid RP, Macintyre IG, Browne KM, Steneck RS, Miller T (1995) Modern marine stromatolites in the Exuma Cays, Bahamas: uncommonly common. Facies 33:1–17

Reid RP, Visscher PT, Decho AW, Stolz JF, Bebout BM, Dupraz C, Macintyre IG, Paerl HW, Pinckney JL, Prufert-Bebout L, Steppe TF, DesMarais DJ (2000) The role of microbes in accretion, lamination and early lithification of modern marine stromatolites. Nature 406:989–992

Risk MJ (1999) Paradise lost: how marine science failed the world's coral reefs. Mar Freshwater Res 50:831–837

Roberts HH (1974) Variability of reefs with regard to changes in wave power around an island. Proceedings of the 2nd International Coral Reef Symposium 2, pp 497–512

Rogers AD (1999) The biology of *Lophelia pertusa* (Linnaeus 1758) and other deep-water reef-forming corals and impacts from human activities. Int Rev Hydrobiol 84:315–406

Sarnthein M, Walger E (1973) Classification of modern marl sediments in the Persian Gulf by factor analysis. In: Purser BH (ed) The Persian Gulf Holocene Carbonate sedimentation and diagenesis in a shallow epicontinental sea. Springer-Verlag, New York, 471

Schlager W, Ginsburg RN (1981) Bahama carbonate platforms – the deep and the past. Mar Geol 44:1–24

Seibold E, Diester L, Fütterer D, Lange H, Muller P, Werner F (1973) Holocene sediments and sedimentary processes in the Iranian part of the Persian Gulf. In: Purser BH (ed) The Persian Gulf: Holocene carbonate sedimentation and diagenesis in a shallow epicontinental sea. Springer-Verlag, New York, pp 57–80

Smith CL (1940) The Great Bahama Bank I. General hydrogeographical and chemical features. J Mar Res 3:147–169

Wagner CW, van der Togt C (1973) Holocene sediment types and their distribution in the Southern Persian Gulf. In: Purser BH (ed) The Persian Gulf: Holocene Carbonate sedimentation and diagenesis in a shallow epicontinental sea. Springer, New York, 471

Wantland KF, Pusey WCI (1975) Belize Shelf – carbonate sediments, clastic sediments, and ecology, AAPG Studies in Geology 2, p 599

Westphal H, Halfar J, Freiwald A (in press) Heterozoan carbonates in subtropical to tropical settings in the present and past. Int J Earth Sci

Wilkinson BH, Diedrich NW, Drummond CN, Rothman ED (1998) Michigan hockey, meteoric precipitation, and rhythmicity of accumulation on peritidal carbonate platforms. GSA Bull 110:1075–1093

Wood R (1993) Nutrients, predation and the history of reef-building. Palaios 8:526–543

Index

A

Abu Dhabi, 151, 158, 167, 170, 179–183, 185, 186, 190, 192, 194, 200, 203

Accomodation space, 46, 66–67, 128, 131, 132

Acropora, 27, 29, 31, 89, 96, 98, 100, 111, 115, 116, 188, 190, 191, 197, 201, 202, 226

Airborne sediments, 155

Air temperature, 11, 82–85, 160, 166, 223

Algae, 24, 26, 30–32, 37, 38, 57, 58, 61, 94, 96, 99–101, 108, 111, 114, 119–121, 123, 167, 173, 177, 178, 180, 184, 195–203, 205, 221, 224

Anhydrite, 181, 182

Anticline, 149, 171, 187

Arabian, 145–151, 153, 156, 160–162, 167, 168, 171, 174, 177, 179, 180, 187, 191, 194, 199, 201, 203, 216, 218

Arabian Gulf, 145, 147–149, 151, 156, 161, 168, 171, 174, 177, 191, 199, 216, 218

Arabian Peninsula, 149, 191

Aragonite, 24, 26, 45, 57, 59, 63, 65, 95, 123, 157, 169–171, 186, 205, 224

Architecture, 1, 10, 11, 67, 124, 138, 215–217

Arid, 2, 6, 145, 148, 152, 180, 181, 216, 219, 222

Atoll, 9, 81, 87, 100, 104, 106, 112–117, 119, 120, 123, 124, 126, 132, 137, 138, 217, 221, 222

B

Bahamas/Bahamian, 1, 2, 5–69, 145, 180, 189, 198, 200–202, 215–226

Bahrain, 160, 161, 187, 190

Barrier, 11, 27, 29, 40, 43, 44, 56, 60, 81, 82, 85, 87–89, 91–102, 104–112, 115–118, 120, 122–124, 126, 129–133, 137, 138, 151, 165, 178, 179, 183–185, 191, 201–204, 217, 218, 220, 222

Barrier island, 56, 123, 151, 165, 179, 183–185, 202, 203

Barrier reef, 27, 29, 60, 81, 82, 85, 87–89, 91–96, 99–102, 104–112, 115–118, 120, 123, 126, 129, 131–133, 137, 138, 191, 201, 202, 217, 218, 220, 222

Bathymetric high, 56, 129, 150, 151, 204

Beachrock, 170

Belize, 2, 38, 81–138, 145, 189, 198, 200–203, 205, 215–226

Bioerosion, 27, 31, 37–38, 196, 224

Biogenic sediment, 174

Biological evolution, 215, 225

Bivalve, 26, 37, 184, 186, 194–198

Bryozoa, 173

C

Calcarenite, 173, 177

Calcareous algae, 24, 57, 58, 167, 173, 197–198, 202, 205

Calcite, 24, 26, 59, 95, 123, 154, 157, 171, 182, 205, 224

Caprock, 189

Carbonate ramp, 2, 145–205, 217, 218

Carbonate saturation, 2

Cenozoic, 149

Central Swell, 148, 157, 164

Chenier, 158, 179, 204

Circulation, 17, 19, 22, 29, 69, 89, 92, 93, 112, 137, 152, 161, 163, 164, 185, 199, 224

Clay, 24, 119, 122, 173, 174, 218, 221, 223

Climate, 1, 2, 11, 22, 59, 68, 81–91, 145, 148, 152, 181, 188, 202, 204, 215, 219, 222–224

CO_2, 24, 25, 173, 215, 225

Cold-water, 2, 219, 225

Compound grain, 151, 177–179, 184
Cool-water, 2, 219, 225
Coral, 19, 27–29, 31–33, 37, 44, 59, 60,
     87–89, 91, 96, 98, 100, 106, 108, 111,
     114–117, 119, 120, 123, 132–135,
     149–151, 160, 161, 172, 178, 179,
     184, 187–192, 197–199, 201–203,
     218, 221, 224, 225
Coral buildup, 29, 160, 161
Coralgal, 29, 40, 59–60, 121, 138, 196
Coral reef, 22, 28, 29, 32, 59, 60, 91, 115, 116,
     191, 192, 197, 198, 202, 203, 221, 224
Crustacaea, 173
Current, 2, 9–15, 17, 19, 21, 29, 36, 37, 43, 45,
     46, 49–52, 56, 57, 61–66, 82, 87,
     89–93, 96, 110, 145, 151, 153,
     155–157, 161–166, 169, 174, 178, 183,
     193, 197, 202–204, 218, 220–223
Current regime, 21, 29, 161, 162

D
Deep sea, 2, 32
Delta, 8, 49, 52, 53, 56, 57, 129, 130, 149,
     164, 165, 174, 176, 183, 185, 187,
     218, 220
Density cascading, 29, 63
Diversity, 33, 102, 132, 187, 191, 193,
     202, 221
Dolimitization, 95, 181, 182, 204
Dolomite, 24, 59, 95, 145, 150, 158, 180–182,
     219, 222
Dune, 37, 40, 46, 48, 49, 53, 55, 56, 154–156,
     180, 204, 220
Dust, 7, 154, 155, 217

E
Echinoderm, 26, 33–34, 37, 157, 176, 197,
     202, 205
Environmental parameters, 1–3, 5–69
Eolian, 5, 8, 40, 48, 49, 51, 55, 56, 155, 176,
     178, 180, 203, 204, 220
Eolianite, 1, 40, 48, 55, 56
Eolian sedimentation, 56, 176
Euphrates, 149, 164, 187
Evaporation, 22, 29, 45, 49, 59, 101, 148, 152,
     153, 162, 166, 170, 179, 181, 182, 203,
     204, 219, 221–224
Evaporative pumping, 181, 182
Evaporite, 2, 23, 24, 145, 146, 149, 180, 184,
     203, 204, 219, 222
Evolution, 6, 67–69, 124, 128, 129, 132–138,
     172, 215, 225

Extratropical, 225–226
Exuma, 7, 8, 12, 19, 22, 24, 27, 28, 35, 37, 42,
     45–47, 49, 52, 53, 55, 57, 60–62, 64, 220

F
Facies, 3, 5–69, 82, 94, 100–131, 137, 138,
     145–205, 215, 217–220
  association, 204, 215
  mosaic, 38, 218
  prosaic, 38, 217
Fish, 24, 37, 167, 173, 174
Florida, 7, 8, 10–15, 21, 22, 31–33, 37, 38,
     63–66, 137, 145
Foraminifer, 26, 32–33, 58, 63, 94, 95,
     101–104, 108, 114, 119, 120, 122, 123,
     137, 146, 150, 151, 167, 173, 174,
     176–178, 184, 185, 188, 193–197,
     203–205, 221, 224
Foraminifera, 94, 95, 101–103, 108, 114, 119,
     120, 122, 123, 146, 150, 151, 167, 173,
     174, 176–178, 184, 185, 188, 193–197,
     203–205, 221
Foreland basin, 2, 147, 148, 177, 222
Fore-reef, 19, 28, 32, 60, 88, 106, 108–111,
     114, 117–120, 223
Framework, 27, 33, 88, 116, 128, 172, 189,
     191–192, 197, 200–203, 219, 226
Fringing reef, 5, 27–29, 35, 37, 46, 60, 105,
     115, 116, 137, 138, 177, 179, 191

G
Gastropod, 26, 37, 59, 135, 178, 184–186,
     188, 195, 197, 200, 224
Geometry, 3, 5–69, 130, 172–187, 216–218
Geomorphology, 106, 146
Grain, 3, 21, 45, 50, 51, 54, 57, 101, 108–110,
     119–123, 149, 155, 172, 179,
     184, 196, 219
Green algae, 26, 30, 32, 38, 45, 96, 98–99,
     118, 120, 167, 198–199, 202
Gulf, 8, 11–13, 25, 65, 81, 82, 93, 94, 96, 98,
     101, 106, 128, 132, 145, 146, 149,
     152–154, 156, 157, 160–162, 166–169,
     171–181, 183, 185, 187–191, 194–197,
     199, 201–204, 215–226
Gypsum, 24, 59, 178–182

H
Halimeda, 22, 30, 32, 45, 61, 63, 95, 96,
     99–101, 104–106, 108, 111, 114, 115,
     117–123, 197, 199, 202, 218

Halite, 24, 181, 182
Hardground, 27, 40, 171, 172, 189,
    196–199, 203
Holocene, 8, 29, 40, 43, 45, 46, 48, 49, 55, 56,
    60, 63, 66, 67, 69, 95, 98, 108,
    114–116, 122, 123, 126, 129–138, 145,
    151, 156, 163, 171, 174, 181, 187, 191,
    200, 204
Honduras, 81, 82, 93, 94, 96, 98, 101,
    106, 132
Humid, 2, 6, 11, 59, 81, 82, 216,
    222, 223
Humidity, 1, 82, 219, 222
Hydrology, 146, 181, 182, 204, 223

I
Infratidal, 185
Intertidal, 34, 35, 37, 38, 40, 46, 48, 54, 58,
    146, 154, 160, 170, 179, 180, 182, 184,
    194, 195, 197, 199, 204
Iran, 147, 149, 153, 154, 157, 160, 162,
    164–166, 168, 169, 172–174, 177,
    187, 190, 191, 193–195, 204, 216,
    218, 226
Iraq, 149, 154, 155, 163
Island, 5, 7, 8, 12, 13, 17–21, 23, 24, 27–29,
    33–52, 54–60, 63, 66, 67, 69, 89, 96,
    101, 108, 112, 114, 122–124, 132, 136,
    137, 149, 151, 153, 161, 164, 165, 170,
    173, 179, 180, 182–187, 189–192, 194,
    201–203, 218–220

K
Kaus, 153
Kelvin wave, 160
Kuwait, 155, 160, 163, 164, 166, 167, 179,
    180, 190, 199

L
Lagoon, 2, 17, 27, 28, 32–35, 38, 44, 60,
    81–85, 87, 89, 91, 93–105, 108,
    112–116, 120–124, 126–133, 135, 137,
    138, 149, 153, 165, 178–180, 183–188,
    197, 199, 203, 205, 215–226
Lag time, 66
Landbreeze, 154, 162, 204
Light, 22, 27, 29, 30, 32, 34, 81, 91, 153, 161,
    167, 221–223
Light penetration, 22, 29, 153, 167,
    221, 223
Lithification, 35, 169–172, 184, 188

M
Mangrove, 12, 49, 59, 96, 98, 108, 112, 123,
    124, 132, 136, 185, 200
Marl, 68, 82, 104, 106, 121, 122, 149, 172,
    173, 176, 177, 216, 218
Meteor, 146, 166
Mexico, 8, 11, 81, 98, 112, 116, 128
Microbial mat, 26–27, 34–37, 59, 199
Modern analog, 1, 215–216
Modern environment, 1–3, 5, 33, 37, 199, 215
Mollusk, 114, 121, 123, 146, 173, 175, 176,
    178, 203, 221
Monospecific, 198, 221
Morphology, 1, 6–9, 28, 36, 43, 47, 49, 67, 68,
    81, 82, 96, 97, 100, 106, 108, 114, 115,
    117, 120, 130, 137, 138, 146, 151, 159,
    170, 184, 199, 200, 203, 204, 216, 217,
    219, 220, 222
Musandam, 148, 149, 179, 180, 190

N
Nutrient, 1, 2, 26, 27, 32, 43, 91, 98, 168–169,
    204, 221, 223, 224, 226

O
Ocean chemistry, 2
Oman, 146, 148, 149, 153, 173, 174, 191
Ooid, 6, 17, 21, 34, 40, 42–44, 46, 50–57, 66,
    67, 166, 167, 170, 178, 179, 182, 185,
    203, 218, 221
Ostracod, 104–106, 121, 123, 173, 176, 178, 188
Oxygen, 29, 168–169

P
Paleozoic, 149, 150
Patch reef, 27, 28, 38, 40, 44, 60, 97–99,
    106, 108, 112–116, 126, 184, 191,
    201, 202, 222
Pelagosite, 170
Persian, 145, 151, 222–224
Persian Gulf, 2, 8, 11–13, 65, 82, 93, 94, 96,
    98, 106, 132, 145–205, 215–226.
    See also Arabian Gulf; The Gulf
Photic zone, 2, 152, 167, 204, 221
Physical parameters, 2, 3, 10–22, 30, 145–205,
    219, 222, 223
Pinnacle, 97, 108, 115, 116, 137, 222
Platform, 2, 5–69, 81, 87–89, 91, 94, 97, 100,
    104, 106, 108, 110, 112, 114, 115, 118,
    120, 122–126, 130, 132, 133, 138, 148,
    149, 187, 188, 201, 202, 216–223

Pleistocene, 8, 39, 40, 43, 46–49, 55, 68, 69,
      95, 98, 108, 114–116, 123, 126,
      128–131, 137, 138, 149, 156, 172, 174,
      185, 191, 216, 217, 220
Prediction, 21, 215
Productivity, 1, 10, 26, 65, 167, 194,
      200, 204, 225
Progradation, 6, 10, 59, 63, 67, 68, 180

Q
Qatar, 146, 156, 158, 160, 166, 167, 170,
      178–180, 185, 188, 190, 193, 203
Quartz, 104–106, 120–122, 138, 158, 170,
      176, 178, 203

R
Radiation, 22
Ramp, 2, 145–205, 217–219, 222
Red algae, 26, 30–32, 38, 96, 101, 114,
      195–198, 201, 202, 221
Reef, 5, 81, 150, 217
Reworking, 222
Rock record, 1, 2, 215

S
Sabkha, 146, 149, 152, 179–187, 203, 204
Salinity, 2, 19, 22–25, 27, 29, 30, 49, 91–96,
      98, 101, 104, 137, 148, 152, 166–169,
      184, 187, 188, 194, 196, 199, 203–205
Salt flat, 146, 149, 179
Sand belt, 17, 41–43, 45, 46, 49–55,
      180, 217, 218
Sand body/Sand bodies, 6, 11, 17, 40, 43–46,
      49–57, 120, 182–184, 187,
      218–220, 222
Sand shoal, 11, 46, 123, 160, 192
Saudi Arabia, 146, 153, 155, 156, 160, 167,
      185, 186, 190, 191, 196, 220
Scending brine, 182
Scleractinian, 191, 201
Seabreeze, 154, 162, 204
Seagrass, 38, 89, 94, 98, 101, 135, 195, 200
Sea level, 1, 6, 8, 40, 42, 46–49, 51, 52, 55,
      56, 60, 64–69, 81, 88, 91, 94, 97–99,
      104, 106–108, 115, 116, 127–129,
      132–138, 145, 180, 181, 187, 188,
      191, 200, 204
Sea temperature, 21–22, 84, 166, 223
Seawater pumping, 182
Sedimentary record, 2, 67
Shakki, 153

Shamal, 152–154, 156–159, 162, 166, 179,
      204, 220, 222
Shatt el Arab, 152, 156, 160, 162, 164, 165,
      167, 169, 185, 193, 195
Siliciclastic, 2, 65, 67, 81–138, 145, 150, 154,
      203, 205, 217, 218
Slope, 3, 6, 8, 22, 32, 33, 37, 38, 40, 43, 45,
      51, 56, 60–65, 68, 69, 98, 111,
      115, 117–120, 147, 164, 173,
      177, 216, 218
Sponge, 26, 33, 37, 61, 99, 111, 119, 120,
      173, 195
Starved, 174, 177
Steep-sided platform, 8, 219, 220
Straits of Hormuz, 147, 152, 154, 160, 162,
      166–168, 179, 193, 195, 203
Stringer reef, 191, 192
Stromatolite, 34–38, 96, 199–221
Supratidal, 34, 48, 58, 59, 170, 179, 180, 182,
      184–186, 199

T
Tectonic, 1, 2, 112, 124–128, 138, 145, 203,
      204, 218, 219
Tectonic setting, 2, 8–10, 147–151,
      215–217, 222
Temperature, 1, 2, 10, 11, 13, 21–22, 27, 29,
      30, 32, 34, 45, 82–85, 91–93, 98, 148,
      152, 153, 160, 166–167, 169, 187, 199,
      201, 202, 204, 219, 222, 223
Tertiary, 10, 67, 124, 128, 147, 149, 170, 182,
      186, 216, 218
The Gulf, 2, 145–205, 216
Tidal delta, 49, 52, 53, 56–57, 165, 183, 185,
      187, 220
Tidal flat, 6, 21, 23, 40, 46, 49, 58–59, 66,
      67, 149, 171, 178–182, 184, 199, 204,
      218, 219
Tidal range, 2, 10, 20, 45, 50, 52–54, 56, 57,
      89, 90, 158, 160, 161, 181, 220, 223
Tidal regime, 161, 162
Tides, 7, 19, 20, 44, 46, 50–52, 54, 58, 59, 67,
      89–91, 111, 150, 153, 158–161, 165,
      181, 186, 203, 204, 217, 219–221
Tigris, 149, 164, 165, 187
Tongue of the Ocean (TOTO), 7, 8, 10, 17,
      19, 22, 27, 41, 43, 45, 46, 49–55,
      60–64, 220
Topography, 5, 8–10, 21, 29, 38, 40–44, 47,
      49, 56, 67, 98, 105, 114, 115, 123,
      124, 126–132, 137, 138, 149–151,
      174, 187, 189, 192, 201, 203, 204,
      215, 217

Transport, 12, 13, 15, 17, 19, 22, 32, 42–46,
     51, 52, 56, 58, 63, 64, 82, 87, 89, 111,
     122, 138, 152, 154–158, 163–165, 172,
     174, 179, 182, 184, 194, 203, 204, 218,
     220, 221
Turbidity, 9, 27, 61–64, 68, 164, 167,
     204, 221

U
United Arab Emirates, 145, 154, 165, 199, 220

W
Wall, 36, 60, 61, 117, 119, 120
Water clarity, 2
Water composition, 215, 223, 224
Water energy, 1, 10, 11, 29, 32, 150, 199, 215,
     219, 221

Water temperature, 2, 29, 82–85, 92, 93, 153,
     160, 166
Wave energy, 2, 10, 17–19, 29, 36, 43, 44,
     87–89, 91, 108, 123, 138, 156–158,
     185, 199, 202, 203, 220, 222
Whitings, 24–26, 58, 95, 169, 185, 203,
     205, 226
Wind, 2, 6–8, 10, 13, 19, 27–29, 39–57, 60,
     61, 63, 67, 84–89, 91, 93, 105, 111,
     114, 116, 138, 149, 151–158, 161–164,
     177, 181, 185, 186, 192, 201, 203, 204,
     217–220, 222, 223
Wind energy, 1, 15–17, 55, 204, 218
Winnowing, 17, 111, 179, 180

Z
Zagros, 147, 149, 187
Zagros Mountains, 148, 149, 153, 163

Printed by Printforce, the Netherlands